Statistical Methods in
Engineering and
Quality Assurance

Statistical Methods in Engineering and Quality Assurance

PETER W. M. JOHN

A Wiley-Interscience Publication
JOHN WILEY & SONS, INC.
New York • Chichester • Brisbane • Toronto • Singapore

In recognition of the importance of preserving what has been written, it is a policy of John Wiley & Sons, Inc. to have books of enduring value published in the United States printed on acid-free paper, and we exert our best efforts to that end.

Copyright © 1990 by John Wiley & Sons, Inc.

All rights reserved. Published simultaneously in Canada.

Reproduction or translation of any part of this work beyond that permitted by Section 107 or 108 of the 1976 United States Copyright Act without the permission of the copyright owner is unlawful. Requests for permission or further information should be addressed to the Permissions Department, John Wiley & Sons, Inc.

Library of Congress Cataloging-in-Publication Data
John, Peter William Meredith.
    Statistical methods in engineering and quality assurance / Peter W. M. John.
      p.   cm. -- (Wiley series in probability and mathematical statistics. Applied probability and statistics section)
   Includes bibliographical references (p.   ).
   ISBN 0-471-82986-2
    1. Engineering--Statistical methods. 2. Quality control--Statistical methods. 3. Quality assurance. I. Title. II. Series.
TA340.J65   1990
620'.0072--dc20   90-33718
                                                     CIP

Printed in the United States of America

10 9 8 7 6 5 4 3 2

Contents

Preface xv

1. Quality Assurance and Statistics 1

    1.1   Quality and American Business, 1
    1.2   Competition, 2
    1.3   The Reaction, 2
    1.4   Total Quality, 3
    1.5   *Kaizen*, 4
    1.6   Why Statistics?, 4
    1.7   The Future, 5
    1.8   Gathering Information, 6
    1.9   This Book, 7
    1.10 Instructional Considerations, 7
    1.11 Reprise, 8

2. Descriptive Statistics 9

    2.1   Introduction, 9
    2.2   The Mean and Standard Deviation, 9
    2.3   The Median and Quartiles, 11
    2.4   Ishikawa's Seven Tools, 13
    2.5   Histograms, 14
    2.6   An Example of a Nonsymmetrical Distribution, 16
    2.7   Dotplots, 19
    2.8   Stem-and-Leaf Diagrams, 20
    2.9   Splitting Stems, 21
    2.10 Box-and-Whisker Plots, 22
    2.11 Pareto Diagrams, 24

2.12 Summary, 25
Exercises, 26

3. **Discrete Variables, or Attributes**   29

   3.1   Introduction, 29
   3.2   Random Digits, 29
   3.3   A Mathematical Model, 30
   3.4   Some Basic Probability Theory, 31
   3.5   The Coin Tossing Model, 33
   3.6   The Binomial Distribution, 33
   3.7   Unequal Probabilities, 35
   3.8   The Binomial Distribution (General Case), 36
   3.9   Mathematical Expectation, 37
   3.10 The Variance, 38
   3.11 Sampling without Replacement, 39
   3.12 The Hypergeometric Distribution, 40
   3.13 Conditional Probabilities, 41
   3.14 Bayes' Theorem, 42
   3.15 Bivariate Distributions, 44
   3.16 The Geometric Distribution, 45
   3.17 The Poisson Distribution, 46
   3.18 Properties of the Poisson Distribution, 47
   3.19 The Bortkiewicz Data, 48
   3.20 Other Applications, 48
   3.21 Further Properties, 48
         Exercises, 49

4. **Continuous Variables**   52

   4.1   Introduction, 52
   4.2   Probability Density Functions, 52
   4.3   The Cumulative Distribution Function, 54
   4.4   The Uniform Distribution, 55
   4.5   The Exponential Distribution, 56
   4.6   The Quartiles of a Continuous Distribution, 58
   4.7   The Memoryless Property of the Exponential Distribution, 58
   4.8   The Reliability Function and the Hazard Rate, 59
   4.9   The Weibull Distribution, 60
   4.10 The Gamma Distribution, 61

4.11 The Expectation of a Continuous Random Variable, 62
4.12 The Variance, 64
4.13 Moment-Generating Functions, 64
4.14 Bivariate Distributions, 66
4.15 Covariance, 67
4.16 Independent Variables, 68
Exercises, 69

5. The Normal Distribution   72

5.1 Introduction, 72
5.2 The Normal Density Function, 72
5.3 The Tables of the Normal Distribution, 74
5.4 The Moments of the Normal Distribution, 75
5.5 The Moment-Generating Function, 76
5.6 Skewness and Kurtosis, 76
5.7 Sums of Independent Random Variables, 77
5.8 Random Samples, 79
5.9 Linear Combinations of Normal Variables, 80
5.10 The Central Limit Theorem, 80
5.11 The Chi-Square Distribution, 81
5.12 The Lognormal Distribution, 82
5.13 The Cumulative Normal Distribution Curve and Normal Scores, 83
Exercises, 86

6. Some Standard Statistical Procedures   88

6.1 Introduction, 88
6.2 The Three Experiments, 88
6.3 The First Experiment, 89
6.4 The Second Experiment, 89
6.5 The Third Experiment, 90
6.6 The Behavior of the Sample Average, 90
6.7 Confidence Intervals for the Normal Distribution, 90
6.8 Confidence Intervals for the Binomial Distribution, 94
6.9 Confidence Intervals for the Exponential Distribution, 96
6.10 Estimating the Variance, 97
6.11 Student's t Statistic, 100
6.12 Confidence Intervals for the Variance of a Normal Population, 101

- 6.13 Chi-Square and Exponential Variables, 102
- 6.14 Another Method of Estimating the Standard Deviation, 104
- 6.15 Unbiased Estimators, 105
- 6.16 Moment Estimators, 106
- 6.17 Maximum-Likelihood Estimates, 108
  Exercises, 110

7. Hypothesis Testing   113

- 7.1 Introduction, 113
- 7.2 An Example, 114
- 7.3 The Decisions, 115
- 7.4 Null and Alternate Hypotheses, 116
- 7.5 Acceptance and Rejection Regions, 116
- 7.6 One-Sided Tests, 117
- 7.7 The Choice of Alpha, 118
- 7.8 The z-Test, 119
- 7.9 The Type II Error, 120
- 7.10 The t-Test, 122
- 7.11 Tests about the Variance, 123
- 7.12 The Exponential Distribution, 124
- 7.13 Tests for the Percent Defective, 124
- 7.14 Chi-Square and Goodness of Fit, 125
- 7.15 Multinomial Goodness of Fit, 126
  Exercises, 127

8. Comparative Experiments   129

- 8.1 Introduction, 129
- 8.2 Comparing Two Normal Means with Known Variance, 129
- 8.3 Unknown Variances, 131
- 8.4 Unequal Variances, 132
- 8.5 The Paired t-Test, 133
- 8.6 Wilcoxon's Two-Sample Test, 134
- 8.7 The Duckworth Test, 136
- 8.8 Comparing Variances, 137
- 8.9 Confidence Intervals for the Variance Ratio, 138
- 8.10 Comparing Exponential Distributions, 139
- 8.11 Comparing Binomial Populations, 140
- 8.12 Chi-Square and 2×2 Tables, 141
  Exercises, 142

9. Quality Control Charts — 144

- 9.1 Introduction, 144
- 9.2 Quality Control Charts, 145
- 9.3 The x-Bar Chart, 146
- 9.4 Setting the Control Lines, 147
- 9.5 The R Chart, 148
- 9.6 An Example, 149
- 9.7 Another Example, 151
- 9.8 Detecting Shifts in the Mean, 153
- 9.9 Alternative Stopping Rules, 154
- 9.10 Average Run Lengths, 154
- 9.11 s Charts, 156
- 9.12 Setting the Control Limits for an s Chart, 156
- 9.13 Alternative Control Limits, 157
- 9.14 Nonnormality, 158
- 9.15 Process Capability, 158
- 9.16 I Charts, 160
- Exercises, 161

10. Control Charts for Attributes — 164

- 10.1 Introduction, 164
- 10.2 Binomial Charts, 164
- 10.3 Binomial Charts for a Fixed Sample Size, 165
- 10.4 Variable Sample Sizes, 167
- 10.5 Interpreting Outlying Samples, 169
- 10.6 The Arcsine Transformation, 171
- 10.7 c-Charts, 172
- 10.8 Demerits, 174
- Exercises, 175

11. Acceptance Sampling I — 177

- 11.1 Introduction, 177
- 11.2 The Role of Acceptance Sampling, 177
- 11.3 Sampling by Attributes, 178
- 11.4 Single Sampling Plans, 179
- 11.5 Some Elementary Single Sampling Plans, 180
- 11.6 Some Practical Plans, 182
- 11.7 What Quality is Accepted? AQL and LTPD, 183

- 11.8 Choosing the Sample Size, 183
- 11.9 The Average Outgoing Quality, AOQ and AOQL, 184
- 11.10 Other Sampling Schemes, 186
- 11.11 Rectifying Inspection Plans, 186
- 11.12 Chain Sampling Plans, 187
- 11.13 Skip Lot Sampling, 187
- 11.14 Continuous Sampling, 187
- 11.15 Sampling by Variables, 188
  Exercises, 188

12. Acceptance Sampling II   189

- 12.1 The Cost of Inspection, 189
- 12.2 Double Sampling, 189
- 12.3 Sequential Sampling, 191
- 12.4 The Sequential Probability Ratio Test, 191
- 12.5 The Normal Case, 192
- 12.6 An Example, 193
- 12.7 Sequential Sampling by Attributes, 194
- 12.8 An Example of Sequential Sampling by Attributes, 194
- 12.9 Two-Sided Tests, 195
- 12.10 An Example of a Two-Sided Test, 196
- 12.11 Military and Civilian Sampling Systems, 197
- 12.12 MIL-STD-105D, 197
- 12.13 The Standard Procedure, 198
- 12.14 Severity of Inspection, 199
- 12.15 Tightened versus Normal Inspection, 199
- 12.16 Reduced Inspection, 200
- 12.17 Switching Rules, 201
- 12.18 Acceptance Sampling by Variables, 201
  Exercises, 202

13. Further Topics on Control Charts   203

- 13.1 Introduction, 203
- 13.2 Decision Rules Based on Runs, 204
- 13.3 Combined Rules, 204
- 13.4 An Example, 206
- 13.5 A Table of Values of ARL, 206
- 13.6 Moving Averages, 206

13.7	Arithmetic-Average Charts, 207	
13.8	Cumulative Sum Control Charts, 209	
13.9	A Horizontal V-Mask, 212	
13.10	Geometric Moving Averages, 213	
13.11	EWMA and Prediction, 214	
	Exercises, 215	

14. Bivariate Data: Fitting Straight Lines   217

- 14.1   Introduction, 217
- 14.2   The Correlation Coefficient, 218
- 14.3   Correlation and Causality, 219
- 14.4   Fitting a Straight Line, 220
- 14.5   An Example, 221
- 14.6   The Method of Least Squares, 223
- 14.7   An Algebraic Identity, 225
- 14.8   A Simple Numerical Example, 225
- 14.9   The Error Terms, 226
- 14.10   The Estimate of the Slope, 227
- 14.11   Estimating the Variance, 228
- 14.12 t-Tests, 228
- 14.13   Sums of Squares, 229
- 14.14   Standardized Residuals, 231
- 14.15   Influential Points, 231
- 14.16   Confidence Intervals for a Predicted Value, 232
- 14.17   Lines through the Origin, 232
- 14.18   Residual Plots, 233
- 14.19   Other Models, 235
- 14.20   Transformations, 237
-         Exercises, 238

15. Multiple Regression   243

- 15.1   Introduction, 243
- 15.2   The Method of Adjusting Variables, 244
- 15.3   The Method of Least Squares, 245
- 15.4   More Than Two Predictors, 247
- 15.5   Properties of the Estimates, 248
- 15.6   More about the Example, 248
- 15.7   A Geological Example, 249

- 15.8 The Printout, 249
- 15.9 The Second Run, 252
- 15.10 Fitting Polynomials, 253
- 15.11 A Warning about Extrapolation, 255
- 15.12 Testing for Goodness of Fit, 256
- 15.13 Singular Matrices, 257
- 15.14 Octane Blending, 258
- 15.15 The Quadratic Blending Model, 259
  Exercises, 260

16. The Analysis of Variance   264

- 16.1 Introduction, 264
- 16.2 The Analysis of Variance, 264
- 16.3 The One-Way Layout, 265
- 16.4 Which Treatments Differ?, 267
- 16.5 The Analysis of Covariance, 267
- 16.6 The Analysis of Covariance and Regression, 270
- 16.7 Quantitative Factors, 271
- 16.8 The Two-Way Layout, 273
- 16.9 Orthogonality, 274
- 16.10 Randomized Complete Blocks, 276
- 16.11 Interaction, 276
- 16.12 Interaction with Several Observations in Each Cell, 278
- 16.13 Three Factors, 279
- 16.14 Several Factors, 280
  Exercises, 280

17. Design of Experiments: Factorial Experiments at Two Levels   284

- 17.1 Introduction, 284
- 17.2 Experimenting on the Vertices of a Cube, 285
- 17.3 An Example of a 2^2 Experiment, 286
- 17.4 The Design Matrix, 287
- 17.5 Three Factors, 288
- 17.6 The Regression Model, 290
- 17.7 Yates' Algorithm, 292
- 17.8 An Example with Four Factors, 293
- 17.9 Estimating the Variance, 294
- 17.10 Normal Plots, 295

CONTENTS   xiii

    17.11   Fractions, 296
    17.12   Three Factors in Four Runs, 296
    17.13   Fractions with Eight Runs, 297
    17.14   Seven Factors in Eight Runs, 298
    17.15   The $L(8)$ Lattice, 299
    17.16   The $L(16)$ Lattice, 300
    17.17   Screening Experiments, 301
    17.18   Foldover Designs, 303
    17.19   Resolution V Fractions, 304
    17.20   Fractions with 12 Points, 304
              Exercises, 305

18. Design of Experiments: Factorial Experiments at Several Levels   **311**

    18.1   Factors with Three Levels, 311
    18.2   Two Factors, 311
    18.3   Fractions: Four Factors in Nine Points, 312
    18.4   Three Factors in 27 Runs, 313
    18.5   Four Factors in 27 Runs, 316
    18.6   Thirteen Factors in 27 Runs, 318
    18.7   The $L(18)$ Lattice, 319
    18.8   The $L(36)$ Lattice, 319
    18.9   A Nonorthogonal Fraction, 321
    18.10   Latin Squares, 323
    18.11   Graeco-Latin Squares, 323
    18.12   Hyper-Graeco-Latin Squares, 324
    18.13   Five Factors at Four Levels in 16 Runs, 324
    18.14   Four-Level Factors and Two-Level Factors, 325
    18.15   Factors at Five Levels, 328
              Exercises, 328

19. Taguchi and Orthogonal Arrays   **331**

    19.1   Introduction, 331
    19.2   Terminology, 333
    19.3   The Three Stages of Process Design, 333
    19.4   A Strategy for Parameter Design, 333
    19.5   Parameter Design Experiments, 335
    19.6   Signal-to-Noise Ratios, 335
    19.7   Inner and Outer Arrays, 337

19.8	An Example of a Parameter Design Experiment, 338	
19.9	The Confirmation Experiment, 340	
19.10	Tolerance Design, 341	
19.11	Finale, 341	
	Exercises, 343	

References   347

Tables   351

    I. The Normal Distribution, 351
   II. The t-Distribution, 353
  III. The Chi-Square Distribution, 354
  IV. The F Distribution, 356

Index   367

Preface

This book is designed to give engineers an introduction to statistics in its engineering applications, with particular emphasis on modern quality assurance. It reflects more than 30 years of teaching statistics to engineers and other scientists, both in formal courses at universities and in specialized courses at their plants, augmented by broad consulting experience. The examples in the text cover a wide range of engineering applications, including both chemical engineering and semiconductors.

My own interest in quality assurance evolved from my long-standing specialization in the design of experiments, which began with my experience as the first research statistician in the laboratories at Chevron Research Corporation, from 1957 to 1961. Since returning to academia in 1961, I have devoted much of my research to fractional factorial experiments. That aspect of the design of experiments has become the basis of off-line quality control, a key element of modern quality assurance. Given that background, you can imagine how satisfying I find American industry's current surge of interest in quality assurance.

Almost equally satisfying is the burst of technical capabilities for the implementation of quality assurance methods by engineers at plant level—a direct result of the wide accessibility and growing power of the personal computer.

Why is it so exciting to have these capabilities available for the PC? Before the 1950s, statistical calculations were a forbidding chore. As a practical matter, we could not perform such tasks as multiple regression. Then came the mainframe, which meant that the engineer had to work through the computing section with its programmers and keypunch operators. Although the mainframes facilitated wonderful progress, using them could be intimidating and often complicated. Under some accounting procedures, it was also costly. People felt inhibited about using the company's computers unless they had a relatively big problem and an expert to help them.

Now, after three decades of ever-accelerating progress in both hardware and software, the personal computers that most engineers have on their desks can do more easily and quickly all that the mainframes did in the late 1950s. Hence, most engineers have little or no occasion to deal with the mainframe that may be miles away from their plants. All of the computations that I have made in writing this book have been carried out on such a PC, using Minitab for calculations and Statgraphics for figures, in addition, of course, to a word processor for composition.

I use Minitab on the mainframe for my teaching at the University of Texas. It has the great advantage of being very easy for the students to use, and most of them use it on our instructional VAX. Moreover, it has lately become available in a PC version, which I find more convenient. I also use Minitab in much of my research. When I have more complicated problems, I can turn to SAS or to GLIM; they, too, are available in versions for the PC with a hard disk.

Not only has the PC freed engineers from the mainframes for everyday computations, but better still, the engineers themselves can keep their data files on their own disks. Then, as a matter of routine, they can apply standard statistical procedures easily and quickly. This is particularly advantageous with graphical procedures. In just a few seconds, today's engineers can make plots that normally would have taken their fathers all day. Now engineers can, and should, use statistical methods and statistical thinking on an everyday basis.

Even though engineers now can and should perform many of the applications independently, it is even better to have convenient access to a statistician with whom to discuss problems and seek advice. This book is only an introduction to a broad field. As to the more complicated problems, I have been able only to touch upon some of them. Parts of the last chapters on the design of experiments may look deceptively easy. However, a few words with a statistician before you get started may save you hours of pain later.

We are at a stage in our technological development when engineering and statistics can walk together hand in hand. I intend this book to help us both to do that. I hope that it will persuade engineers to use statistical thinking as a standard operating procedure, both to collect their data and to analyze them.

PETER W. M. JOHN

Austin, Texas

Statistical Methods in Engineering and Quality Assurance

CHAPTER ONE

Quality Assurance and Statistics

1.1. QUALITY AND AMERICAN BUSINESS

If, 30 years ago, you had asked engineers or executives what quality assurance was, they probably would not have known the expression and would have said something about being sure that we kept on doing things as we had always done them. If you had asked them what quality control was, and they knew the conventional wisdom, the answer would have been something about Shewhart control charts and government sampling plans. That is what quality control and quality assurance were.

Today, quality has beome a vital part of American industry, as we are engaged in a life and death struggle with foreign nations that hardly existed as competitors 30 years ago. Without dedication to total quality in our manufacturing companies, we will be unable to compete in areas that once we dominated. We will be beaten not by price, but by inferior product quality.

Two of the main techniques in quality assurance—control charts and acceptance sampling—date back to the 1930s at Bell Labs. They were the work of such pioneers as Walter Shewhart, H. F Dodge, and Enoch Farrell. During World War II, academic statisticians, of whom there were only a few, worked on the development of acceptance sampling, mainly at Princeton and Columbia. After the war, that group spread out over the country to form the core of the growth of statistics in American universities. But they no longer worked on quality control. The topic was relegated to a few lectures in departments of industrial engineering.

It is not too harsh to say that American industry became interested in quality control during World War II because the government insisted upon it. Industry lost interest as soon as the war was over, except for those companies that were engaged in defense work, where the government still insisted on its use.

In some industries, the production people churned out the parts, good, bad, or indifferent; the inspectors inspected the lots, accepted some, and rejected others. One hoped that they were able to spot most of the defective parts, which were then either scrapped or reworked. Some managers

1

resented the expense and the hassle of this, and would have been happier to ship the product as is and let the buyer beware.

In some companies, the division of functions was simple. Once the production department shoveled the product out of the plant door, it became marketing's problem. Marketing sold the product. Warranty and service were pushed down to the level of the local retailer or the local distributor. This meant two things. It meant that service after the sale was often subject to cavalier neglect. It also meant that there was really very little feedback from the final buyer, the user, to the manufacturing department concerning what was wrong with the items and how they could be improved. So-called new models might appear annually with cosmetic changes, but often little real improvement over the previous year's product.

1.2. COMPETITION

Then came the Japanese shock. The Japanese, who had earned an international reputation for shoddy workmanship before 1940, began producing cars, televisions, chips, etc., that were not only cheaper than ours, but worked better and lasted longer. The number of Japanese cars on our highways and television sets in our homes increased by leaps and bounds. Some Americans tried to attribute the increase in imports from Japan entirely to the alleged fact that Japan's wages and standard of living were lower than ours. Then people began to realize that the explanation was that the Japanese had done some things right. One of those things was to listen to William Edwards Deming, an American statistician who had started the Japanese on the road to quality in 1950, but was being completely ignored in his homeland, and to the American engineer J. M. Juran.

American manufacturers then preferred to focus on cost control rather than quality control. Some did not even bother to control costs. For a time, inflation psychology was rampant and extra costs could be passed through to the customer with no great difficulty—especially if that customer was Uncle Sam and the manufacturer had a cost-plus contract.

1.3. THE REACTION

The reaction came a few years ago. Deming, called back from the wilderness, became a prophet in his own country. Juran had also been preaching quality control in this country and in Japan for many years. Once Americans listened to them, a radical change began to take place in much of this nation's industry. Engineers began to talk of quality *assurance* rather than quality *control*. The term *total quality* entered the jargon of manufacturing.

Deming emphasizes that quality requires the complete dedication of management. He summarizes this requirement in "Fourteen Points for

Management," of which the first two are

"Create constancy of purpose toward improvement of product and service."

and

"Adopt the new philosophy; we are in a new economic age."

The final point is

"Put everybody in the company to work in teams to accomplish the transformation."

1.4. TOTAL QUALITY

The first of Deming's points sets the tone for total quality. Total quality means a dedication to quality by the entire company, from the CEO down to the employee who sweeps the corridors. It demands a realization that quality is the concern of everybody. One of the leading converts to total quality, James R. Houghton, the CEO of Corning Glass Company, summed it up more concisely in 1986:

"Quality is
    knowing what needs to be done,
    having the tools to do it right,
    and then doing it right—the first time."

Then he boiled it down even further to two objectives:

"Quality is everyone's job.
Do it right the first time."

The first of these two objectives has far-reaching social implications. It emphasizes that quality is a matter of dedication of the whole company and all its personnel, and not just a function of a few inspectors stationed one step before the customer shipping area. It emphasizes the old, much maligned idea that the objective is customer satisfaction: to supply the customer with good product. The word "customer" does not just mean the end recipient who pays for the product. You also have your own customers within the plant—the people at the next stage of the production cycle. No step in the process stands alone. No worker is an island. All steps are combined in the process. All workers are part of a team working together to achieve quality.

This whole concept of the workers in a plant or a company as a team is revolutionary to many people. It is certainly not a part of the Frederick Taylor model of scientific management that has dominated managerial styles in American manufacturing since it was introduced at the beginning of this century. The Taylor model is sometimes called management by control. Its essence is the breakdown of the manufacturing process into pieces that are then bolted together in a rigid hierarchical structure. The CEO gives his division heads a profit objective and down the line it goes. The division head sets goals or quotas for each of his department heads: increase sales by 10%, decrease warranty costs by 15%, and so on. These dicta end up as quotas or work standards at the shop level, sometimes with unwise, and perhaps unauthorized, insertions and modifications from middle management.

The new thrust in management requires a fundamental change of style. Simply ordering the workers to do this or that is now seen to be counterproductive. Total quality involves workers in a cooperative effort, with initiative encouraged and responsibility shared from top to bottom.

1.5. KAIZEN

The fifth of Deming's points is

"Improve constantly and forever every activity."

The Japanese call this *kaizen*, meaning continuous searching for incremental improvement. Tom Peters, who writes on excellence in manufacturing, points out that some authorities consider this Japanese trait the single most important difference between Japanese and American business. American manufacturers have tended either to go for the big leap or to stop caring about improving the process when they think it is good enough to do the job. There is an expression "overengineering" that is used disparagingly for wasting time fussing at improving a product once it has reached a marketable state.

1.6. WHY STATISTICS?

How is *kaizen* achieved? It is achieved by integrating the research and development effort with the actual production facility and devoting a lot of time and money to "getting the process right" and then continually making it better. The key factor is *the pervasive use of statistical methods*.

Why the pervasive use of statistical methods? Where does statistics come into this picture other than just in teaching people how to draw Shewhart charts?

The distinguished British physicist Lord Kelvin put it this way:

"When you can measure what you are speaking about and express it in numbers, you know something about it; but when you cannot measure it, when you cannot express it in numbers, your knowledge is of the meager and unsatisfactory kind."

But perhaps the most cogent summation is that of W. G. Hunter of the University of Wisconsin, who said:

"1. If quality and productivity are to improve from current levels, changes must be made in the way things are presently being done.
2. We should like to have good data to serve as a rational basis on which to make these changes.
3. The twin question must then be addressed: what data should be collected, and, once collected, how should they be analyzed?
4. Statistics is the science that addresses this twin question."

Quality assurance pioneer Deming was trained as a statistician. Now Japan's principal leader in quality control is an engineer, G. Taguchi, who, with his disciples, vigorously advocates his program of statistically designed experiments, of which more will be said in the last three chapters.

1.7. THE FUTURE

Some of us who have been involved in the quality revolution have asked each other, in our gloomier moments, whether or not this is perhaps a passing fad. Will American manufacturing, once back on track, then lose its appetite for quality assurance and go back to its old noncompetitive ways until another crisis hits?

There are grounds for increased optimism. Quality assurance now has official recognition. Every year the Japanese present their prestigious Deming Award to the company that makes the greatest impact in quality. It has been won by several of the companies that have become household words in this country. The United States has now established the Malcolm Baldrige award for quality, named after the late Secretary of Commerce. It was awarded for the first time in the fall of 1988. Among the winners was a major semiconductor producer, Motorola.

There have been well-publicized success stories in other industries besides semiconductors. We mentioned Corning Glass earlier. We could add Westinghouse, Ford, and a host of other major companies. Many other, less famous, companies are achieving outstanding results from their emphasis on quality. Key examples may be found in the steel industry, where old dinosaurs are being replaced by small speciality firms that have found niches in the market for high-quality items.

Concern about quality is becoming pervasive. Customers demand and expect quality from their manufacturers. Manufacturers demand quality from their suppliers and train them to produce the quality parts that they need. On the day that this paragraph was drafted, the *Wall Street Journal* carried an advertisement from Ford giving a list of their suppliers in the United States and other parts of the world—Mexico, Brazil, Liechtenstein, and more—who had won Preferred Quality Awards for the quality of the parts that they supplied to the automotive manufacturer. Quality is working its way down to the small subsubcontractor, down to the grass roots of American manufacturing.

Quality pays, and management is learning that. Two statistics can be cited in support of this principle. Juran has said for some time that at least 85% of the failures in any organization are the fault of systems controlled by management. Fewer than 15% are worker-related. Lately, he has tended to increase that earlier figure of 85%. The second, more telling statistic came from James Houghton. In some companies the cost of poor quality—the cost of the twin demons

SCRAP and REWORK

—runs somewhere between 20 to 30% of sales. And that is before taking into account sales lost through customer dissatisfaction. American management has been learning that statistic the hard way. It should be a long while before they forget it.

1.8. GATHERING INFORMATION

If we are to improve our manufacturing processes, we must learn more about them. We cannot just sit in our offices, developing the mathematical theory, and then turn to the computers to solve the differential equations. We are working, whether we like it or not, in the presence of variability. Shewhart's great contribution was to educate engineers about the existence of variability. Agricultural scientists already knew about natural variability. For them it was the reality of variability in weather, in soil fertility, in weights of pigs from the same litter. They necessarily learned to conduct experiments in its presence.

Engineers conduct experiments all the time. They may not realize it. When an engineer nudges the temperature at one stage of a plant by a degree or so, he is doing an experiment and geting some data. When another engineer uses a catalyst bought from another manufacturer, she is doing an experiment. There are data out there. They have to be gathered and analyzed. Better still, the data should be obtained in an organized way so that their yield of information shall be as rich as possible. The statistician can help to extract all the information that is to be found in the data, and, if

INSTRUCTIONAL CONSIDERATIONS 7

given the opportunity, will help to choose a designed experiment to yield good data.

1.9. THIS BOOK

There are 19 chapters in this book. Chapters two through eight can be taken as a course in probability and statistics designed for engineers. Chapters nine through thirteen deal with the traditional topics of quality control—charts and acceptance sampling. Those topics deal with passive quality assurance. The charts tell whether the process is moving along under statistical control. The sampling procedures tell whether the incoming or outgoing batches of product meet the specifications.

The last six chapters point the reader toward active quality assurance. How can the process be improved? Chapters fourteen and fifteen are devoted to regression, i.e., fitting models to the data by the method of least squares. Chapters sixteen through nineteen are concerned with off-line quality control. They consider how to conduct experiments to determine the best conditions for the operation of the process.

The semiconductor industry is full of horror stories about devices that have been designed miles away from the future fabricators and "tossed over the transom." The design people have made one batch of the prototype device under laboratory conditions. Now let the manufacturing engineer make their design manufacturable! But how? The manufacturing engineer has to learn more about the process and to find good, preferably optimum (whatever that really means), operating conditions. That means carrying out experiments at the manufacturing level.

Chapter sixteen is an introduction to the analysis of variance, a procedure that Sir Ronald Fisher introduced 60 years ago to design and analyze agricultural experiments. That procedure has become the basis of modern experimental design. Chapters seventeen and eighteen are about factorial experiments. Although based on the work of Fisher and Frank Yates, these chapters focus on the developments of the past 40 years in industrial applications by George Box, Cuthbert Daniel, and others. The last chapter, nineteen, is mainly devoted to the so-called Taguchi method. This is the procedure for designed experiments developed by G. Taguchi in Japan, where it is widely used. It is now used increasingly in the west as well. The Taguchi method combines the ideas developed by Fisher, Box, and others with some new insights and emphases by Taguchi to form a systematic approach to conducting on-line experiments.

1.10. INSTRUCTIONAL CONSIDERATIONS

There is enough material in this book for a two-semester course. The instructor who has only one semester will necessarily make some choices.

For example, he could omit some of the more advanced and specialized topics in chapters three through eight. If the students already know some statistics, judicious cuts could be made in the assignment of introductory segments.

The emphasis in the later sections would depend on the focus of the course. For example, some instructors may choose to leave out all of the material on acceptance sampling in chapters eleven and twelve, or at least to omit the sections on sequential sampling in the latter chapter. Others may elect to omit the advanced topics on control charts in chapter thirteen.

In the last part of the book, one could postpone the last half of chapter eighteen and some of the specialized topics in the two chapters on regression. Less thorough coverage could be accorded the chapter on analysis of variance. However, any modern course in statistics for engineers should include the material in chapters seventeen and nineteen. Those techniques for designed experiments are fundamental to the improvement of manufacturing processes.

1.11. REPRISE

Why do engineers need to learn statistics? We have mentioned some reasons earlier. You cannot make good decisions about quality without good information, any more than a general can make good decisions on the battlefield without good intelligence. We have to gather good data and interpret it properly. That is what statistics is all about: gathering data in the most efficient way, preferably by properly designed experiments; analyzing that data correctly; and communicating the results to the user so that valid decisions can be made.

CHAPTER TWO

Descriptive Statistics

One picture is worth more than ten thousand words.

2.1. INTRODUCTION

In this chapter, we discuss how one might summarize and present a large sample of data. There are two ways of doing this. One is by words and figures. We can calculate from the sample certain important statistics like the mean and the standard deviation that, in some sense, characterize the data. The other is by using graphical methods. This used to be a chore, calling for hours of tedious tabulation, but nowadays computers have made it easier. Table 2.1.1 shows the daily yields (percent conversion) in 80 runs in a chemical plant. We use this data set to illustrate the ideas throughout this chapter.

As they stand, these are only 80 numbers, which at first glance convey little. In the next sections, we make this set of 80 numbers more intelligible in two ways. First, we derive some numbers (statistics) that summarize the data points. An obvious candidate is the average of the 80 observations. Then, we show how the points can be presented graphically by a histogram and also in several other ways.

2.2. THE MEAN AND STANDARD DEVIATION

The word *statistic* is used to describe a number that is calculated from the data. Two of the most commonly used statistics for describing a sample are the mean and the standard deviation. The name mean is short for the arithmetic mean. It is the simple average of the data. If there are n observations denoted by $x_1, x_2, \ldots, x_n$, the mean, which is written as $\bar{x}$, or x-bar, is defined as

$$\bar{x} = \sum \frac{x_i}{n},$$

Table 2.1.1. Yields of 80 Runs in a Chemical Plant

71.8	71.6	73.0	76.4	67.4	74.6	76.8	73.0	76.8	69.1
77.1	63.9	77.3	66.7	68.5	69.5	70.1	68.1	73.5	72.1
72.7	74.5	69.7	73.7	72.9	72.3	73.2	73.4	69.9	74.5
69.8	77.1	72.0	68.4	71.9	70.5	68.2	73.6	66.6	68.4
71.3	72.0	73.6	66.0	70.7	70.7	70.4	75.7	74.3	70.7
69.5	69.3	74.5	70.9	68.4	69.0	75.5	76.8	68.5	60.7
78.5	72.5	70.4	70.2	68.0	80.0	75.5	76.5	64.4	68.5
65.9	77.1	68.1	68.9	71.4	69.9	69.3	73.8	61.8	75.2

where the summation is over the values $i = 1, 2, \ldots, n$. For the set of data in table 2.1.1. x-bar = 71.438.

The set of all possible daily yields of the plant forms a statistical *population*. Its average is the population mean, denoted by the greek letter μ. The population mean is often unknown and has to be estimated from the data. Our set of data is a sample of 80 observations from that unknown population. If the sample has not been taken under unusual or biased circumstances, the sample mean provides a good estimate of μ. (We say more about the criteria for "good" estimates in later chapters.) The mean represents the center of the data in the same way that the centroid, or center of gravity, represents the center of a distribution of mass along the x-axis.

The variance and standard deviation of x are measures of the spread of the data. We try to use uppercase letters, X, to denote the names of variables and lowercase, x, to denote values that the variables actually take, i.e., observations, but it is very hard to be absolutely consistent in this notation. For any observation x_i, the quantity dev $x_i = x_i - \mu$ is the deviation of the *i*th observation from the population mean. The average of the squares of the deviations is the variance of X. It is denoted by $V(X)$, or σ^2.

When μ is known, the variance is estimated from the sample by

$$\hat{\sigma}^2 = \frac{\sum (x_i - \mu)^2}{n}.$$

(The hat ˆ denotes that the expression is an estimate from the data; we could also have written $\bar{x} = \hat{\mu}$.)

Usually, we do not know μ, so we redefine dev x_i as $x_i - \bar{x}$, the deviation from the sample mean. Then the variance is estimated by the sample variance

$$s^2 = \frac{\sum (x_i - \bar{x})^2}{n - 1}.$$

The divisor is $n - 1$ rather than n because x-bar has been used instead of μ.

We see later that dividing by n would give a biased estimate—an estimate that is, on the average, too low.

The standard deviation, σ, is the square root of the variance. It is usually estimated by s, although we use a different method when we come to set up Shewhart charts. For this set of data, $s = 3.79$.

2.3. THE MEDIAN AND THE QUARTILES

The median is another measure of the center of the data. The sample median is a number with the property that there are as many points above it as below it. If n is odd, the sample median is clearly defined. For example, with only five observations, 1, 8, 4, 9, and 2, we arrange them in ascending order, 1, 2, 4, 8, and 9, and take the middle observation—the third one. If we only have four observations, 1, 8, 4, and 9, any number $4 \leq \tilde{x} \leq 8$ will satisfy the definition. A less ambiguous definition follows. Suppose that the data are written down in order and that $x_{(i)}$ is the ith largest observation. If $n = 2t + 1$, the median is defined as

$$\tilde{x} = x_{(t+1)}.$$

If $n = 2t$,

$$\tilde{x} = \frac{x_{(t)} + x_{(t+1)}}{2}.$$

In the example of table 2.1.1, $\tilde{x} = (x_{(40)} + x_{(41)})/2 = 71.35$.

Just as the median divides the data into two halves, the quartiles divide it into quarters. The median is the second quartile. The first quartile, Q_1, lies to the left of the median. Approximately one-quarter of the points are below it and approximately three-quarters are above it. We define Q_1 and Q_3 in the following way. As before, the points are arranged in ascending order. Let $u = (N+1)/4$. If u is an integer, then

$$Q_1 = x_{(u)} \quad \text{and} \quad Q_3 = x_{(3u)}.$$

If u is not an integer, we use linear interpolation. For example, when $N = 80$, $u = 20.25$, and so we define

$$Q_1 = x_{(20)} + 0.25(x_{(21)} - x_{(20)})$$
$$= \frac{3x_{(20)} + x_{(21)}}{4} = 68.925.$$

Similarly, $3u = 60.75$ and so Q_3 is defined by

$$Q_3 = x_{(60)} + 0.75(x_{(61)} - x_{(60)})$$
$$= \frac{x_{(60)} + 3x_{(61)}}{4} = 74.175.$$

Those are the definitions that the Minitab package uses, but they are not universal. Some other packages interpolate differently when u is not an integer. They place the quartile midway between the observations immediately above and below it. With $N = 80$, they define

$$Q_1 = \frac{x_{(20)} + x_{(21)}}{2} = 68.95,$$

and

$$Q_3 = \frac{x_{(60)} + x_{(61)}}{2} = 74.05.$$

There is another possibility for confusion about quartiles. We have defined a quartile as a particular value that separates one-quarter of the data from the remainder. Some authors refer to all the observations in the bottom quarter as constituting the first quartile. Others interchange Q_1 and Q_3, so that $Q_1 > Q_3$. High school counselors may say that you have to be "in the first quartile" of your high school class if you are to go on to college. They mean that you must be in the top 25% of the class.

These statistics are easily obtained on personal computers. In the Minitab package, the data and the results of calculations are contained on an imaginary work sheet. One sets the data in the first column and types the command DESCRIBE C1. The machine prints out the mean, as is shown in table 2.3.1, the standard deviation, median, quartiles, maximum, and minumum values and two other statistics that are introduced later, the trimmed mean (TRMEAN) and the standard error of the mean (SEMEAN).

Individual statistics can be obtained by typing MAXI C1, or MINI C1, etc. If you are in the habit of storing your data in computer files, it takes only one or two typewritten commands to read a data set into the work sheet.

Statistics such as the median, the quartiles, the maximum, and the minimum are called order statistics. The points that divide the data into ten

Table 2.3.1. Description of the Yield Data

N	MEAN	MEDIAN	TRMEAN	STDEV	SEMEAN
80	71.438	71.350	71.546	3.788	0.423
	MIN	MAX	Q1	Q3	
	60.700	80.000	68.925	74.175	

groups are called deciles, and we can also use percentiles. The ninety-fifth percentile may be defined as the observation in place $95(N+1)/100$, using linear interpolation if needed; it is exceeded by 5% of the data.

2.4. ISHIKAWA'S SEVEN TOOLS

Ishikawa (1976) proposed a kit called "The Seven Tools" for getting information from data. These are simple graphical tools, whose use is enhanced by modern computers. His seven tools are the following:

1. tally sheets (or check sheets),
2. histograms,
3. stratification,
4. Pareto diagrams,
5. scatter plots,
6. cause-and-effect diagrams, and
7. graphs.

Control charts fall under the last category, graphs. They involve no deep mathematics and are rarely mentioned in mathematical statistics courses, but their visual impact is simple and directly useful.

The tally chart has long been used for counting how many objects in a group fall into each of several classes. How many of the components that were rejected failed for which of several defects? How many votes were cast for each of several candidates? How many students in a large class got A, B, C, D, or F? The idea is simple. On a piece of paper, you make rows for A, B, ..., and then you take the exams one by one. If the first student got B, you make a vertical mark in the B row. If the second student failed, you make a mark in the F row, and so on. After you have gone through all the papers, you count the number of marks in each row, and there are the totals for the various grades. The counting is usually made easier by counting in groups of five. The first four marks are vertical; the fifth is a diagonal mark through those four, making the set of five look like a gate.

The tally chart is only one step removed from the histogram, which is the topic of the next section. It would be quite appropriate, after tallying, to present the breakdown of the grades as a histogram, or bar graph. You might also choose to stratify the data by making separate charts for men and women, or for graduates and undergraduates. The Pareto diagram is another type of bar chart; it is discussed in section 2.11.

Scatter plots demonstrate the relationships between two variables. For a batch of incoming students at a university, one can plot the entrance examination scores in English and mathematics with English along the vertical axis and mathematics along the horizontal. Each student is a dot on

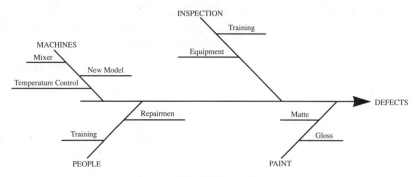

Figure 2.4.1. Fishbone diagram.

the graph. Is there a pattern? Do the points tend to lie on a line, showing that scores in the two subjects are highly correlated, or do they fall into the shotgun pattern that suggests independence? Again, one could use stratification, using one symbol for engineers and another for arts students. Would that suggest a pattern?

A cause-and-effect diagram looks like the diagram that students in English grammar classes used to draw to analyze sentences. Suppose, for example, that we are concerned about the number of defective items being produced. We draw a horizontal line on a piece of paper to denote the defect. Then we draw diagonal lines running to it to denote major areas that lead to that defect, e.g., men, materials, machines, and maintenance. Into each of these diagonal lines we draw lines for subareas and then lines for subsubareas, and so on. The resulting diagram looks like the backbone of a fish, and so these cause-and-effect diagrams are commonly called fishbone diagrams. An example is shown in figure 2.4.1.

2.5. HISTOGRAMS

One way in which a sample can be represented graphically is by using a histogram. The range of values taken by X is divided into several intervals, usually of equal width. A rectangle is then erected on each interval. The area of the rectangle is proportional to the number of observations in the interval.

Minitab prints the histogram sideways. Figure 2.5.1 shows the histogram for the data of table 2.1.1. It was obtained by typing the command HISTOGRAM C1.

The program has arbitrarily elected to use intervals of length 2.0 centered on the even integers. The first interval runs from 59.0 ($=60.0-1$) to 61.0 ($=60.0+1$). It contains one observation, 60.7. The second interval runs from 61.0 to 63.0 and also contains one observation, 61.8. As we have phrased it, the value 61.0 occurs in both intervals. There is no agreement on

HISTOGRAMS

```
MIDDLE OF       NUMBER OF
INTERVAL        OBSERVATIONS
   60.            1      *
   62.            1      *
   64.            2      * *
   66.            4      * * * *
   68.           12      * * * * * * * * * * * *
   70.           19      * * * * * * * * * * * * * * * * * * *
   72.           12      * * * * * * * * * * * *
   74.           14      * * * * * * * * * * * * * *
   76.            9      * * * * * * * * *
   78.            5      * * * * *
   80.            1      *
```

Figure 2.5.1. Histogram for table 2.1.1.

handling this. Minitab puts points on the boundary in the higher interval, so that the intervals are $59.0 < x \leq 61.0$, $61.0 < x \leq 63.0$ . . . . Other programs put them in the lower interval. Some authors suggest counting them as being half a point in each interval.

This difficulty is illustrated by figure 2.5.2, which shows another histogram for the same data set obtained with Statgraphics. It has a slightly different shape. Minitab has 12 observations in the interval centered on 68 and 19 in the interval centered on 70; Statgraphics has 13 in the former interval and 18 in the latter. There is an observation at the boundary, 69.0. Minitab puts it in the upper interval; Statgraphics puts it in the lower interval. Similarly, there are two observations at $x = 73.0$ that are treated differently in the two packages.

The difference between the two is a matter of choice. It is not a question of one writer being correct and the other wrong. The number of intervals is also arbitrary. Most software packages can give you intervals of your own choosing by a simple modification in the histogram command. For example, Statgraphics suggests 16 intervals for this data set, but it gives the user freedom to choose 10.

Several other things should be noted about these histograms. The distribution of the yields appears to follow the bell-shaped curve that is associated with the Gaussian, or normal, distribution. The normal distribution is discussed in later chapters; it is a symmetric distribution. The histograms of samples of normal data like that in table 2.1.1 are usually almost symmetric. The interval with the most observations is called the modal interval. In this case, the modal interval is centered on 70.0. We have already noted that the sample mean and the sample median (71.44 and 71.35, respectively) are close to one another; this too happens with samples of data from symmetric distributions.

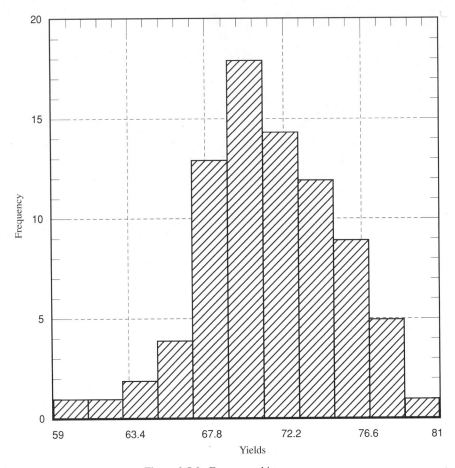

Figure 2.5.2. Frequency histogram.

2.6. AN EXAMPLE OF A NONSYMMETRICAL DISTRIBUTION

Some years ago a realtor listed in the *Austin American-Statesman* newspaper the asking prices of all the houses that he had on his list. They are given in thousands of dollars in table 2.6.1. The prices are listed in descending order. It is obvious that the two highest priced houses are markedly more expensive than the others.

Suppose that we can regard this list as typical of a realtor serving clients in the middle and upper income brackets in those days. The histogram, figure 2.6.1, shows those two houses appearing as outliers. In this case, there is a clear explanation of such outliers—there are always a few expensive houses on the market for the wealthy buyers. If these had been measurements of the yield of a chemical plant, we might have doubted that

AN EXAMPLE OF A NONSYMMETRICAL DISTRIBUTION

Table 2.6.1. House Prices (in Thousands of Dollars)

208.000	275.000	225.000	195.000	178.500	156.500
154.900	149.500	144.950	139.500	134.950	129.500
129.000	125.000	124.500	122.500	119.950	114.500
110.000	104.950	96.000	95.500	94.900	94.500
93.000	92.000	89.500	88.200	87.500	85.000
79.950	79.900	77.950	76.900	74.950	74.950
74.000	73.950	71.500	71.000	70.500	69.500
69.000	67.900	66.500	65.450	62.950	62.500
61.000	60.950	59.900	57.500	56.950	55.000
54.950	53.900	52.950	51.950	51.900	49.950
48.950	39.900				

they were valid and gone back to the records to check whether mistakes had been made in recording the observations. The histogram also shows that the distribution is not symmetric. It is skewed with a long tail to the bottom.

We saw earlier, in the case of a sample from a symmetric distribution, that the sample mean and median are close to one another. Each provides an estimate of a central value of the distribution. For a nonsymmetric distribution, one can argue about what is meant by a central value. A case can be made that, for an ordinary person coming to town, the median is a better estimate of how much one should expect to pay for a house than the mean. This is because the median is much less affected by outlying observations. If one takes a small sample of faculty incomes at an American university, the mean will vary considerably, according to whether or not the head football coach was included in the sample. On the other hand, the coach has less effect on the median.

MIDDLE OF INTERVAL	NUMBER OF OBSERVATIONS	
40.	3	* * *
60.	18	* * * * * * * * * * * * * * * * * *
80.	15	* * * * * * * * * * * * * * *
100.	7	* * * * * * *
120.	8	* * * * * * * *
140.	4	* * * *
160.	2	* *
180.	1	*
200.	1	*
220.	1	*
240.	0	
260.	0	
280.	2	* *

Figure 2.6.1. Histogram of house prices.

The mean and the median of the house prices are

$$\bar{x} = \$97{,}982 \quad \text{and} \quad \tilde{x} = \$79{,}925,$$

respectively.

When the two most expensive houses are removed, the mean drops to $91,998, a decrease of 6%. The median decreases to $78,925, a loss of only 1.25%. The reason is that the calculation of the median never involved the outlying values directly. The median moves only from $(x_{(31)} + x_{(32)})/2$ to $(x_{(30)} + x_{(31)})/2$, a small change in the middle of the data, where the observations are close together. This is why, even when estimating the population mean for a normal distribution, which is symmetric, some statisticians prefer the median over the sample mean. Although the sample mean can be shown to be a more precise estimate mathematically when all the observations are valid, it is more sensitive to outliers, some of which may be erroneous observations. The median is said to be a more robust statistic, or to provide a more robust estimate of μ.

Indeed, some scientists go further and use trimmed means. The DESCRIBE subroutine that was used in section 2.3 gave, among other statistics, the 5% trimmed mean of the yield data. This is obtained by dropping the largest 5% of the observations and the lowest 5% and finding the mean of the remaining 90%. In this example, the four highest and the four lowest observations were dropped; the trimmed mean is 71.546 as opposed to the mean of the whole set, 71.430.

The amount of trimming is arbitrary; some packages trim as much as the upper and lower 20% of the data. It can be argued that the median itself is essentially a 50% trimmed mean because all the observations are trimmed away save one! Sports fans will note that in some athletic contests, such as Olympic gymnastics, the rating of each competitor is a trimmed mean. Each competitor is evaluated by a panel of several judges; her highest and lowest

MIDDLE OF INTERVAL	NUMBER OF OBSERVATIONS	
3.6	1	*
3.8	1	*
4.0	10	* * * * * * * * * *
4.2	12	* * * * * * * * * * * *
4.4	12	* * * * * * * * * * * *
4.6	7	* * * * * * *
4.8	8	* * * * * * * *
5.0	6	* * * * * *
5.2	2	* *
5.4	1	*

Figure 2.6.2. Histogram of natural logarithms of house prices.

scores are discarded (trimmed); her rating is the average of the remaining scores. It has been suggested that this is a precaution against outliers caused by unduly partisan judges. It is, of course, important that the engineer decide on the level of trimming before looking at the data and not after seeing the observations and forming an opinion.

We have no reason to expect that the housing prices should follow the normal bell-shaped curve. Even when we omit the two highest houses, the curve is still skewed. If it is important to have a variable that has a bell-shaped distribution, we can try a transformation. One possibility is to replace the price by its natural logarithm. Figure 2.6.2 shows the histogram of ln(price) with the two highest houses omitted. This second histogram is more like a bell curve than the first.

2.7. DOTPLOTS

The dotplot is a simple graphical device that is rather like a quick histogram. It is very easy to make and is useful when there are fewer than 30 points. It is a simple diagram that can be made on any piece of paper without any fuss. The engineer draws a horizontal scale and marks the observations above it with dots. Figure 2.7.1 shows a dotplot of the last 10 yields in table 2.1.1.

Dotplots are also useful when comparing two samples. Figure 2.7.2 shows dotplots of the first and third rows of table 2.1.1 on the same horizontal scale; we see immediately that the averages of the two rows are about the same, but the third row has less variation than the first row. I selected the third row on purpose to illustrate this point: there may actually have been an improvement in the variance during that period, but it seems to have been temporary.

Figure 2.7.1. Dotplot of row 8.

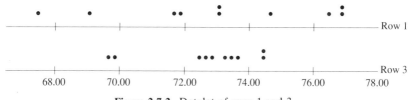

Figure 2.7.2. Dotplot of rows 1 and 3.

2.8. STEM-AND-LEAF DIAGRAMS

A variation of the histogram was introduced as a tool in exploratory data analysis by J. W. Tukey (1977). It is called the stem-and-leaf diagram. We now illustrate the procedure for the house price data in table 2.6.1. We begin by rounding the price of each house down to the number of thousands by trimming off the last three digits. In ascending order, the data are now 39, 48, 49, 51, 51, 52, 53, 54, ... The diagram is shown in figure 2.8.1.

The last digit of each trimmed observation is called a leaf. The other digits form the stems; 39 is leaf 9 on stem 3; 195 is leaf 5 on stem 19; 48 and 49 become stem 4 with leaves 8 and 9, respectively. The Minitab program, with the command STEM C1, separates 225, 275, and 280 at the end as high values.

There are two main differences between the histogram and the stem-and-leaf diagram. In the stem-and-leaf diagram, the digits are written as leaves. The histogram only has diamonds, stars, or rectangles. This can be considered a point in favor of the stem and leaf because it shows more information without adding more clutter to the picture.

On the other hand, the stems have to come away from the trunk at unit intervals. In the example, there has to be a stem every $10,000. This means

```
              STEM-AND-LEAF DISPLAY
              LEAF DIGIT UNIT = 1.0000
              1   2 REPRESENTS 12.

         1     3   9
         3     4   89
        12     5   112345679
        21     6   012256799
       (11)    7   01134446799
        30     8   5789
        26     9   234456
        20    10   4
        19    11   049
        16    12   24599
        11    13   49
         9    14   49
         7    15   46
         5    16
         5    17   8
         4    18
         4    19   5
```

Figure 2.8.1. Stem-and-leaf display of house prices.   HI   225, 275, 280

SPLITTING STEMS 21

that there tend to be more intervals in the stem-and-leaf diagram than in the histogram, which may, or may not, be a good idea.

In this example, the histogram suggests a smooth curve, with a maximum at about $71,000; the stem and leaf draws attention to the fact that the curve may have two maxima. There is only one house in the $100,000 to $110,000 range. This is a matter of sales technique—$99,950 is less alarming to a potential customer than $101,000. If there are not many data points, the stem-and-leaf diagram may have too many intervals, which makes it hard to determine the general shape of the distribution of the observations.

The first column of the figure gives running totals of the number of observations from the ends. By the time we have reached the end of the sixth stem, we have counted 21 observations from the low end of the data. The seventh stem contains the median; (11) indicates that the stem has eleven leaves. This makes it easier to compute the median. Once we have passed the median, we start to count from the high end of the data. At the end of stem 12 (the beginning of stem 13), there are 11 observations to go. Note that stems 16 and 18 have no leaves.

2.9. SPLITTING STEMS

If there are too many leaves on a stem, the diagram may not have room for them. Tukey has a method for splitting stems into five smaller stems. Figure 2.9.1 illustrates this for the chemical data in table 2.1.1.

Each yield is rounded down to an integer by trimming the last digit. The seventh stem has been split into five shorter stems. 7* contains observations 70 and 71; 7T contains 72 and 73; 7F contains 74 and 75; 7S contains 76 and

```
STEM-AND-LEAF DISPLAY
LEAF DIGIT UNIT = 1.0000
 1   2 REPRESENTS 12.

        LO   60

   2    6*   1
   3    6T   3
   5    6F   45
   9    6S   6667
  30    6.   888888888889999999999
 (14)   7*   00000000011111
  36    7T   2222222333333333
  20    7F   444445555
  11    7S   666667777
   2    7.   8
   1    8*   0
```

Figure 2.9.1. Stem-and-leaf display for table 2.1.1.

77; 7. contains 78 and 79. Notice Tukey's choice of symbols for the last digit of the integer: * with 0 and 1, T with 2 and 3, F with 4 and 5, S with 6 and 7, and, finally, . with 8 and 9.

2.10. BOX-AND-WHISKER PLOTS

Another of Tukey's graphical representations is the box-and-whisker plot. This plot emphasizes the spread of the data. It shows the median, the quartiles, which Tukey calls hinges, and the high and low values. The box-and-whisker plot of the chemical data is shown in figure 2.10.1.

The box is the rectangle in the middle of the diagram. The two vertical ends of the box are the quartiles, or hinges. The median is denoted by +. The asterisk at the far left denotes the lowest value, 60.7, which is considered to be unusually low by a criterion described in the next paragraph. $Q_3 - Q_1$ is called the interquartile range, or H-spread. $Q_3 + 1.5(Q_3 - Q_1)$ and $Q_1 - 1.5(Q_3 - Q_1)$ are called the upper and lower fences, respectively. Recall that for this data set, the median is 71.35, $Q_1 = 68.925$, and $Q_3 = 74.175$. The H-spread is $74.175 - 68.925 = 5.25$, and so the fences are $68.925 - 7.875 = 61.05$ and $74.175 + 7.875 = 82.05$.

The horizontal lines coming out from either side of the box are called whiskers. The whiskers protrude in either direction until they reach the next observed value before the fence. In this example, the right-hand whisker goes as far as the largest observation, which is 80.0. The left-hand whisker goes as far as the second lowest observation, 61.8. One observation, 60.7, lies beyond the fence; it is classified as a low value and marked with an asterisk.

In the box plot of the houses data, which is shown in figure 2.10.2, the three outliers are all on the high side.

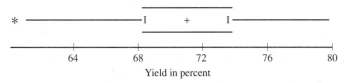

Figure 2.10.1. Box-and-whisker plot of chemical yields in table 2.1.1.

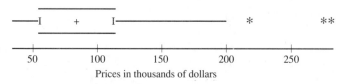

Figure 2.10.2. Box-and-whisker plot for houses data in table 2.6.1.

BOX-AND-WHISKER PLOTS

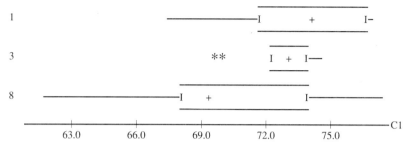

Figure 2.10.3. Multiple box-and-whisker plots for three samples.

We recall that the median is $79,925. The hinges are $Q_1 = \$62,840$, $Q_3 = \$123,000$, and the fences are at $\$62,840 - \$90,240 = -\$27,400$ and $\$123,000 + \$90,240 = \$213,240$. The lower fence is negative and we have no observations below it; the left whisker, therefore, stops at $39,900. The right whisker stops at the fourth highest observation, $195,000, leaving the three highest values marked by asterisks.

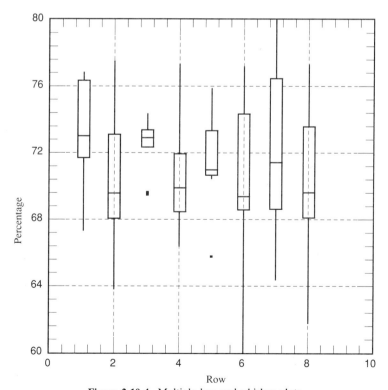

Figure 2.10.4. Multiple box-and-whisker plots.

It is sometimes convenient to make box and whisker plots of several samples on the same scale. Figure 2.10.3 shows such a multiple plot for rows 1, 3, and 8 of the data in table 2.1.1, considered as three separate samples.

With small samples, multiple box-and-whisker plots may raise more questions than they can answer. One should be careful not to read too much into the appearance that the variance for sample 3, as manifested by distance between the hinges, is much less than the variance for the other two samples. Remember that there are only ten observations in each of these samples.

Figure 2.10.4 goes a step further and shows, in a figure, box-and-whisker plots for each of the eight rows considered as separate samples.

2.11. PARETO DIAGRAMS

One of the main purposes of a quality assurance program is to reduce the cost of poor quality. An obvious step is to reduce the number of items produced that do not conform to specifications. We inspect some of the items produced and set aside those that are nonconforming or defective. In most processes, there are several types of defect that could cause an item to be rejected as nonconforming. We are not going to be able to eliminate every type of defect at once. Which should we work on first?

J. M. Juran, one of the pioneers of quality assurance, argued that only a few types of defect account for most of the nonconforming items. He called this phenomenon

"the rule of the vital few and the trivial many,"

and introduced Pareto analysis, which he named in honor of the Italian economist and sociologist Alfredo Pareto. The essence of the analysis is to record the number of each type of defect found in the nonconforming items that were rejected and then to make a chart rather like a histogram that will emphasize clearly the important types. Table 2.11.1 gives a set of data from a group of 97 nonconforming items from a production run. Note that there

Table 2.11.1. Analysis of Defects

Type of Defect	Number Reported	Percent	Cumulative Percent
Insulating varnish	54	40	40
Loose leads	39	29	68
Solder joint A	20	15	83
Solder joint B	9	7	90
Resistor 1	7	5	95
Resistor 2	5	4	99
Capacitor	2	1	100

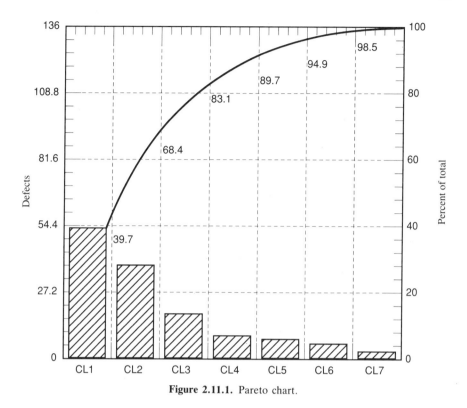

Figure 2.11.1. Pareto chart.

are more than 97 defects recorded because some items had more than one. The table shows that insulating varnish, loose leads, and the solder on joint A between them account for over 80% of the defects. The corresponding chart is figure 2.11.1.

2.12. SUMMARY

In this chapter, we have discussed some methods of presenting data sets graphically. The sets consisted of observations on a single variable, such as percent conversion or housing prices. The methods used in the main part of the chapter were histograms, dotplots, stem-and-leaf plots, and box-and-whisker plots. In later chapters, we mention other types of plots. Control charts appear in chapters nine, ten, and thirteen, followed by x–y plots for bivariate data and residual plots in the two chapters on regression. With a few easy commands on a personal computer, each of these plots gives the viewer a simple visual summary of the data that provides a picture of "how the data looks."

Not many years ago, allowing for false starts and interruptions along the way, it took an hour or more to calculate the mean and variance and to locate the maximum and minimum and potential outliers in a large set of data, and even longer to make a histogram. Perhaps that was an excuse for a busy engineer not taking the time to do it. Now that these simple graphics take only a few minutes at the terminal that is conveniently situated on one's desk, there is no excuse not to make them a routine first step in any analysis.

Always have a good look at your data set before you try to do anything complicated with it!

EXERCISES

2.1. During World War II, thin slices of mica were used as dielectric material in condensers for radios. The following data set is made up of measurements of the thickness of 100 strips of mica in one-thousandths of an inch. Make a histogram of the data and calculate the mean, median, and standard deviation.

15.7	13.5	15.6	15.3	14.0	16.0	16.2	14.1
16.9	16.5	16.2	13.7	14.8	15.2	10.9	15.1
17.4	14.5	15.5	17.5	15.9	16.7	15.8	12.5
13.3	11.0	14.2	11.8	15.6	14.4	13.6	12.8
14.3	14.7	14.8	16.4	15.8	19.0	13.6	16.5
13.7	18.0	13.6	14.4	17.2	15.9	13.4	16.3
16.3	13.5	15.1	16.6	14.5	15.1	14.5	18.2
16.4	15.0	14.0	17.2	15.0	15.6	13.4	13.6
15.4	14.8	12.6	16.6	12.1	17.6	14.7	16.8
15.5	12.6	13.1	15.4	13.4	14.8	15.1	16.2
13.6	13.9	15.5	14.3	13.8	13.4	15.0	14.2
15.7	12.7	15.8	18.3	16.1	14.3	18.0	17.2
15.0	17.2	14.9	15.5				

2.2. Verify that you would get the same results in exercise 2.1 if, for convenience, you subtracted 10.0 from each observation before making your calculations (and added it back to obtain the mean and median).

2.3. Verify that in exercise 2.1, there would be little change in the values of the mean, median, and standard deviation if you rounded off each observation to the nearest integer.

2.4. An engineer takes a sample of 19 wafers and measures the oxide thickness. The reported thicknesses, in angstroms, are

1831	1828	1831	1813	1816	1781	1874	1876	1847	1848
1839	1800	1850	1817	1832	1832	1578	1821	1823	

Obtain the mean, standard deviation, median, the trimmed mean, and the quartiles of the data. Make a dotplot and a box-and-whisker plot. In both plots, the observation 1578 is obviously an outlier.

2.5. Suppose that after investigation, the engineer decides, in exercise 2.4, that 1578 is an invalid observation and that he is justified in throwing it out. Do that and recalculate the mean, standard deviation, median, trimmed mean, and the quartiles for the remaining 18 wafers; make a new dotplot and a new box-and-whisker plot. The observation 1781 is the only observation below 1800; why is it not starred as an outlier in the box-and-whisker plot?

Note that when we drop the outlier, the median is unchanged and the trimmed mean and quartiles increase slightly. On the other hand, there are appreciable changes in both the mean and the standard deviation.

2.6. Make a stem-and-leaf diagram for the 18 observations in exercise 2.4 (after 1578 has been excluded) and use it to find the median and the quartiles.

2.7. Make a histogram of the following set of 50 observations and use it to find the median and the quartiles. Does the data set seem to have the traditional bell curve (the normal distribution)?

180	222	183	182	160	209	285	218	228	155
215	189	219	184	207	241	258	214	231	199
179	262	193	171	208	185	264	183	187	202
193	202	188	201	249	243	223	211	219	189
247	211	187	185	231	189	175	257	204	205

2.8. Make a box-and-whisker plot of the data in exercise 2.7. Find the hinges and fences. Explain why the maximum observation is flagged as a possible outlier, but the minimum observation is not.

2.9. Make a histogram of the following set of observations and use it to find the median and the mean. Does this set of data approximate a bell-shaped curve? The observations are the reciprocals of the observations in exercise 2.7 multiplied by 10^4 and rounded.

56	45	55	55	63	48	35	46	44	65
47	53	46	54	48	41	39	47	43	50
56	38	52	58	48	54	38	55	53	50
52	50	53	50	40	41	45	47	46	53
40	47	53	54	43	53	57	39	49	49

2.10. Make a box-and-whisker plot of the data in exercise 2.9.

2.11. An electric component is specified to have an average life of 1000 hours. Sixty of the components are put on test and their lifetimes are given below. Make a histogram of the data. Find the mean and median lifetimes of the components tested, together with the standard deviation and the quartiles. The average exceeds the specification by 27%, and 25% of the components lasted more than twice as long as specified. This sounds encouraging, but what fraction of the data fails to meet the specification? This suggests that it is not enough just to look at the sample average.

15	72	74	99	100	127	195	195
213	241	270	285	302	361	366	430
434	452	464	504	526	652	666	704
722	723	877	959	1024	1051	1075	1087
1108	1171	1316	1328	1410	1451	1463	1529
1576	1601	1631	1657	1893	1975	2059	2076
2217	2252	2316	2684	2858	2862	2873	3088
3159	3215	3832	4295				

2.12. A plant has ten machines working a single shift. In a week, the following numbers of defects were noted for the ten machines, with an average of 20 defects per machine

A 2	B 5	C 54	D 12	E 8
F 67	G 15	H 13	I 10	K 14

Make a Pareto diagram that provides a graphical illustration of the importance of focusing on machines C and F.

CHAPTER THREE

Discrete Variables, or Attributes

3.1. INTRODUCTION

Suppose that we have a process for manufacturing metal rods and that we are concerned about the uniformity of the rod lengths. An inspector takes a sample of rods from the production line and measures the lengths. The inspection can now proceed in two ways.

The inspector can record the actual length, Y, of each rod. Because the rods will not have exactly the same length, Y will not take the same value every time. Thus, Y is a random variable. It may, perhaps, take values anywhere in a certain range, in which case it is said to be a continuous random variable. In the jargon of quality control, the inspector is sampling by variables. Continuous variables are discussed in the next chapter.

A simpler alternative is sampling by attributes. A rod may be usable if its length lies in the interval $109 \text{ mm} \leq Y \leq 111 \text{ mm}$, but unusable if its length is outside that interval, either too long or too short. Hence, the inspector may classify each rod as either good or bad, acceptable or not acceptable. The customary terms are defective or nondefective. The inspector may record the number of defective rods, X, in the sample; X is said to be a discrete random variable. It takes only integer values $0, 1, 2, \ldots$, as opposed to values over a continuous range. We now develop some theoretical background for discrete variables.

3.2. RANDOM DIGITS

Table 3.2.1 shows 320 digits taken from the Austin phone book. I took the right-hand column from several pages (for convenience), and then I took the third digit after the prefix in each number, i.e., the last digit but one. The reason that I chose the third digit rather than the last was that I was not sure whether the phone company held back some numbers ending in zero for test purposes.

Table 3.2.1. 320 Digits Taken from the Phone Book

										Number of Zeros
1620	3076	9104	2476	4741	8350	0546	7957	1057	1301	7
6773	6761	2140	6699	0722	1449	9131	0229	8404	5198	4
4114	4255	8518	5270	2024	5518	5659	2440	0037	4590	6
2269	7063	6807	3992	3073	7781	0158	1042	8534	6081	6
7510	9485	6089	6298	7673	5272	0863	2953	8008	5253	5
4305	1573	0325	6370	3733	1663	2118	8593	2191	5343	3
1516	1589	8966	5308	6653	3902	1085	9706	1938	8764	4
8796	4879	6136	4241	8068	3472	0480	5054	8227	1013	5

It is reasonable to think that the digits ought to be evenly distributed, in the sense that each of the ten values is equally likely to occur. This implies that each value 0, 1, ..., 9 has probability 0.1.

3.3. A MATHEMATICAL MODEL

Suppose that we regard this data as a sample of the output from a process for producing digits. Let X denote the value of a digit. X takes the values 0, 1, 2, ..., 9 according to a probability model, or law; it is, therefore, called a *random variable*. We may also say that X has a particular probability distribution.

In this example, the probability law is simple. We write $P(X=0)$, or $P(0)$, for the probability that X takes the value zero, and, more generally, $P(X=i)$, or $P(i)$, for the probability that X takes the value i. The probability model for the process is given by

$$P(0) = P(1) = P(2) = \cdots = P(9) = 0.1.$$

The probability distribution of the random variable X in this example is called the multinomial distribution. It is a generalization of the binomial distribution that is discussed later. Some go further and call it a *uniform discrete distribution* because all the probabilities are equal.

We do not claim that, as a consequence of this probability law, exactly one-tenth of the digits in our sample must be 0, one-tenth must be 1, another tenth must be 2, and so on. That is clearly not so. Table 3.3.1 shows the actual frequencies of occurrence of the digits. A plot of them appears as

Table 3.3.1. Frequencies of Occurrences of Digits

0	1	2	3	4	5	6	7	8	9
40	36	29	33	29	34	32	30	31	26

SOME BASIC PROBABILITY THEORY

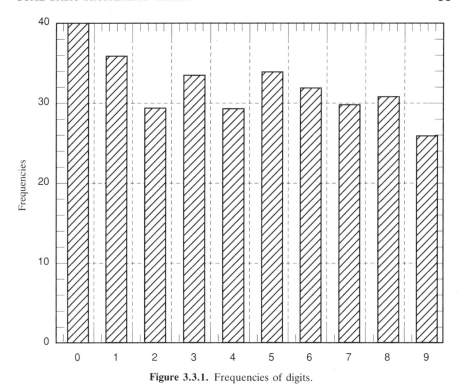

Figure 3.3.1. Frequencies of digits.

figure 3.3.1. On the other hand, we expect that each digit should appear about one-tenth of the time.

Out of the total of 320 digits in the sample, each digit should have occurrred about 32 times, and yet 0 occurred 40 times and 9 appeared only 26 times. Is this unreasonable? The answer to that question can be shown to be no. The deviations of the frequencies from 32 that we see in this particular set of data are not unreasonable. In section 7.15, we present a statistical test for the hypothesis that this set of data did indeed come from a process with equal probabilities.

We should also expect that out of the 40 digits in each row, the number 0 should occur four times. It occurs more often in the first row than in any of the others. In only one of the rows are there fewer than four zeros. Is that unreasonable?

3.4. SOME BASIC PROBABILITY THEORY

We can also look at this from the point of view of an experiment. The experiment consists of picking a digit. It has ten possible results, or outcomes. Similarly, tossing a die can be regarded as an experiment with six outcomes. In the general case, the set of possible results of the experiment is called the sample space and is denoted by S. For a discrete variable, S is a set of discrete points. The number of points in S may be either finite or infinite. An example in which S has an infinite number of points appears in section 3.16.

A probability, or, more precisely, a probability measure, is assigned to each point of S. These probabilities satisfy two requirements. They are nonnegative and their sum is unity. One can picture a 1-pound jar of probability jam being ladled out, not necessarily in equal portions, among the various points. Some might get no jam at all; those points have zero probability.

An event, E, is a set of points in S. For example, the event E_1 might be the event $X \leq 3$, which consists of the points 0, 1, 2, and 3. An event is said to occur if the result, or outcome, of the experiment is one of the points in the set. The probability that the event occurs is the sum of the probabilities of the points in the set. Thus, in the digit example, $P(E_1) = P(X \leq 3) = P(0) + P(1) + P(2) + P(3) = 0.4$.

The event E^c is the complement of E; it is the set of points of S that are not included in E and corresponds to the event that E does *not* occur. It follows that $P(E^c) = 1 - P(E)$. In this case, E^c is the event $X > 3$. It contains six points. Its probability is $0.6 = 1 - P(E)$.

We can now define unions and intersections of events. Suppose that E_2 is the event that X is odd, i.e., $X = 1$ or 3 or 5 or 7 or 9. The union of E_1 and E_2 is written $E_1 \cup E_2$. It is the set of all points that are in either E_1 or E_2, or both; it occurs if either E_1 *or* E_2 occurs. The intersection, or product, of E_1 and E_2 is written as $E_1 \cap E_2$. It is the set of points that are in both E_1 and E_2; it occurs only if both E_1 *and* E_2 occur. In this example, $E_1 \cup E_2$ is the set (0, 1, 2, 3, 5, 7, 9) with probability 0.7; $E_1 \cap E_2$ is the set (1, 3) with probability 0.2.

In computing the probability of $E_1 \cup E_2$, we must be careful not to count twice points 1 and 3, which occur in both sets. As another example, we can consider an ordinary deck of playing cards. It consists of 52 cards, of which 4 are aces and 13 are spades. Let E_1 be the event of drawing a spade, and E_2 be drawing an ace. $P(E_1) = 13/52$ and $P(E_2) = 4/52$, but $P(E_1 \cup E_2) = 16/52$, not $(13 + 4)/52$. The latter answer is wrong because the ace of spades has been counted twice. The ace of spades is the only card in $E_1 \cap E_2$; $P(E_1 \cap E_2) = 1/52$. The following formula expresses this:

$$P(E_1 \cup E_2) = P(E_1) + P(E_2) - P(E_1 \cap E_2) \qquad (3.4.1)$$

Example 3.4.1. A simple part in an electronic system can have two kinds of defect, either an input defect or an output defect. Of the parts offered by a supplier, 20% have an input defect, 30% have an output defect, and 10% have both defects. What is the probability that a part will have no defects?

Let E_1 be the event that a part has the input defect, and let E_2 be the event that a part has the output defect. We are given that $P(E_1) = 0.20$,

THE BINOMIAL DISTRIBUTION

$P(E_2) = 0.30$, and $P(E_1 \cap E_2) = 0.10$. The probability that a part will have at least one defect is, from equation 3.4.1,

$$P(E_1 \cup E_2) = 0.20 + 0.30 - 0.10 = 0.40.$$

It follows that the probability that a part will be free of defects is $1.00 - 0.40 = 0.60$. □

3.5. THE COIN TOSSING MODEL

Table 3.5.1 is based on the data in Table 3.2.1. This time the digits have been classified as odd (1, 3, 5, 7, 9) or even (0, 2, 4, 6, 8). The random variable recorded in Table 3.5.1 is the number of even digits in each set of four. The first entry in the table is 3, since the first set of four digits in Table 3.2.1 is 1620, which contains three even numbers.

The probability that a digit from the set 0, 1, 2, 3, 4, 5, 6, 7, 8, 9 will be even is one-half, the same as the probability that the digit will be odd.

The mathematical model for drawing a digit at random from that set of possible values is the same as the model for tossing a fair coin. Let X take the value 1 if the digit is even, and the value 0 if the digit is odd. Alternatively, toss a fair coin once, and let X denote the number of heads that appear; $X = 1$ if the coin shows a head, and $X = 0$ if the coin shows a tail.

In both examples, the behavior of X follows the probability law

$$P(0) = P(1) = 0.5.$$

This is the simple coin tossing model.

3.6. THE BINOMIAL DISTRIBUTION

The random variable, X, that is observed in table 3.5.1 is the number of even digits in a group of 4. This is equivalent to observing the number of

Table 3.5.1. 320 Digits Taken in Groups of 4 (Number of Even Digits)

																				Number of Twos
3	2	2	3	2	1	3	0	1	1	1	2	3	2	3	2	0	4	3	1	6
2	2	2	1	4	1	1	4	2	2	4	2	3	1	1	1	2	3	2	3	8
1	2	3	3	1	2	3	1	4	1	2	0	2	2	0	2	2	1	1	1	7
1	0	2	2	2	2	3	2	1	2	2	2	2	3	4	2	4	2	3	1	11

Table 3.6.1. Frequencies of Occurrence of Values of X

X =	0	1	2	3	4
Observed frequency	5	21	32	15	7
We expect	5	20	30	20	5

heads that occur when a coin is tossed four times; X takes the five values 0, 1, 2, 3, 4, but not with equal probabilities. The numbers of times that these five values occurred are shown in table 3.6.1 and in figure 3.6.1.

The probability, $P(0)$, that X takes the value 0 is the same as the chance of tossing four tails in succession. We multiply the probabilities of tails on the individual tosses.

$$P(0) = (\tfrac{1}{2})(\tfrac{1}{2})(\tfrac{1}{2})(\tfrac{1}{2}) = \tfrac{1}{16} = P(4).$$

The sample space, S, has 16 points. They are equally likely; each has probability 1/16. The points are

HHHH, HHHT, HHTH, HHTT, HTHH, HTHT, HTTH, HTTT

THHH, THHT, THTH, THTT, TTHH, TTHT, TTTH, TTTT

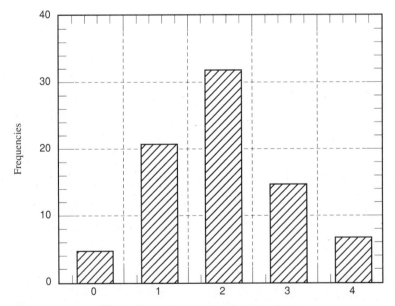

Figure 3.6.1. Frequencies of values of X.

$X = 0$ for the point TTTT. Exactly one head in four tosses occurs at each of the following four points:

$$\text{HTTT}, \quad \text{THTT}, \quad \text{TTHT}, \quad \text{TTTH},$$

while $X = 3$ at

$$\text{HHHT}, \quad \text{HHTH}, \quad \text{HTHH}, \quad \text{THHH}.$$

Thus, $P(1) = P(3) = 4/16$.

Finally, exactly two heads occur at six points:

$$\text{HHTT}, \quad \text{HTHT}, \quad \text{HTTH}, \quad \text{THHT}, \quad \text{THTH}, \quad \text{TTHH},$$

and so $P(2) = 6/16$.

An alternative version of the sample space would be the five points, 0, 1, 2, 3, 4, corresponding to the values taken by X with probabilities 1/16, 4/16, 6/16, 4/16, 1/16, respectively. In a set of 80 observations, we should expect $X = 2$ to occur $(6/16)(80) = 30$ times.

The random variable X is said to have a binomial distribution. This topic is developed further in the next two sections.

Example 3.6.1. A fair coin is tossed five times. What are the probabilities that $X = 0$, $X = 3$, and $X = 5$?

There are $2^5 = 32$ points in the sample space; $X = 0$ occurs only at the point TTTTT; $P(0) = 1/32 = P(5)$; $X = 3$ requires three heads and two tails; three places out of five for the heads can be found in $5!/(3!2!) = 10$ ways. Hence, $P(3) = 10/32$. □

3.7. UNEQUAL PROBABILITIES

The binomial distribution was introduced in the previous section from the coin tossing model. The probability of a head at each toss is 0.5. That is a special case of a more general model. Again, we use the data in table 3.2.1. This time, replace all the nonzero digits by ones. Now we have a sample of data from a process that produces 10% zeros and 90% ones. We can think of it as representing the output from a production process that produces 10% defective parts, represented by zeros, and 90% that are nondefective. One might ask the following question. If the data were given in the form of a sequence of zeros and ones, how would you decide whether the process is operating under control at a 10% defective rate?

In the general case, the statistician thinks of a process, or a "sequence of trials," in which each trial has exactly two outcomes, called success or failure. Associated with each trial is a random variable X, which takes the value one if the trial is a success and the value zero if the trial is a failure. The probability of success is the same for each trial and is denoted by p. The probability of failure at each trial is $1 - p$, which is often denoted by q.

In the coin tossing model with "heads" as the successful outcome, $p = q = 0.50$. If we were to classify a defective as a failure, the example in the first paragraph of this section would have $p = 0.90$, and $q = 0.10$. As a practical matter, quality control engineers are more interested in focusing upon the number of defectives produced by a process than on the number of nondefectives. They apply this model with defectives classified as successes, and with the probability of any part being defective denoted by p. Thus, they would have $p = 0.10$ in the example and $q = 0.90$.

We have described what is known to statisticians as a sequence of Bernoulli trials, named after the Swiss mathematician James Bernoulli (1654–1705). The idea that the trials are independent with constant probability of success is important to the argument. If we toss a fair coin nine times and observe tails each time, the probability of a head on the tenth throw is still 0.5. It would not be reasonable to argue that, since the probability of one head and nine tails in ten throws is greater than the probability of ten tails, the chance of a head on the tenth is greater than a half. The system is postulated to have no memory. It is not reasonable to bet in favor of heads on the tenth throw.

Indeed, a case could be made for the argument that a sequence of nine tails in succession is strong evidence that the original assumption of $p = 0.5$ is false. This would lead one to conclude that p is considerably less than 0.5 and to bet on tails for the tenth toss. That argument takes us into the area of hypothesis testing and decision making, which is the topic of chapter seven.

3.8. THE BINOMIAL DISTRIBUTION (GENERAL CASE)

Suppose that we have a series of n independent Bernoulli trials with probability p of success at each trial, and let X denote the total number of successes in all n trials. X is said to have a binomial distribution with parameters n and p.

We saw in table 3.5.1 an example of a binomial variable with $n = 4$ and $p = 0.5$. We argued, for example, that $X = 1$ occurs for four sequences, each of which has the letter H once and the letter T three times. Each sequence has probability $1/16$, and so $P(1) = 4/16$.

By generalizing this for $p \neq 0.5$, the probability of a sequence with one head and three tails is pq^3, and so

$$P(1) = 4pq^3.$$

To compute $P(2)$, we multiplied p^2q^2 by 6, which is the number of sequences with H twice and T twice, i.e., the number of ways of choosing two places out of four for the successes. In the language of permutations and combinations, this is the number of combinations of two objects out of four. If we wish to have x successes in n trials, the number of sequences available is

$$C(n, x) = \frac{n!}{x!(n-x)!}$$

and the probability that $X = x$ is

$$P(x) = \frac{n! p^x q^{n-x}}{x!(n-x)!} .$$

In quality control work, the values of n may be rather large, and it is hoped that the values of p are small. Under those circumstances, it is easier, and adequate, to use the Poisson distribution as an approximation to the binomial. The Poisson distribution is discussed in sections 3.17–3.19.

3.9. MATHEMATICAL EXPECTATION

We said earlier that out of 320 random digits, we should expect to have 32 zeros. In practice, we should not be surprised to see a few more or a few less. The mathematical expectation, or expected value, of a random variable X is the long-term average of X, weighted according to the probability law. Mathematically, it is defined by the center of gravity of the probability distribution. It is denoted by $E(X)$. For a discrete variable, the formula for $E(X)$ is

$$E(X) = \sum x P(x) , \qquad (3.9.1)$$

where the summation is made over all values x that X takes.

Consider throwing a single die. Let X be the number that shows on the top face. X takes the values 1, 2, 3, 4, 5, 6, each with probability 1/6. Then

$$E(X) = \tfrac{1}{6} + \tfrac{2}{6} + \tfrac{3}{6} + \tfrac{4}{6} + \tfrac{5}{6} + \tfrac{6}{6} = 3.5 .$$

We do not believe that we shall ever see the die showing three and one-half points. We do state that if we were to throw a fair die for hours on end, then the average number of points per throw would be approximately 3.5.

In a series of Bernoulli trials, let X_i denote the number of successes on the ith trial:

$$P(X_i = 1) = p \quad \text{and} \quad P(X_i = 0) = q ,$$
$$E(X_i) = (1)(p) + (0)(q) = p . \qquad (3.9.2)$$

For a binomial distribution, X is the sum of n independent variables X_1, $X_2, \ldots$, and so

$$E(X) = E(X_1) + E(X_2) + \cdots = np . \qquad (3.9.3)$$

The following properties of expectations are easily verified.
If a is some constant,

$$E(aX) = aE(X) . \qquad (3.9.4)$$

This is merely a matter of change of scale. If X has been measured in yards and if Y is the corresponding measurement in feet, then $Y = 3X$, and so $E(Y) = 3E(X)$.

If X_1, $X_2, \ldots$ are random variables with expectations m_1, $m_2, \ldots$, respectively, and if $Y = a_1 X_1 + a_2 X_2 + \cdots$, then

$$E(Y) = a_1 m_1 + a_2 m_2 + \cdots = \sum a_i m_i . \qquad (3.9.5)$$

3.10. THE VARIANCE

We are now able to give a more formal definition of the variance of a random variable. This definition is compatible with the definition that was given in chapter two.

Definition. The variance of a random variable X is

$$V(X) = \sigma^2 = E\{[X - E(X)]^2\} . \qquad (3.10.1)$$

In practice, we note that

$$\begin{aligned} E\{[X - E(X)]^2\} &= E(X^2) - 2E(X) \cdot E(X) + [E(X)]^2 \\ &= E(X^2) - [E(X)]^2 . \end{aligned} \qquad (3.10.2)$$

For a single toss of a coin,

$$E(X^2) = 1^2 \cdot p + 0^2 \cdot q = p ,$$

and

$$E(X) = p ,$$

whence

$$V(X) = p - p^2 = p(1 - p) = pq .$$

For independent variables, we can develop properties of the variance that parallel equations (3.9.4) and (3.9.5).

If a and b are constants,

$$V(aX + b) = a^2 V(X), \qquad (3.10.3)$$

and if $X_1, X_2, \ldots$ are independent random variables with expectations $m_1, m_2, \ldots$, respectively, and if $Y = a_1 X_1 + a_2 X_2 + \cdots$, then

$$V(Y) = a_1^2 V(X_1) + a_2^2 V(X_2) + \cdots = \sum a_i^2 V(X_i). \qquad (3.10.4)$$

Applying equation (3.10.4), we see that if X is a binomial variable with n trials and probability p, then

$$V(X) = V(X_1) + V(X_2) + \cdots = npq. \qquad (3.10.5)$$

□

3.11. SAMPLING WITHOUT REPLACEMENT

In acceptance sampling by attributes, an inspector draws a sample of n items from a lot, or batch, of N items and observes the number, X, of defectives in the sample. We have argued until now that the probability of drawing a defective part at any stage of the sampling is a constant, p, which is independent of the parts that have been sampled so far. This was the basis for deriving the binomial distribution.

However, the sample can be drawn in two ways. If the inspector returns each item to the batch after examination and then draws the next item from all N members of the batch as before, we have *sampling with replacement*. Clearly, the probability of drawing a defective item remains constant because the composition of the batch from which the inspector is choosing is unchanged. With this scheme, it is quite possible for the same item to be counted more than once before the inspector has reached the quota.

However, if the testing of the items is destructive, as, for example, stretching a piece of yarn until it breaks in order to determine its strength, sampling with replacement is impossible. The alternative is *sampling without replacement*. The inspector sets each item aside after it has been tested, and the next item is drawn from the remainder of the batch. Suppose that a batch consists of only six items, of which three are defective. When the first item in the sample is chosen, $p = 3/6$. If the first item chosen were defective, there would now be five items left in the batch, of which only two are defective; for the second item, we should have $p = 2/5$. If the first item were nondefective, we should have, for the second item, $p = 3/5$. In computing p for the second item, the denominator is $N - 1$, where N is the size of the original sample; the numerator depends on the result of the first drawing. The correct model would not be the binomial model, but a model based on the change in batch size and content. That model is called the hypergeometric law, or the hypergeometric distribution.

On the other hand, suppose that a large university has 45,000 students, of whom 30,000 are men and 15,000 are women. The probability that a student chosen at random will be female is 1/3. When a second student is chosen, the probability of selecting a woman will be either $14,999/44,999 = 0.3333185$, or $15,001/44,999 = 0.333363$, both of which are close enough to 1/3 for practical purposes. When the batch size, N, and the number of defectives are not very small, the binomial model is still a good enough approximation to the real world to justify its use, even if we sample without replacement.

3.12. THE HYPERGEOMETRIC DISTRIBUTION

Example 3.12.1. Returning to the batch of six items, of which three are defective, suppose that a random sample of three items is drawn. What is the probability that it will contain two defectives and one nondefective?

We can choose two defectives out of three in three ways, which is the number of combinations of two items out of three. Similarly, we can choose one nondefective out of three in three ways. Altogether, we can choose a sample of three out of six in 20 ways, the number of combinations of three out of six. Of these 20 ways, $3 \times 3 = 9$ give us the sample that we seek, and so the desired probability is 9/20. □

That is a simple example of the hypergeometric distribution. In the general case, suppose that we have N items of which d are defective and $N - d$ are nondefective, and that we draw a sample of n items. Let x be the number of defective items in the sample. We wish to find the probability law, or distribution, for $P(x)$.

The sample of n items contains x out of d defectives and $n - x$ of $N - d$ nondefectives. We can choose x defectives out of d in $C(d, x)$ ways, $n - x$ nondefectives out of $N - d$ in $C(N - d, n - x)$ ways, and a sample of n items out of N in $C(N, n)$ ways, so that

$$P(x) = \frac{C(d, x)C(N - d, n - x)}{C(N, n)} \qquad (3.12.1)$$

The algebraic expression in the numerator resembles an expression that appears in classical mathematics in a term in the hypergeometric series, hence, the name hypergeometric for this distribution. It can be shown that the sum of these probabilities for the various values of X is indeed unity and that $E(X) = nd/N$.

3.13. CONDITIONAL PROBABILITIES

Suppose that in the previous problem, A denotes the event that the first item drawn is defective, and B is the event that the second item is defective. Let A^c denote the complement of A, i.e., the event that A did *not* occur; in this case, A^c is the event that the first item is nondefective.

If the first item is defective, the probability that the second item is also defective is $2/5$. We write this as $P(B|A) = 2/5$. The notation $P(B|A)$ means the conditional probability of event B, given that event A has occurred. Similarly, $P(B|A^c) = 3/5$, and $P(A|B) = 2/5$, $P(A|B^c) = 3/5$.

Event B can occur in two distinct ways; either it is preceded by A or else it is preceded by A^c. Hence,

$$P(B) = P(A)P(B|A) + P(A^c)P(B|A^c). \tag{3.13.1}$$

Substituting the values given above, we calculate

$$P(B) = (\tfrac{1}{2})(\tfrac{2}{5}) + (\tfrac{1}{2})(\tfrac{3}{5}) = \tfrac{1}{2}.$$

Let AB denote the joint event that both items are defective (we drop the symbol $\cap$). We see that $P(AB) = 1/5 = P(A)P(B|A)$.

This is an example of the basic law of conditional probability:

$$P(AB) = P(A)P(B|A) = P(B)P(A|B). \tag{3.13.2}$$

Mathematically, we define $P(B|A)$ as the ratio $P(AB)/P(A)$; the ratio has no meaning if $P(A) = 0$. If $P(B|A) = P(B)$, then $P(A|B) = P(A)$ and $P(AB) = P(A)P(B)$. In that case, we say that the events A and B are (statistically) *independent*.

Example 3.13.1. A supplier is manufacturing electronic components. The components each contain two resistors. In 12% of the components, the first resistor is defective. In 40% of the components, the second resistor is defective. In 8% of the components, both resistors are defective. An inspector examines a component and finds that the first resistor is defective. What is the probability that the second resistor will be defective also?

Let A be the event that the first resistor is defective and B the event that the second is defective. We see from the data that $P(A) = 0.12$, $P(B) = 0.40$, and $P(AB) = 0.08$.

$$P(B|A) = \frac{P(AB)}{P(A)} = \frac{0.08}{0.12} = 0.67.$$

The two events are not independent. They would have been independent only if we had $P(AB) = (0.12)(0.40) = 0.048$. □

3.14. BAYES' THEOREM

Example 3.13.1 leads us in the direction of Bayes' theorem. It was proved by the Rev. Thomas Bayes, and published in 1763, after his death. The theorem seems at first to be just a routine exercise in manipulating formulas. Its importance has been realized in the last half century. It enables us to make use of prior knowledge in our probabilistic calculations. It is called by various authors Bayes' theorem, Bayes' rule, and Bayes' formula.

Suppose that there are two experiments. The first has outcomes A and A^c. The second has outcomes B and B^c. We notice that B has occurred. What is the probability that A has occurred also? We wish to calculate the conditional probability $P(A|B)$. From equation (3.13.2),

$$P(A|B) = \frac{P(AB)}{P(B)} = \frac{P(A)P(B|A)}{P(B)},$$

and applying equation (3.13.1),

$$P(A|B) = \frac{P(A)P(B|A)}{P(A)P(B|A) + P(A^c)P(B|A^c)}. \quad (3.14.1)$$

More generally, if the first experiment has a set of outcomes $A_1, A_2, \ldots$, the theorem becomes

$$P(A_i|B) = \frac{P(A_i)P(B|A_i)}{\sum P(A_h)P(B|A_h)}. \quad (3.14.2)$$

In equation (3.14.1), $P(A)$ is the prior probability of A. It is the probability of the occurrence of A before we know whether or not B has occurred. On the other hand, $P(A|B)$ is the posterior probability of A. It is our revised value of the probability of the occurrence of A in the light of the additional information that B has occurred. If A and B are independent events, the knowledge that B has, or has not, occurred would provide no information about the occurrence of A, and that probability would remain unchanged.

Example 3.14.1. Your company uses a large number of a certain electronic component. You purchase most of them from a supplier in Dallas who is on record as producing only 5% defectives. Unfortunately, he can only supply 80% of your need, and so you have to buy the remaining 20% from another supplier in Houston who makes 10% defectives. You pick a component at random. It is defective, but you cannot read the serial number. What is the probability that it came from the supplier in Dallas?

Let A denote the event, or outcome, that the component came from Dallas. Let B be the event that the component is defective. If you did not know whether the component was defective, you would argue that $P(A) = 0.80$.

By equation (3.13.1),

$$P(B) = (0.80)(0.05) + (0.20)(0.10) = 0.06 .$$

This is the overall rate of defectives for the supply of components entering your plant. Applying Bayes' theorem,

$$P(A|B) = \frac{(0.80)(0.05)}{0.06} = 0.67 .$$

Because of the comparatively good quality record of the Dallas supplier, the knowledge that the component is defective reduces the odds that it came from Dallas and increases the chance that it came from Houston. □

Another example occurs in medical statistics. Some diagnostic tests are not perfect. They may give a small percentage of false positives. In medicine, this means that the test may indicate, in some cases, that a patient has a disease when that is not so. This can lead to unfortunate consequences.

Example 3.14.2. Suppose that one young male in a thousand has a certain disease that has no obvious unique symptoms, and that the routine test has probability 0.005% (one-half of one percent) of false positives. A young man, chosen at random from the population, takes the test and is rated positive. What is the probability that he actually has the disease?

Let A be the event that the patient has the disease; in this example, the prior probability is $P(A) = 0.001$. Let B be the event that he tests positive. If we assume that there are no false negatives, $P(B|A) = 1.000$, $P(B|A^c) = 0.005$.

By equation (3.13.1),

$$P(B) = (0.999)(0.005) + (0.001)(1.000) = 0.005995 ;$$

then, applying Bayes' theorem,

$$P(A|B) = \frac{(0.001)(1.000)}{0.005995} = 0.1668 .$$

Only one-sixth of the men who test positive actually have the disease! □

$P(A|B)$ is strongly influenced by the rate of false positives and the rareness of the disease. If $P(A)$ were 0.0001 (one in ten thousand) and the rate of false positives were the same, $P(A|B)$ would fall to about 2%. It is important to notice the assumption that the young man was chosen at random from the whole population. Every man in the population had the same chance of being chosen, whether or not he had the disease. This might be an appropriate model for routine medical exams for all the potential inductees to the armed forces under conscription or of a large company that routinely gives medical examinations to all potential employees as a condition of employment. It would not be so appropriate if he was chosen because he was already in a doctor's office or a clinic as a consequence of feeling ill. In that case, $P(A)$ might be larger. The reader with experience of test procedures can think of similar examples in industrial work.

3.15. BIVARIATE DISTRIBUTIONS

An object can sometimes be classified in two ways. In section 3.14, we saw an example in which a component could be classified on the one hand as having come from Dallas or Houston, and on the other as being defective or nondefective. This is an example of a bivariate distribution. Now there are two variables: location and quality. Each variable can take two "values." We usually prefer to say that each has two levels. In this section, we use data from a similar example to illustrate some of the basic ideas of bivariate distributions.

These data show the output of a plant that makes rods and is not very good at it. The data have been modified to make the arithmetic simpler. Each rod is subjected to two tests: length and strength. Each test classifies the rod as either OK or defective. The data can be arranged in a 2×2 table: two rows and two columns. A sample of 150 rods gives the results that are shown in table 3.15.1.

The totals have been written at the rightmost column of the table and along the bottom row—"in the margins." The corresponding probabilities are called the marginal probabilities.

Denote the two variables by X_1 and X_2. In our example, X_1 pertains to length and X_2 pertains to strength. They take the values 0, 1: zero if the test result is OK, and one if the rod is defective. The table has four cells. The ijth cell corresponds to $X_1 = i$, $X_2 = j$; it contains n_{ij} observations. There are N observations altogether. The probability that X_1 takes the value i and X_2

Table 3.15.1

Length	Strength		Total
	OK	Defective	
OK	90	10	100
Defective	30	20	50
Total	120	30	150

takes the value j is $p_{ij} = n_{ij}/N$. In this example, $p_{00} = 90/150$; $p_{01} = 10/150$; $p_{10} = 30/150$; and $p_{11} = 20/150$.

The marginal totals are denoted by $n_{i.}$ and $n_{.j}$; $n_{i.}$ is the total for the ith row of the table; $n_{.j}$ is the total for the jth column. The corresponding marginal probabilities are

$$p_{i.} = \frac{n_{i.}}{N} \quad \text{and} \quad p_{.j} = \frac{n_{.j}}{N}. \tag{3.15.1}$$

In the example, $p_{1.} = 0.2$ is the probability that a rod is defective in length.

The conditional probabilities are defined in terms of the cell probabilities and the marginal probabilities:

$$P(X_2 = j | X_1 = i) = \frac{p_{ij}}{p_i}. \tag{3.15.2}$$

It is sometimes appropriate to ask whether the two variables are independent. Are the two types of defects independent or do they tend to go hand in hand? Let A be the event that a bar passes the length test, and B be the event that it passes the strength test. The events are independent if $P(AB) = P(A)P(B)$. We rephrase that and say that the variables are independent if, and only if,

$$p_{ij} = p_{i.} p_{.j} \quad \text{for every } i \text{ and every } j. \tag{3.15.3}$$

In the example, we have

$$p_{00} = 0.60 \neq (\tfrac{2}{3})(\tfrac{4}{5}).$$

With independence, we should have expected to have had $(2/3)(4/5)(150) = 80$ parts passing both tests. This suggests that there might be a tendency for parts that fail one test to fail the other also. We return to this point when we discuss the chi-square test for 2×2 tables later, in section 8.12.

3.16. THE GEOMETRIC DISTRIBUTION

Suppose that we wish to obtain an estimate of the percentage of nonconforming, or defective, items produced by a manufacturing process. An obvious procedure is to take a batch of items, inspect them, and then argue that the percentage of defectives in the batch is a good estimate of the actual process percentage. We discuss this type of problem in more detail in chapters six and seven.

The mathematical argument is that we inspect N items and find X defectives. X has a binomial distribution with parameters p and N; its expected value is $E(X) = Np$, and so we estimate p by X/N. This procedure fixes the number of items inspected and, thus, the cost of obtaining the estimate.

Another approach to the problem would be to inspect items one at a time until we found a defective. Let X be the number of items inspected (including the defective). We may then estimate p by $1/X$. The probability distribution of X is called the geometric distribution. The sample space, S, consists of all the positive integers 1, 2, 3, ..., and is infinite.

Mathematically, we derive the geometric distribution from a sequence of Bernoulli trials with the probability of success being p, and let X be the number of trials until the first success. This means that we have to have a sequence of $X - 1$ failures followed by the lone success, and so

$$P(x) = q^{x-1}p . \qquad (3.16.1)$$

The infinite series $P(1) + P(2) + P(3) + \cdots$ is a geometric series with first term p and common ratio q. Its sum is

$$\frac{p}{1-q} = \frac{p}{p} = 1 .$$

We can show that $E(X) = 1/p$ and $V(X) = q/p^2$.

The negative binomial distribution is an extension of the geometric distribution. We continue until we have had r successes. In that case,

$$P(x) = C(x - 1, r - 1)p^r q^{x-r} \qquad (3.16.2)$$

with $E(X) = r/p$ and $V(X) = rq/p^2$. The negative binomial variable is merely the sum of r independent geometric variables.

3.17. THE POISSON DISTRIBUTION

The last discrete distribution that we shall discuss is the Poisson distribution, named after the famous French mathematician D. L. Poisson (1781–1840). It is featured in the c charts in chapter ten.

Poisson's mathematical contribution was to obtain an approximation to the binomial distribution when p is small. It is known to some mathematicians as the Poisson limit law. The Poisson approximation is commonly used by engineers in acceptance sampling. This aspect of the Poisson distribution is one of the topics of the chapters on acceptance sampling (chapters 11 and 12).

In 1898, Bortkiewicz found a use for Poisson's distribution in his research. He was investigating the death by accident of soldiers in the Prussian cavalry. He observed ten corps of cavalry over a period of 20 years and

PROPERTIES OF THE POISSON DISTRIBUTION 47

recorded for each of the 200 corps-years the number of men who died from being kicked by a horse. What probability law did his data follow?

One may make this kind of loose argument. Each corps contained about the same number, n, of soldiers, each of whom had a small probability, p, of dying from such an accident. The number of deaths, X, has roughly a binomial distribution, with $E(X) = np$, and $V(X) = npq$. If p is small, we can replace q by unity and write $V(X) = np = E(X)$. We do not know either n or p, but we can write $np = \mu$ and use Poisson's approximation to the binomial. This says that X has a discrete distribution with $X = 0, 1, 2, 3, \ldots$, and

$$P(X = x) = p(x) = e^{-\mu} \frac{\mu^x}{x!}. \tag{3.17.1}$$

We recall that $0! = 1$, and so $p(0) = e^{-\mu}$.

This is the Poisson distribution. It has a single parameter, μ. Since $E(X) = \mu$, we can estimate μ by taking several samples and letting

$$\hat{\mu} = \bar{x}.$$

3.18. PROPERTIES OF THE POISSON DISTRIBUTION

We now confirm that equation (3.17.1) defines a probability distribution. Summing equation (3.17.1) from $x = 0$ to infinity, we have

$$\sum p(x) = e^{-\mu} \frac{\sum x^\mu}{x!} = e^{-\mu} e^\mu = 1.$$

$$E(X) = \sum x p(x) = e^{-\mu} \left(0.1 + 1 \cdot \mu + \frac{2 \cdot \mu^2}{2!} + \frac{3 \cdot \mu^3}{3!} + \cdots \right)$$

$$= e^{-\mu} \mu \left(0 + 1 + \mu + \frac{\mu^2}{2!} + \frac{\mu^3}{3!} + \cdots \right)$$

$$= \mu. \tag{3.18.1}$$

Similarly,

$$E[X(X-1)] = \mu^2. \tag{3.18.2}$$

It follows that

$$E(X^2) = \mu^2 + \mu,$$

and so

$$V(X) = E(X^2) - [E(X)]^2 = \mu^2 + \mu - \mu^2 = \mu. \tag{3.18.3}$$

Table 3.19.1. Poisson Distribution of Deaths

x	Frequency of Observations	fx	Fitted Value
0	109	0	108.7
1	65	65	66.3
2	22	44	20.2
3	3	9	4.1
4	1	4	0.6
	200	122	

3.19. THE BORTKIEWICZ DATA

The data are given in table 3.19.1. The last column of fitted values is explained in the following.

There were 200 observations with a total of 122 deaths. We estimate μ by $x = 0.61$. To make the fit to the data, we substitute 0.61 for μ. The $p(0) = e^{-0.61} = 0.54335$. Out of 200 observations, we should expect $200(0.54335) = 108.7$ to be zero. Similarly, the expected frequency of $x = 1$ is $(200)(0.61 e^{-0.61}) = 66.3$, and so on. The Poisson model gives a very good fit to this set of data.

3.20. OTHER APPLICATIONS

For quite a while, this model seemed to be used only for rare events, such as the number of deaths of centenarians in any month in a big city, and had no place in industry. Eventually, it began to be a popular and useful model for situations in which we count the number of occurrences of some event that we regard as happening at random with a certain average frequency. Examples are the number of cancerous cells per unit area of a section of tissue in a biopsy, or the number of misprints per page in a book, or the number of ships arriving per day in a seaport, or the number of chromosome interchanges in cells. Industrial examples include the number of meteorites that collide with a satellite during an orbit, the number of breaks in a spool of thread, and the number of failures of insulation in a length of electric cable. Other examples in the literature have inspectors taking sheets of pulp and counting the number of dirt specks greater than 0.1 mm^2 in area or counting knots in sheets of plywood.

3.21. FURTHER PROPERTIES

It can be shown that if X_1, X_2 are independent Poisson variables with parameters μ_1 and μ_2, their sum, $X_1 + X_2$, has a Poisson distribution with parameter $\mu_1 + \mu_2$. This is the additive property of the Poisson distribution.

It follows that if $X_1, X_2, \ldots, X_k$ are independent Poisson variables, each with parameter μ_i, and Y is their sum, then Y has a Poisson distribution with parameter $k \Sigma (\mu_i)$.

EXERCISES

3.1. A device consists of six lamps in parallel. There is light as long as at least one of the lamps works. The probability that any given lamp works is 0.95. What is the probability that at least one of the lamps works?

3.2. When two die are thrown, there are 36 possible results $(1, 1)$, $(1, 2), \ldots, (6, 6)$, each with probability $1/36$. Let Y denote the sum of the numbers on the two die. What are the probabilities of the following events?
(a) $Y = 7$,
(b) $Y = 11$,
(c) Y is odd,
(d) the two numbers are the same,
(e) $Y > 8$,
(f) both (d) and (e) occur simultaneously,
(g) both (c) and (d) occur simultaneously.

3.3. A certain electronic device is rather difficult to make. Two-thirds of the devices that we make are defective. What is the probability that, in a sample of six devices, (a) exactly two are defective? (b) At least two are defective?

3.4. Ten judges are each given two brands of coffee, A and B, to taste. If there is no real difference between the two brands, what is the probability that exactly five judges will prefer A and exactly five will prefer B?

3.5. The following results are reported by the registrar of a college. Last year, 40% of the freshmen failed mathematics and 30% failed English; 80% failed at least one of the two subjects. What is inconsistent in that report? The amended report from the registrar said that 50% of the freshmen failed at least one of the two subjects. What percentage of the students passed English but failed mathematics?

3.6. Suppose that the thicknesses of slices of mica made by a machine are symmetrically distributed about the median thickness, and that the manufacturer specifies a median thickness of 0.0015 in. A quality control engineer measures ten slices. Eight of them are thicker than specified. What is the probability that as many as seven slices out of ten will be thicker than the amount specified if the slicing machine is working properly?

3.7. A batch of eight devices contains three defectives. An inspector chooses two of them at random without replacement and tests them. What is the probability that both are defective?

3.8. Another batch of eight devices also contains three defectives. An inspector tests one device. If it is defective, he throws it out; if it is OK, he puts it back. A second inspector comes along later and he, too, tests one device. What is the probability that he chooses a defective device? Note that this is not the same as sampling with replacement, because the first inspector only replaces the device that he tests if it is good.

3.9. A device has two critical components, A and B. Failures in the device are accompanied by failures in one or both components. Component A is found to have failed in 6% of the devices; component B in 8%. Both components have failed in 4% of the devices. Are the events (a) A has failed and (b) B has failed independent? What is the conditional probability that if A has failed, B has failed also? Suppose that the results came from examining 1000 devices. Construct a table like table 3.15.1 to present the data.

3.10. A piece of test equipment detects 99% of the defective parts that it sees. Unfortunately, it also erroneously declares 1% of the good parts to be defective. Our process produces 4% defective items. Suppose that an item is declared to be defective after testing. What is the probability that it actually is defective? Repeat the exercise assuming that the tester catches only 95% of the defectives and misclassifies 8% of the good items.

3.11. A plant has three machines. Machine A makes 200 items a day with 4% defective, B makes 300 items a day with 5% defective, and C makes 400 a day with 2% defective. At the end of the day, the outputs of the three machines are collected in one bin. An item is taken from the bin and tested. What is the probability that it is defective? If it is defective, what is the probability that it came from machine A?

3.12. A batch of ten items is known to contain exactly one defective. Items are tested one at a time until the defective is found. Let X be the number of items tested. Calculate $E(X)$. Generalize your result to the case where there is one defective in a batch of n items.

3.13. A quality control engineer takes two samples each day from a production line and computes the average of some characteristic. If the process is on specification, there is a probability $p = 0.50$ that the value will be above the specified value, which is the center line on his chart, and that the probability will be 0.50% of a value below the line. He decides that he will flag the machine for overhaul over the weekend if there are either seven (or more) out of ten points above the line that week or seven (or more) out of ten below the line. What is the probability that he will unnecessarily flag the machine for overhaul if the process is on specification?

3.14. A plan for examining lots of incoming items calls for an inspector to take a sample of six items from the lot. He will reject the lot if there is more than one defective item in the sample. If the vendor is producing items with 10% defectives, what is the probability that a lot will pass inspection? You may assume that the lot size is so large that it is appropriate to use the binomial distribution rather than the hypergeometric distribution.

3.15. Let X denote the number that is shown when a single die is thrown. X takes the values 1, 2, 3, 4, 5, and 6, each with probability 1/6. Calculate $E(X)$ and $V(X)$.

3.16. Let Y denote the sum of the numbers on two die as in exercise 3.1. Calculate $E(Y)$ and $V(Y)$.

3.17. Calculate $E(X)$ and $V(X)$ for the geometric distribution.

CHAPTER FOUR

Continuous Variables

4.1. INTRODUCTION

In the previous chapter, we considered discrete variables whose sample spaces are sets of discrete points. We turn now to continuous variables whose sample spaces are intervals. A continuous random variable, X, might, for example, take any value in the interval $4 < x < 8$. The picture of the contents of a 1-lb jar of probability jam being allotted in lumps changes. Now we picture the probability being spread over the interval in a layer whose thickness may vary, as one might spread jam unevenly over a piece of bread. One can also have hybrid variables that take values over both intervals and discrete points, sometimes continuous and sometimes discrete. To consider them would take us beyond the scope of this book.

The most important of the continuous probability laws, or distributions, is the normal, or Gaussian, distribution. That is the topic of the next chapter. In this chapter, we discuss the exponential and gamma distributions. The reader who is pressed for time may prefer to postpone some of the sections at the end of the chapter and then refer back to them as needed when reading the later chapters.

4.2. PROBABILITY DENSITY FUNCTIONS

The thickness of the layer of probability jam in our analogy is formally called the *probability density function*. The density at the value $X = x$ is denoted by $f(x)$. The sample space, S, is defined in terms of intervals. We shall usually consider random variables for which S is either the whole line from $-\infty$ to $+\infty$ or the half line from 0 to $+\infty$. In theory, we could have variables for which S consists of a number of disjoint intervals.

Events are subintervals or collections of subintervals of S. The probability that X takes a value in the interval $a < x < b$ is defined by the integral

$$P(a < x < b) = \int_a^b f(x)\, dx. \qquad (4.2.1)$$

PROBABILITY DENSITY FUNCTIONS

Since the probability of an event has to be nonnegative and the probability that X takes a value somewhere in the whole space S is unity, the following conditions are imposed on the density function:

(i) $$f(x) \geq 0 \tag{4.2.2}$$

for all x in S;

(ii) $$\int_S f(x)\,dx = 1, \tag{4.2.3}$$

where the integral is taken over the whole sample space.

It is convenient to adopt the convention that $f(x)$ is zero for values of X that are not in S. One advantage is that formulas such as that in the condition of equation (4.2.3) can be written as the integral from $-\infty$ to $+\infty$. The portions where $f(x) = 0$ make zero contribution to the value of the integral.

Example 4.2.1. A random variable takes values in the interval $0 \leq x \leq +2$. The probability density function is proportional to x. Find $f(x)$ and calculate the probability that $x > 1.0$.

We have $f(x) = kx$, where k is a constant to be calculated. From equation (4.2.3)

$$\int_0^2 kx\,dx = 1,$$

where $k = 1/2$.

$$P(x > 1.0) = \int_1^2 \frac{x}{2}\,dx = \frac{3}{4}.$$

Note that it would have made no difference had we asked for $P(x \geq 1.0)$. The integral that evaluates the probability is the same in both cases. If, in equation (4.2.1), we take the limit as b approaches a, the integral shrinks to zero. This implies that the probability that X takes any single value, as opposed to the set of values in an interval, is zero. □

The probability density functions usually contain constants. It is customary to denote them by Greek letters, such as θ in the uniform and exponential distributions that follow in later sections of this chapter, and μ and σ in the normal distribution. They are called the parameters of the distribution. In practice, their values are unknown, and one of the main problems in statistics is to obtain estimates of the parameters from a set of data.

4.3. THE CUMULATIVE DISTRIBUTION FUNCTION

In general, the probability density function defines a graph $y = f(x)$ that lies on or above the x-axis. (There may be cases when the graph of y is a collection of pieces of curves or a step function.) The defintion of equation (4.2.1) defines probabilities in terms of integrals, or areas under the curve.

The cumulative distribution function of X (we usually omit the word cumulative) is defined by

$$F(x) = \int_{-\infty}^{x} f(x)\, dx. \tag{4.3.1}$$

It is the area under the curve between the lowest value in S and x. Alternatively, we may define $F(x)$ by

$$F(x) = P(X \le x), \tag{4.3.2}$$

and equation (4.2.1) may be modified to read

$$P(a < X < b) = F(b) - F(a).$$

The distribution function, $F(x)$, satisfies the following conditions:

(i) $F(x) \ge 0$ for all x; (4.3.3)
(ii) $F(x)$ is a nondecreasing function, i.e., if $b > a$, $F(b) \ge F(a)$; (4.3.4)
(iii) $F(-\infty) = 0$, $F(+\infty) = 1.0$. (4.3.5)

Example 4.3.1. A random variable has density function

$$\begin{aligned} F(x) &= x & 0 \le x \le 1, \\ &= 2 - x & 1 \le x \le 2, \\ &= 0 & \text{elsewhere}. \end{aligned}$$

Find $F(x)$ and evaluate $P(0.5 \le x \le 1.5)$.

The graph of $y = f(x)$ is an isosceles triangle with a vertex at the point $(1, 1)$. For the left-hand portion of the triangle,

$$F(x) = \int_0^x x\, dx = \frac{x^2}{2}.$$

We may take the lower limit as zero because $f(x) = 0$ for $x \le 0$. For the right-hand portion,

$$F(x) = \int_0^1 x\,dx + \int_1^x (2-x)\,dx = 0.5 + 0.5 - \frac{(2-x)^2}{2}$$

$$= 1 - \frac{(2-x)^2}{2}.$$

$$P(0.5 \le X \le 1.5) = F(1.5) - F(0.5) = \tfrac{7}{8} - \tfrac{1}{8} = \tfrac{3}{4}. \qquad \square$$

Some writers prefer to develop the theory by starting with the distribution function and then introducing the density function as its derivative.

4.4. THE UNIFORM DISTRIBUTION

The simplest continuous distribution is the uniform distribution on the interval $(0, 1)$. It is immaterial whether we use the open interval or the closed interval in the definition. For this distribution,

$$\begin{aligned} f(x) &= 1 \quad 0 < x < 1, \\ &= 0 \quad \text{elsewhere}. \end{aligned} \qquad (4.4.1)$$

This density function obviously satisfies the conditions of section 4.2. The distribution function is

$$\begin{aligned} F(x) &= 0 \quad x < 0, \\ &= x \quad 0 < x < 1, \\ &= 1 \quad x \ge 1. \end{aligned} \qquad (4.4.2)$$

We will see later that if X_1 and X_2 are two independent random variables, each of which has a uniform distribution on $(0, 1)$, their sum, $X = X_1 + X_2$, has the distribution that we used before in example 4.3.1.

There are two common generalizations of the uniform distribution on $(0, 1)$.

The uniform distribution on the interval $(0, \theta)$ has

$$\begin{aligned} f(x) &= \frac{1}{\theta} \quad 0 < x < \theta, \\ &= 0 \quad \text{elsewhere}. \end{aligned} \qquad (4.4.3)$$

More generally, the uniform distribution on the interval $\theta_1 < x < \theta_2$ with $\theta_2 > \theta_1$ has

$$\begin{aligned} f(x) &= \frac{1}{\theta_2 - \theta_1} \quad \theta_2 < x < \theta_1, \\ &= 0 \quad \text{elsewhere}. \end{aligned}$$

A special case occurs when the interval is symmetric about zero, i.e., $\theta_1 = -\theta$ and $\theta_2 = +\theta$.

4.5. THE EXPONENTIAL DISTRIBUTION

The exponential distribution is commonly used in reliability work. It is a good model for the lifetimes of electronic components that do not receive the unusual stresses of sudden surges in current or voltage and have a constant failure rate. Indeed, it is sometimes called the constant failure rate (CFR) model.

The density function is defined as

$$f(x) = \left(\frac{1}{\theta}\right) e^{-x/\theta} \qquad x \geq 0. \qquad (4.5.1)$$

The probability density function for an exponential distribution with $\theta = 100$ is shown in figure 4.5.1. The condition of equation (4.2.2) is clearly satisfied, because, when $x \geq 0$, we have $f(x) > 0$.

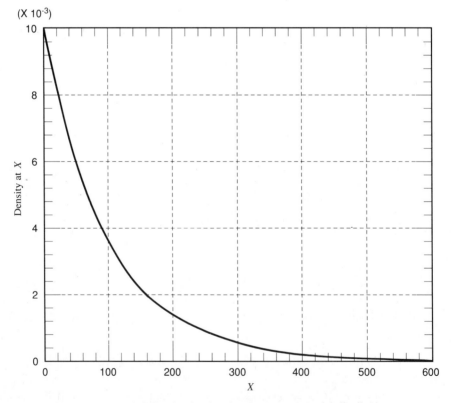

Figure 4.5.1. Probability density function for an exponential distribution.

THE EXPONENTIAL DISTRIBUTION

We confirm that equation (4.2.3) is satisfied by showing that $F(+\infty) = 1$. Integrating $f(x)$ gives us

$$F(x) = 1 - e^{-x/\theta}. \qquad (4.5.2)$$

The limit, as x tends to infinity, of $F(x)$ is indeed unity as required.

It will be shown later that the parameter θ has a physical meaning. It is the population average lifetime of the components. One can rewrite equation (4.5.1) as follows:

$$f(x) = \alpha e^{-\alpha x} \qquad x > 0. \qquad (4.5.3)$$

The only difference between equation (4.5.1) and equation (4.5.3) is that we have written α instead of $1/\theta$. The new parameter α is the death rate, i.e., the rate, measured in components per unit time, at which the components are failing. Whether one uses equation (4.5.1) or equation (4.5.3) to define the distribution is a matter of personal preference. We will stay with equation (4.5.1).

Example 4.5.1. The lifetimes of electrical components made to a certain set of specifications have an exponential distribution with an average lifetime (θ) of 1000 hours. What percentage of them will fail before 500 hours? What percentage will last for more than 1500 hours?

We have

$$F(x) = 1 - e^{-x/1000}.$$

$F(500) = 1 - e^{-0.5} = 0.393$; 39.3% of the components will fail before 500 hours;
$1 - F(1500) = e^{-1.5} = 0.223$; 22.3% will last longer than 1500 hours. □

Example 4.5.2. Specifications call for 95% of the center sections of the wings of a new type of fighter plane to last for 6000 hours. If the lifetimes have an exponential distribution, what should θ be? It was reported in the press that the average lifetime appeared to be only 2000 hours. If that report is correct, what percentage of the wings will last for the desired 6000 hours?

We have $\theta = 2000$, so that $F(6000) = 1 - 0.04978$. By 6000 hours, 95% of the wings will have failed. Only 5% of the wings will last for the required length of time! □

There is a difficulty with this example. How do you estimate the average lifetime of the wings? One possibility is to wait until all the wings have failed and then calculate the average. That is not a very practical solution because by the time that the last wing has failed, there are no planes left. Even if one were to take a sample of the planes and wait until they all failed, the last one might take quite some time. Perhaps the reporter was confusing the median with the mean. We return to this example in the next section.

4.6. THE QUARTILES OF A CONTINUOUS DISTRIBUTION

The quartiles are simply defined in terms of the distribution function. The median of X is the value for which $F(x) = 0.5$. The quartiles, Q_1 and Q_3, are the values for which $F(x) = 0.25$ and 0.75, respectively.

Example 4.6.1. Find the median and the quartiles of the exponential distribution.

The median is obtained by solving the equation

$$1 - e^{-x/\theta} = 0.5,$$

or

$$e^{-x/\theta} = 0.5.$$

Taking natural logarithms of both sides,

$$-\tilde{x}/\theta = \ln(0.5) = -\ln(2),$$

whence $\tilde{x} = \theta \ln(2) = 0.6932\theta$. Similar calculations show that $Q_1 = 0.2877\theta$ and $Q_3 = 1.3863\theta$.

We see here that the parameter θ is a scale factor. When θ is doubled, the median and the quartiles are doubled too. □

Example 4.5.2 (cont.). What if the reporter had said that after 2000 hours, half the wings had failed? This statement implies that the median is 2000, so that we can estimate θ as $2000/0.6932 = 2885$ hours. Using this value for θ, we can calculate that about 12% of the wings would last for 6000 hours. □

4.7. THE MEMORYLESS PROPERTY OF THE EXPONENTIAL DISTRIBUTION

An important property of this probability model is that the system has no memory. If a lamp has lasted until now, the probability that it will last for another 100 hours is the same as the probability of lasting another 100 hours was last week, or the probability that it would have lasted for 100 hours from the time that it was first switched on. Formally,

$$P(X > (a_1 + b|X > a_1) = P(X > (a_2 + b|X > a_2)$$

for all values of a_1 and a_2.

This can be established by using some of the probability concepts that were introduced in the previous chapter. Let X denote the lifetime of a component. Let A be the event that the component lives at least until time a, and B be the event that it lives at least until time $a + b$. Thus, we have A: $X > a$, and B: $X > a + b$. Events A and B can occur if, and only if, B occurs, so that AB is the same as B. We have to show that $P(B|A) = P(X > b)$.

$$P(A) = P(X > a) = 1 - F(a) = e^a.$$
$$P(AB) = P(B) = e^{a+b}.$$

Then, by equation (3.12.2),

$$P(B|A) = \frac{P(AB)}{P(A)} = e^b = 1 - F(b).$$

4.8. THE RELIABILITY FUNCTION AND THE HAZARD RATE

Two other probability functions are commonly used by reliability engineers. The reliability function, $R(x)$, is the probability that an item will survive to time x. It is, therefore, defined by

$$R(x) = 1 - F(x). \qquad (4.8.1)$$

For the exponential distribution (constant failure rate),

$$R(x) = e^{-x/\theta}; \qquad (4.8.2)$$

$R(\theta) = 0.368$, and so only 36.8% of the items will survive past time θ.

The hazard rate $h(x)$ is the instantaneous failure rate. It is defined in the following way: $h(x) dx$ is the incremental probability that the component will fail in the time period x to $x + dx$, given that it has lasted until time x. The computation of $h(x)$ uses the conditional probabilities. Going through all the stages of the calculation and taking the limit is similar to computing a derivative using the step process. We end up with the formula

$$h(x) = \frac{f(x)}{1 - F(x)} = \frac{f(x)}{R(x)}. \qquad (4.8.3)$$

The exponential distribution has a constant hazard rate:

$$h(x) = \frac{(1/\theta)e^{-x/\theta}}{e^{-x/\theta}} = \frac{1}{\theta}.$$

4.9. THE WEIBULL DISTRIBUTION

We have just seen that the exponential distribution has a constant failure rate and a constant hazard rate. We now modify the hazard rate to make it time-dependent. If we take the simplest form of the exponential distribution,

$$f(x) = e^{-x} = \exp(-x),$$

and replace x by t^β, the element of probability becomes

$$dP = f(x)\,dx = \exp(-t^\beta)\beta t^{\beta-1}\,dt,$$

and the probability density function of t is

$$f(t) = \beta t^{\beta-1}\exp(-t^\beta), \tag{4.9.1}$$

and $F(t) = 1 - \exp(-t^\beta)$;

$$R(t) = \exp(-t^\beta)\,;\, h(t) = \beta t^{\beta-1}.$$

This is the basic Weibull distribution.

If we write t/α instead of just t, we obtain a more general density function:

$$f(t) = \frac{\beta}{\alpha}\left(\frac{t}{\alpha}\right)^{\beta-1}\exp\left[-\left(\frac{t}{\alpha}\right)^\beta\right] \tag{4.9.2}$$

and

$$R(t) = \exp\left[-\left(\frac{t}{\alpha}\right)^\beta\right]\,;\, h(t) = \frac{\beta}{\alpha}\left(\frac{t}{\alpha}\right)^{\beta-1}.$$

If we set $t = \alpha$, we have $R(\alpha) = e^{-1} = 0.368$; α is called the characteristic life. Like the parameter θ in the exponential distribution, it is the lifetime by which 63.2% of the devices have failed.

Readers who are interested in reliability will also want to read about the lognormal distribution in section 5.12.

Example 4.9.1. Suppose that the lifetime of a piece of equipment has a Weibull distribution with $\alpha = 15$ years and $\beta = 2$. These items are warranted to last for 3 years. What percentage is still alive at the end of the warranty period?

$$R(t) = \exp\left[-\left(\frac{t}{\alpha}\right)^\beta\right];$$

$$R(3) = \exp\left[-\left(\frac{3}{15}\right)^2\right] = 0.9608.$$

Only 4% of the items do *not* survive the warranty period. □

4.10. THE GAMMA DISTRIBUTION

Suppose that we put on test n items that have exponential lifetimes. The sum of the lifetimes and also the average lifetime have gamma distributions. The chi-square distribution, which is very important in later chapters, is a special case of a gamma distribution.

The name gamma comes from the gamma function and is defined by

$$\Gamma(m) = \int_0^\infty x^{m-1} e^{-x}\, dx \qquad m > 0. \tag{4.10.1}$$

Integrating by parts, we have the reduction formula

$$\Gamma(m) = (m-1)\Gamma(m-1). \tag{4.10.2}$$

When m is an integer, we keep applying the reduction formula until we arrive at

$$\Gamma(m) = (m-1)!\,\Gamma(1) = (m-1)!, \tag{4.10.3}$$

where $(m-1)!$ denotes the factorial

$$(m-1)(m-2)(m-3)\cdots(3)(2)(1).$$

That is the usual definition of the factorial. When m is not an integer, it is conventional to define $(m-1)!$ as $\Gamma(m)$. When m is not an integer, we cannot evaluate the gamma function by the methods of elementary calculus.

The gamma distribution is defined for a continuous random variable that takes nonnegative values. The density function is

$$f(x) = \frac{x^{\alpha-1} e^{-x/\beta}}{\beta^\alpha (\alpha-1)!}, \tag{4.10.4}$$

with $\alpha > 0$ and $\beta > 0$.

When $\alpha = 1$, the gamma distribution reduces to the exponential distribution with $\theta = \beta$. When $\alpha > 1$, the distribution is skewed with a maximum

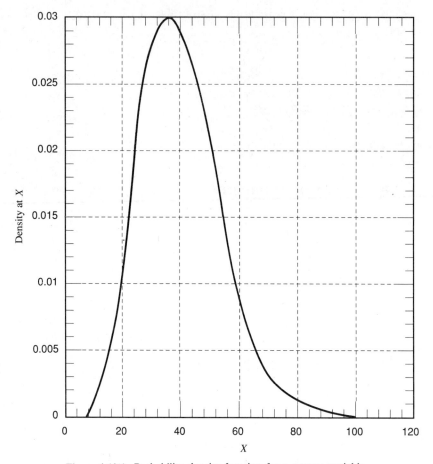

Figure 4.10.1. Probability density function for a gamma variable.

(mode) at $x = \beta(\alpha - 1)$ and a long tail on the right as x goes to infinity. The probability density function for a gamma variable with $\alpha = 8.0$ and $\beta = 5.0$ is shown in figure 4.10.1.

4.11. THE EXPECTATION OF A CONTINUOUS RANDOM VARIABLE

In section 3.9, we defined the expectation of a discrete random variable by the formula of equation (3.9.1),

$$E(X) = \sum xP(x).$$

For a continuous variable, the corresponding definition is

$$E(X) = \int_{-\infty}^{\infty} xf(x)\, dx. \tag{4.11.1}$$

THE EXPECTATION OF A CONTINUOUS RANDOM VARIABLE

This, too, can be thought of as the long-term average or the center of gravity of the probability distribution. It is sometimes called the mean of the distribution, and is often denoted by μ.

Example 4.11.1. Find the expectation of a variable with the exponential distribution.

$$E(X) = \frac{1}{\theta} \int_0^\infty x e^{-x/\theta} \, dx$$

$$= \int_0^\infty e^{-x/\theta} \, dx$$

$$= \theta . \qquad (4.11.2)$$

□

Example 4.11.2. Let X have a gamma distribution with parameters α and β. Find $E(X)$.

$$E(X) = \int_0^\infty \frac{x^\alpha e^{-x/\beta}}{\beta^\alpha (\alpha - 1)!} .$$

The numerator has the form of the gamma function with α instead of $\alpha - 1$, so that

$$E(X) = \frac{\beta^{\alpha+1} \alpha!}{\beta^\alpha (\alpha - 1)!} = \alpha \beta . \qquad (4.11.3)$$

□

More generally, the expectation of a function $g(X)$ is defined by

$$E[g(X)] = \int_{-\infty}^{+\infty} g(x) f(x) \, dx . \qquad (4.11.4)$$

The moments of a distribution are defined in two ways. The rth moment about the origin is $E(X^r)$. We write

$$m'_r = \int_{-\infty}^{+\infty} x^r f(x) \, dx . \qquad (4.11.5)$$

The rth moment about the mean, or the rth central moment, is defined by

$$m_r = \int_{-\infty}^{+\infty} (x - \mu)^r f(x) \, dx . \qquad (4.11.6)$$

The second central moment is the variance. It is the topic of the next section.

4.12. THE VARIANCE

The variance of a continuous random variable, denoted by $V(X)$ or by σ^2, is defined by

$$V(X) = \int_{-\infty}^{+\infty} (x - \mu)^2 f(x)\, dx. \qquad (4.12.1)$$

It corresponds to the moment of inertia of a distribution of mass. The positive square root of the variance, σ, is called the standard deviation.

It is often easier to use the following formula:

$$V(X) = E(X^2) - [E(X)]^2$$
$$= m_2' - \mu^2. \qquad (4.12.2)$$

Example 4.12.1. Find the variance of an exponential variable.

$$m_2' = \frac{1}{\theta} \int_0^\infty x^2 e^{-x/\theta}\, dx$$
$$= 2 \int_0^\infty x e^{-x/\theta}\, dx = 2\theta^2.$$

We saw in example 4.11.1 that $E(X) = \theta$. Then

$$V(X) = 2\theta^2 - \theta^2 = \theta^2. \qquad (4.12.3)$$

□

Example 4.12.2. Find the variance of a random variable that has a gamma distribution with parameters α and β.

Following the method of example 4.11.2, we calculate

$$m_2' = \beta^2 \alpha(\alpha + 1).$$

Then

$$V(X) = \beta^2 \alpha(\alpha + 1) - \alpha^2 \beta^2 = \alpha\beta^2. \qquad (4.12.4)$$

□

4.13. MOMENT-GENERATING FUNCTIONS

A useful technique for finding the moments of a distribution is to take derivatives of the moment-generating function. That is, as the name suggests, a function that generates the moments.

MOMENT-GENERATING FUNCTIONS

The moment-generating function (or m.g.f.) of a random variable, X, is defined as

$$M_x(t) = E(e^{tx}) = \int_{-\infty}^{+\infty} e^{tx} f(x) \, dx. \tag{4.13.1}$$

We recall that e^{tx} can be expanded as the infinite series

$$e^{tx} = 1 + tx + \frac{(tx)^2}{2} + \frac{(tx)^3}{3!} + \frac{(tx)^4}{4!} + \cdots.$$

When we evaluate the integral of equation (4.13.1) and expand it as a series in powers of t, the coefficient of t^r in the series is $m'_r/r!$. It follows that we can obtain the rth moment about the origin by taking the rth derivative of $M(t)$ with respect to t and evaluating it at $t = 0$.

Engineers will notice the similarity between the moment-generating function and the Laplace transform.

Example 4.13.1. Suppose that X has a uniform distribution on the interval $(0, \theta)$. Find the rth moment about the origin.

This is a simple example, but it illustrates the basic idea. We have $f(x) = 1/\theta$, and we can see immediately that the rth moment is

$$m'_r = \frac{1}{\theta} \int_0^\theta x^r \, dx = \frac{\theta^r}{r+1}.$$

The moment-generating function is

$$M(t) = \frac{1}{\theta} \int_0^\theta e^{tx} \, dx$$

$$= \frac{1}{\theta t} (e^{t\theta} - 1)$$

$$= \frac{\theta t}{2} + \frac{(\theta t)^2}{3!} + \cdots + \frac{(\theta t)^r}{(r+1)!} + \cdots,$$

and the coefficient of $t^r/r!$ is

$$m'_r = \frac{\theta^r}{r+1}.$$

$\square$

Example 4.13.2. For the exponential distribution

$$M(t) = \frac{1}{\theta} \int_0^\infty e^{tx} e^{-x/\theta} \, dx$$

$$= \frac{1}{\theta} \int_0^\infty \exp\left[\frac{-x(1-\theta t)}{\theta}\right] dx$$

$$= \frac{1}{1-\theta t}.$$

This is the geometric series

$$1 + (\theta t) + (\theta t)^2 + (\theta t)^3 + \cdots,$$

and so

$$m'_r = \theta^r r!.$$

□

Example 4.13.3. For the gamma distribution, the m.g.f. is

$$M(t) = [(\alpha - 1)!\beta^\alpha]^{-1} \int_0^\infty x^{\alpha-1} e^{-x} e^{tx}\, dx$$

$$= [(\alpha - 1)!\beta^\alpha]^{-1} \int_0^\infty x^{\alpha-1} \exp\left[\frac{-x(1 - \beta t)}{\beta}\right]$$

$$= (1 - \beta t)^{-\alpha}.$$

□

In later chapters, we use the moment-generating function in an inverse way. For example, if we are investigating a random variable and we find that its m.g.f. is $(1 - 2t)^{-3}$, we can say that the variable has a gamma distribution with parameters $\alpha = 3$ and $\beta = 2$. This is useful when we wish to talk about the sums, or the averages, of random variables. We will see that the sum of n independent exponential variables is a gamma variable with $\alpha = n$ and $\beta = \theta$.

4.14. BIVARIATE DISTRIBUTIONS

We discussed bivariate discrete distributions in section 3.15. The same ideas carry forward to continuous variables. Summations are replaced by integrals. The probability density is spread over an area rather than along a line. Suppose that we denote two continuous random variables by X and Y. We have a bivariate density function $\phi(x, y)$. The probability that both $a < X < b$ and $c < Y < d$ is given by the double integral

$$\int_a^b \int_c^d \phi(x, y)\, dx\, dy. \tag{4.14.1}$$

The density function satisfies these two requirements:

(i) $\quad\quad\quad \phi(x, y) \geq 0 \quad$ for all x, y;

(4.14.2)

(ii) $\quad\quad\quad \displaystyle\int_{-\infty}^{+\infty} \int_{-\infty}^{+\infty} \phi(x, y)\, dx\, dy = 1.$

The distribution function is defined by

$$P(X < x, Y < y) = F(x, y) = \int_{-\infty}^{x} \int_{-\infty}^{y} \phi(x, y) \, dx \, dy \, . \quad (4.14.3)$$

It has to be a nondecreasing function of both x and y. It must be zero when both variables are at $-\infty$ and one when they are at $+\infty$.

In the discrete case, we obtained the marginal probability of one variable by summing over all the values taken by the other. Here we obtain the marginal density of X by integrating out the other variable, Y:

$$f(x) = \int_{-\infty}^{+\infty} \phi(x, y) \, dy \, . \quad (4.14.4)$$

The conditional density function for Y, given that $X = x$, is obtained by dividing the joint density by the marginal density of X, evaluated at the value x:

$$h(y|x) = \frac{\phi(x, y)}{f(x)} \, . \quad (4.14.5)$$

In this function, x is a given value; it is no longer a variable; $h(y|x)$ is a function of y alone. The conditional density is not defined when $f(x) = 0$.

4.15. COVARIANCE

The covariance of two random variables, X and Y, is a bivariate generalization of the concept of variance. It is defined by

$$\text{Cov}(X, Y) = E\{[X - E(X)][Y - E(Y)]\} \, . \quad (4.15.1)$$

The definition holds whether the variables are discrete or continuous. For purposes of calculation, equation (4.15.1) can be rewritten as

$$\text{Cov}(X, Y) = E(XY) - E(X)E(Y) \, . \quad (4.15.2)$$

As the name implies, the covariance measures how the variables vary together. In the simplest case, if $Y > E(Y)$ whenever $X > E(X)$, and if $Y < E(Y)$ whenever $X < E(X)$, the product $[X - E(X)][Y - E(Y)]$ is never negative, and so the covariance is positive.

Example 4.15.1. Suppose that the joint density function for X and Y is

$$\phi(x, y) = x + y, \quad 0 < x < 1, 0 < y < 1 \, ;$$

and zero elsewhere.

The marginal density of X is

$$f(x) = \int_0^1 (x+y)\,dy = x + \frac{1}{2},$$

which is similar to the marginal density of Y. Then

$$E(X) = \int_0^1 x\left(x + \frac{1}{2}\right) dx = \frac{7}{12} = E(Y),$$

and

$$E(X^2) = \int_0^1 x^2\left(x + \frac{1}{2}\right) dx = \frac{5}{12}.$$

It follows that

$$V(X) = \frac{5}{12} - \left(\frac{7}{12}\right)^2 = \frac{11}{144} = V(Y),$$

and

$$E(XY) = \int_0^1 \int_0^1 xy(x+y)\,dx\,dy = \frac{1}{3},$$

whence

$$\text{Cov}(X, Y) = \frac{1}{3} - \left(\frac{7}{12}\right)\left(\frac{7}{12}\right) = -\frac{1}{144}.$$

The conditional density of Y, given that $X = 1/3$, is obtained by substituting $x = 1/3$ in equation (4.14.5). This gives

$$h\left(y|x = \frac{1}{3}\right) = \frac{2(1+3y)}{5}.$$

The conditional density of Y when $x = 1/2$ is $(2y + 1)/2$, which is the same as the marginal distribution of Y. This fact does not make X and Y independent. For independence, we need $h(y|x) = f(y)$ for all values of x; we have just seen that this is not true for $x = 1/3$. □

4.16. INDEPENDENT VARIABLES

In section 3.15, we said that two discrete random variables are independent if

$$p_{ij} = p_i \cdot p_{\cdot j} \quad \text{for each } i \text{ and each } j,$$

i.e., the joint probability function is the product of the two marginals.

EXERCISES

The corresponding statement for continuous random variables is this:
Let $f(x)$ and $g(y)$ be the marginal density functions of X and Y respectively. The two variables are independent if, and only if,

$$\phi(x, y) = f(x)g(y). \tag{4.16.1}$$

When the condition of equation (4.16.1) holds, we have

$$E(XY) = \int_{-\infty}^{+\infty} \int_{-\infty}^{+\infty} xy\phi(x, y) \, dx \, dy$$

$$= \int_{-\infty}^{+\infty} \int_{-\infty}^{+\infty} xyf(x)g(y) \, dx \, dy$$

$$= \int_{-\infty}^{+\infty} xf(x) \, dx \int_{-\infty}^{+\infty} yg(y) \, dy$$

$$= E(X)E(Y).$$

Thus, if X and Y are independent, their covariance is zero.

The converse is true when both the variables are normally distributed. When the variables are not normal, zero covariance may not imply independence. This is illustrated in example 4.16.1.

Example 4.16.1. Suppose that X takes the values -1, 0, and $+1$, each with probability $1/3$, and that $Y = X^2$. The sample space consists of the three points $(-1, +1)$, $(0, 0)$, $(+1, +1)$ with equal probabilities. The marginal probabilities for Y are

$$p_{\cdot 0} = \tfrac{1}{3} \quad \text{and} \quad p_{\cdot 1} = \tfrac{2}{3}.$$

The variables are not independent because $p_{00} \neq p_{0 \cdot} p_{\cdot 0}$; $p_{00} = \tfrac{1}{3}$, and $p_{0 \cdot} p_{\cdot 0} = (\tfrac{1}{3})(\tfrac{1}{3}) = \tfrac{1}{9}$. However, $E(X) = 0$ and

$$E(XY) = (-1)(+1)(\tfrac{1}{3}) + (0)(0)(\tfrac{1}{3}) + (+1)(+1)(\tfrac{1}{3}) = 0.$$

It follows from equation (4.15.2) that Cov(X, Y) = 0, even though the variables are not independent. □

EXERCISES

4.1. The lifetime of a certain component has an exponential distribution. If $\theta = 1500$ hours, what percentage of the components will last for more than 400 hours?

4.2. If, in exercise 4.1, a buyer wishes to have a 95% probability that a component will last for 5000 hours, what should the buyer specify the average life to be?

4.3. An engineer puts 100 lamps on test. After 1000 hours of use, 710 of the lamps have failed. What would the engineer estimate the value of θ to be?

4.4. A device is designed to have a mean lifetime of 20,000 hours. Suppose that the lifetime is exponentially distributed. What is the median lifetime? What percentage fails in the first year (8760 hours)? What percentage lasts for 5 years? What should be the mean lifetime if we wish 90% of the devices to last for 40,000 hours?

4.5. The life of a certain device has a Weibull distribution with $\beta = 2$; 63.2% of them last for 8000 hours. What fraction lasts for 10,000 hours?

4.6. Suppose that in exercise 4.5, we did not know β, but that we did know that 25% of the devices lasted until 9000 hours. What fraction lasts for 10,000 hours?

4.7. Lifetimes of field windings for a generator have a Weibull distribution with $\alpha = 12$ years and $\beta = 2$. They are warranted for a life of only 2 years. What percentage is still operating at the end of the warranty period?

4.8. Derive the following identities for obtaining central moments from moments about the origin:

$$m_3 = m_3' - 3m_2'\mu + 2\mu^3$$

$$m_4 = m_4' - 4m_3'\mu + 6m_2'\mu^2 - 3\mu^4.$$

4.9. Compute m_3 and m_4 for the exponential distribution. (See exercise 4.8.)

4.10. Show that the moment-generating function for a Bernoulli variable is $q + pe^t$. Use it to calculate the mean, variance, and m_3.

4.11. A continuous variable has the density function

$$f(x) = kx^2 \quad 1 < x < 3.$$

Calculate the constant, k, and find the mean and variance.

EXERCISES

4.12. The Beta function is defined by

$$B(m, n) = \int_0^1 x^{m-1}(1-x)^{n-1}\, dx = \frac{(m-1)!(n-1)!}{(m+n-1)!}.$$

A variable has a beta distribution with parameters α and β if its density function is

$$f(x) = \frac{x^{\alpha-1}(1-x)^{\beta-1}}{B(\alpha, \beta)} \qquad 0 < x < 1.$$

Calculate the mean and variance of X.

4.13. Two variables, X and Y, have joint density

$$\phi(x, y) = kx(x + y) \qquad 0 < x < 1, \quad 0 < y < 2.$$

Calculate the constant, k. Find the marginal densities of X and Y, and calculate $E(X)$, $V(X)$, and $\text{Cov}(X, Y)$.

4.14. A variable has density function

$$f(x) = k \sin x \qquad 0 < x < \pi.$$

Calculate the constant, k, and find $E(X)$ and $V(X)$.

4.15. A random variable, X, has a uniform distribution on the interval $(-\theta, +\theta)$:

$$f(x) = \frac{1}{2\theta} \qquad -\theta < x < +\theta,$$
$$= 0 \qquad \text{elsewhere}.$$

Calculate $E(X)$, $E(X^2)$, and $V(X)$.

4.16. It is stated in the text that the sum of two independent random variables that have uniform distributions on the interval $(0, 1)$ has the distribution of example 4.3.1. Confirm that.

4.17. Derive the moment-generating function for a Poisson variable.

CHAPTER FIVE

The Normal Distribution

5.1. INTRODUCTION

The normal distribution, or probability law, is the cornerstone of statistical practice. It is perhaps too extreme a simplification to say that it is called the normal distribution because it is the distribution that naturally occurring random variables normally have. However, it is true that many random phenomena have probability distributions that are well approximated by the normal.

This distribution is sometimes called the Gaussian distribution after K. F. Gauss, who, in his early nineteenth-century work with astronomical data, used it as a law for the propagation of errors. At about the same time, Laplace published the same distribution in his book on probability theory, so there are, of course, arguments about priority between supporters of the two men. Actually, Abraham DeMoivre derived the normal distribution as early as 1718 as the limit of the binomial distribution as n becomes large.

5.2. THE NORMAL DENSITY FUNCTION

The simplest form of the normal density function is

$$f(z) = \frac{1}{\sqrt{2\pi}} \exp\left(\frac{-z^2}{2}\right). \qquad (5.2.1)$$

Its graph is the familiar bell curve shown in figure 5.2.1. The curve is symmetric about the value $z = 0$, which is also the maximum value. The mean, the median, and the mode of the distribution are all at $z = 0$.

The normal density function cannot be integrated by the ordinary methods of calculus. It is, however, proved in advanced calculus texts that

$$\int_{-\infty}^{+\infty} \exp\left(\frac{-t^2}{2}\right) dt = \sqrt{2\pi}. \qquad (5.2.2)$$

THE NORMAL DENSITY FUNCTION

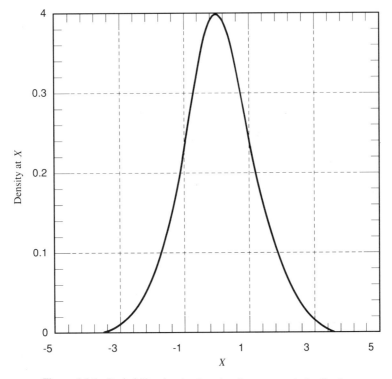

Figure 5.2.1. Probability density function for a normal distribution.

The general form of the normal density is obtained by introducing the two parameters, μ and σ^2, and setting

$$z = \frac{x - \mu}{\sigma}.$$

Then, writing dP for the element of probability, we have

$$dP = f(z)\, dz = \phi(x)\, dx,$$

where $dz = dx/\sigma$. The density function for X is

$$\phi(x) = \frac{1}{\sigma\sqrt{2\pi}} \exp\left[\frac{-(x-\mu)^2}{2\sigma^2}\right]. \tag{5.2.3}$$

The graph of $\phi(x)$ looks exactly like figure 5.2.1, except that it is centered at $x = \mu$ and that the scale is changed. We show later that the parameter σ is the scale factor. It is the standard deviation of X; $V(X) = \sigma^2$.

We write $X \sim N(\mu, \sigma^2)$ to denote that X is normally distributed with mean, or expectation, $E(X) = \mu$, and variance, $V(X) = \sigma^2$. Z is said to have a standard normal distribution or to be a standard normal variable. We write $Z \sim N(0, 1)$. We have seen that if $X \sim N(\mu, \sigma^2)$, then $(X - \mu)/\sigma \sim N(0, 1)$.

5.3. THE TABLES OF THE NORMAL DISTRIBUTION

Table I (see Appendix) lists values of $F(z)$. From this table, we can, given $E(X)$ and $V(X)$, find values of $F(x)$.

Entering the table at $z = 1.96$, we see that $F(1.96) = 0.9750$, so that $P(Z > 1.96) = 0.025$. From the symmetry of the curve, we also have $P(Z < -1.96) = 0.025$ and $P(-1.96 \leq Z \leq 1.96) = 0.95$. For 95% of the time, a standard normal variable will lie in the interval $(-1.96, +1.96)$; 2.5% of the time it will be in the upper tail, $Z > +1.96$, and 2.5% of the time in the lower tail, $Z < -1.96$.

By writing $X = \mu + z\sigma$, this translates to the following two statements that are each true with probability 0.95:

$$\mu - 1.96\sigma \leq x \leq \mu + 1.96\sigma ; \tag{5.3.1}$$

$$P(X > \mu + 1.96\sigma) = P(X < \mu - 1.96\sigma) = 0.025 . \tag{5.3.2}$$

If we enter the table at $z = 2.0$ rather than 1.96, we see that the values of $F(z)$ change from 0.0250 and 0.9750 to 0.0228 and 0.9554. The change is small enough for us to take as a rule of thumb that 95% of the members of a normal population lies within the two sigma limits, i.e., within a distance $\pm 2\sigma$ of the mean.

Several other useful values of z are listed in table 5.3.1. We see that 99% of the population lies within the 2.6 sigma limits. Only one observation in 400 lies outside the bounds set by $\mu \pm 3\sigma$, one observation in 800 on the high side, and one in 800 on the low side.

Example 5.3.1. Let $X \sim N(89.0, 36.00)$. Find (a) $P(x > 93.5)$, (b) $P(X < 81.5)$, and (c) $P(86.0 < X < 98.0)$.

Table 5.3.1. Probabilities for Special Values of z

z	$P(-z \leq Z \leq +z)$	$P(Z > z)$
1.00	0.6832	0.1584
1.645	0.9000	0.0500
1.96	0.9500	0.0250
2.00	0.9544	0.0228
2.575	0.9900	0.0050
2.60	0.9906	0.0047
3.00	0.9974	0.0013

(a) The mean of X is 89.0; the standard deviation is 6.0. When $X = 93.5$, $Z = (93.5 - 89.0)/6.0 = 0.75$, and so $P(X > 93.5) = P(Z > 0.75) = 1 - F(0.75) = 1 - 0.7734 = 0.2266$.

(b) When $X = 81.5$, $Z = -7.5/6.0 = -1.25$. From symmetry, $P(Z < -1.25) = P(Z > +1.25) = 1 - F(1.25) = 1 - 0.8944 = 0.1056$.

(c) When $X = 86.0$, $Z = -0.5$; when $X = 98.0$, $Z = +1.5$. We need to calculate

$$F(1.5) - F(-0.5) = F(1.5) - 1 + F(0.5)$$
$$= 0.9332 - 1 + 0.6915$$
$$= 0.6247 .$$

□

5.4. THE MOMENTS OF THE NORMAL DISTRIBUTION

In this section, we calculate the expectation, the variance, and the third and fourth central moments of the normal distribution. Our procedure is to calculate the moments for the standard normal distribution. We then argue that since $X = \sigma Z + \mu$, $E(X) = \sigma E(Z) + \mu$ and the rth central moment of X is σ^r times the corresponding moment of Z.

Since the standard normal density function is symmetric about the line $z = 0$, we have

$$f(z) = f(-z) ,$$

so that

$$(-z)f(-z) = -(z)f(z)$$

and

$$\int_{-\infty}^{0} zf(z) \, dz = -\int_{0}^{+\infty} zf(z) \, dz .$$

Hence,

$$E(Z) = \int_{-\infty}^{+\infty} zf(z) \, dz = \int_{-\infty}^{0} zf(z) \, dz + \int_{0}^{+\infty} zf(z) \, dz = 0 ,$$

and $E(X) = \mu$.

A similar argument shows that all the odd moments of Z are zero. It follows that the third central moment of X is also zero.

Since $E(Z) = 0$,

$$V(Z) = \int_{-\infty}^{+\infty} z^2 f(z) \, dz = \int_{-\infty}^{+\infty} z[zf(z)] \, dz \, .$$

By integrating by parts,

$$V(Z) = \frac{1}{2\pi} \left[-z \exp\left(\frac{-z^2}{2}\right) \right]_{-\infty}^{+\infty} + \int_{-\infty}^{+\infty} f(z) \, dz \, .$$

Application of L'Hôpital's rule shows that the first term on the right side is zero. The second term is unity. Thus,

$$V(Z) = 1 \quad \text{and} \quad V(X) = \sigma^2 \, .$$

A similar argument shows that the fourth moment of Z is $m_4 = 3$, and the fourth moment of X is $3\sigma^4$.

5.5. THE MOMENT-GENERATING FUNCTION

The results of the previous section can also be obtained from the moment-generating function. It can be shown that the moment-generating function for the normal distribution is

$$M(t) = \exp\left(\mu t + \frac{\sigma^2 t^2}{2}\right). \quad (5.5.1)$$

We actually want the generating function for the central moments, which is

$$M^*(t) = M(t) \exp(-\mu t) = \exp\left(\frac{\sigma^2 t^2}{2}\right).$$

Differentiating $M^*(t)$ as many times as necessary and putting $t = 0$ gives us the desired central moments.

5.6. SKEWNESS AND KURTOSIS

Scale-free versions of the third and fourth moments are used as measures of the shape of density functions. The skewness is measured by

$$g_3 = \frac{m_3}{\sigma^3} \, .$$

For a symmetric density function, meaning that the curve $f(x)$ is symmetric about the line $x = E(X)$, the odd central moments are all zero, and so the skewness is zero.

The kurtosis of a density function is measured by

$$g_4 = \frac{m_4}{\sigma^4}.$$

This a measure of the peakedness of the density curve. The kurtosis of the normal distribution is $m_4(Z) = 3$. Some use the normal distribution as a reference distribution. They subtract 3 from the kurtosis and call the difference the excess. Although the standard deviation is a measure of the spread of a density curve, it is possible for two densities to have the same mean and variance and still differ markedly in shape.

Example 5.6.1. Suppose that X has a uniform distribution over the interval $(-\theta, \theta)$. The density is $1/2\theta$. It is symmetric about the line $X = E(X) = 0$. The odd moments are zero. The even moments are

$$V(X) = \frac{\theta^2}{3} \quad \text{and} \quad m_4 = \frac{\theta_4}{5}.$$

If $\theta = \sqrt{3}$, this density has the same mean and the same variance as the standard normal distribution. They clearly do not have the same shape. The normal distribution has tails going off to infinity. The graph of the uniform distribution is a rectangle. The difference is manifested in the kurtosis. The kurtosis of the uniform distribution is $9/5 = 1.8$, which is less than the kurtosis of the more widely spread normal distribution. □

5.7. SUMS OF INDEPENDENT RANDOM VARIABLES

In the next chapter, we begin our investigation of statistical procedures. One of the most important statistics we use is the mean, or average, of a random sample of observations, which is their sum divided by n. In the next three sections, we discuss the behavior of sums of random variables. In section 5.10, we introduce the important central limit theorem.

We now consider how the sum of two independent random variables behaves. Let X_1 and X_2 be two independent random variables with density functions $f_1(x)$ and $f_2(x)$, respectively, and moment-generating functions $M_1(t)$ and $M_2(t)$, respectively. Let $U = X_1 + X_2$. Denote the moment-generating function of U by $M(t)$. The result of this section is expressed as a theorem.

Theorem 5.7.1

$$M(t) = M_1(t) M_2(t).$$

We wish to find $E(tx_1 + tx_2)$. Because X_1 and X_2 are independent, their joint density function is $f_1(x_1)f_2(x_2)$, and so

$$M(t) = \int_{-\infty}^{+\infty} \exp(tx_1 + tx_2) f_1(x_1) f_2(x_2) \, dx_1 \, dx_2$$

$$= \int_{-\infty}^{+\infty} \exp(tx_1) f_1(x_1) \, dx_1 \int_{-\infty}^{+\infty} \exp(tx_2) f_2 x_2 \, dx_2$$

$$= M_1(t) M_2(t). \tag{5.7.1}$$

□

More generally, we can show that if U is the sum of several independent random variables, its moment-generating function is the product of the individual moment-generating functions. The details of the generalization are left to the reader.

To find $E(U)$, we differentiate $M(t)$ with respect to t and equate t to zero. Then

$$E(U) = M_1'(0) M_2(0) + M_1(0) M_2'(0)$$
$$= E(X_1) + E(X_2). \tag{5.7.2}$$

Similarly,

$$E(U^2) = M_1''(0) M_2(0) + 2M_1'(0) M_2'(0) + M_1(0) M_2''(0)$$
$$= E(X_1^2) + 2E(X_1)E(X_2) + E(X_2^2)$$

and

$$V(U) = E(U^2) - [E(U)]^2 = V(X_1) + V(X_2). \tag{5.7.3}$$

The generalizations of equations (5.7.2) and (5.7.3) are that the expectation of the sum of independent random variables is the sum of their expectations, and the variance of the sum is the sum of the variances. Furthermore, if $w_1, w_2, \ldots$ are constants, and

$$W = \sum w_i X_i,$$

then

$$E(W) = \sum w_i E(X_i) = \sum w_i \mu_i \tag{5.7.4}$$

and

$$V(W) = w_i^2 V(X_i) = \sum w_i^2 \sigma_i^2. \tag{5.7.5}$$

W is said to be a linear combination of the random variables.

An important special case involves the difference between two random variables. If $W = X_1 - X_2$, we have $w_1 = +1$ and $w_2 = -1$. Substituting in equations (5.7.4) and (5.7.5), we see that the expectation of the difference is the *difference* between the individual expectations, but because $w_1^2 = w_2^2 = +1$, the variance of the difference is the *sum* of the individual variances.

5.8. RANDOM SAMPLES

A random sample of n observations from a population is a set of n independent observations. Independence is essential. Every member of the population must have the same chance of being included in the sample. Basically, it is a "fair" sample that is typical of the population being sampled. If we want a random sample of the student population, we do not just go into the student union at lunch time, because that will bias the sample against students who take their lunch in residence halls, fraternities, and sororities. Nor do we just set up a booth outside the engineering building or the mathematics department, because that will bias the sample against the arts students. It is not easy to obtain a random sample. There are numerous examples, especially in political polling, in which the pollsters reached wrong conclusions because their samples were not random, but had biases built into them.

Suppose that we do have a random sample of n observations: $x_1, \ldots, x_n$. We can treat them as realizations of n independent random variables, each of which has the same probability distribution. Then the formulas of equations (5.7.4) and (5.7.5) become

$$E(W) = \sum w_i E(X) \tag{5.8.1}$$

and

$$V(W) = \sum w_i^2 V(X). \tag{5.8.2}$$

If we substitute $1/n$ for w_i in the formulas of the previous section, we obtain

$$E(\bar{x}) = E(X) \tag{5.8.3}$$

and

$$V(\bar{x}) = \frac{V(X)}{n}. \tag{5.8.4}$$

These equations show that the sample mean provides a good estimate of $E(X)$ inasmuch as, on the average, $\bar{x} = E(X)$, and its precision (the reciprocal of the variance) improves as we take more observations. We say more about this in the next chapter.

5.9. LINEAR COMBINATIONS OF NORMAL VARIABLES

If X is a random variable with moment-generating function $M(t)$, the moment-generating function of wX is $M(wt)$. It follows from the results of section 5.7 that the moment-generating function of W is

$$M(t) = M_1(w_1 t) M_2(w_2 t) \cdots M_n(w_n t). \tag{5.9.1}$$

If the X_i are all normal variables, we take equation (5.5.1) and add the exponents to get

$$M(t) = \exp\left(\sum w_i \mu_i t + \frac{\sum w_i^2 \sigma_i^2 t^2}{2}\right), \tag{5.9.2}$$

which has the same form as equation (5.5.1). We have shown that if W is a linear combination of independent normal variables, then W too is a normal variable. In particular, the sample mean has a normal distribution.

5.10. THE CENTRAL LIMIT THEOREM

We saw in the last section that the mean of a sample of independent normal variables itself has a normal distribution. What if the variables do not have a normal distribution? The central limit theorem tells us that for all practical purposes, the distribution of the sample mean approaches normality for large enough n, no matter what the parent distribution of the population may be. How large is large enough depends on the shape of the parent distribution. For a uniform distribution, the distribution of the mean is well approximated by the normal when n is as few as six. For a skewed distribution such as the exponential distribution, a larger sample is needed.

The proof of the central limit theorem takes us further into mathematical theory than we want to go in this book. It consists essentially of showing that as n increases, $M(t)$, as given in equation (5.9.1), converges to the form of equation (5.9.2). One approach is to look at the logarithm of $M(t)$ as a power series and to note that the coefficient of t^r contains a multiplier, $(1/n)^r$. As n increases, the higher-order terms tend to zero, and we are left with

$$\ln[M(t)] = \mu t + \sigma t^2,$$

which is the logarithm of a normal moment-generating function.

The importance of the central limit theorem is that we can quite reasonably act as if the mean of a random sample of observations is normally distributed. We lean heavily on this in our applications.

5.11. THE CHI-SQUARE DISTRIBUTION

Let $Z_1, Z_2, \ldots, Z_n$ be independent standard normal variables and let $W = \Sigma Z_i^2$. The distribution of W becomes important when we investigate the behavior of the variance of a sample from a normal distribution in the next chapter. The random variable W is said to have a chi-square distribution with n degrees of freedom. Table IV (see Appendix) gives tabulated values of the chi-square distribution for various values of degrees of freedom. The expression "degrees of freedom" is usually abbreviated as d.f.

The chi-square distribution is a special case of the gamma distribution that was introduced in section 4.10, and we can use this property to derive the density function.

The moment-generating function of Z^2 is

$$M(t) = E[\exp(z^2 t)]$$
$$= \frac{1}{\sqrt{2\pi}} \int_{-\infty}^{+\infty} \exp\left(\frac{-z^2}{2} + tz^2\right) dz$$
$$= \frac{1}{\sqrt{2\pi}} \int_{-\infty}^{+\infty} \exp\left[\frac{-z^2(1-2t)}{2}\right] dz . \tag{5.11.1}$$

This has the same form as the integral for a normal distribution with zero expectation and variance $(1 - 2t)^{-1}$ except that the factor $1/\sigma = (1 - 2t)^{1/2}$ is missing. Hence, the value of the integral is

$$M(t) = (1 - 2t)^{-1/2} .$$

It follows that the moment-generating function of W is

$$M_W(t) = (1 - 2t)^{-n/2} . \tag{5.11.2}$$

We recall from example 4.13.3 that the moment-generating function for a gamma variable with parameters α and β is

$$(1 - \beta t)^{-\alpha} . \tag{5.11.3}$$

Comparing equation (5.11.3) with $M_W(t)$, we see that W has a gamma distribution with parameters $n/2$ and 2. We can now write the density function of W as

$$f(w) = \frac{w^{(n-2)/2} e^{-w/2}}{2^{n/2}(n/2 - 1)!} \qquad w \geq 0 . \tag{5.11.4}$$

Fortunately, we shall have no occasion in this book to use this formula or to quote it again!

An important consequence of the mathematical result that we have just obtained is that we can establish a connection between the exponential and chi-square distributions. We showed in example 4.13.2 that the moment-generating function of an exponential variable is

$$M(t) = \frac{1}{1 - \theta t}.$$

Let Y be the sum of n independent exponential variables with the same θ. The moment-generating function of Y is

$$M_Y(t) = \frac{1}{(1 - \theta t)^n}.$$

The moment-generating function of $2Y/\theta$ is obtained by replacing t by $2t/\theta$. It is

$$\frac{1}{(1 - 2t)^n},$$

which we recognize as the m.g.f. of chi-square with $2n$ d.f.

We sum up the result of the previous paragraph as follows: Let $x_1, \ldots, x_n$ be exponential variables with the same parameter, θ, and let $Y = \Sigma x_i$. Then, $2Y/\theta$ has a chi-square distribution with $2n$ d.f.

Example 5.11.1. An engineer has a supply of ten lamps whose lifetimes are exponentially distributed with an expected life of 700 hours. As soon as one lamp fails, a new one is plugged in. What is the probability that the supply of ten lamps will last for a year (8760 hours)?

$$P(Y > 8760) = P\left(\frac{2Y}{700} > 25.03\right).$$

Entering the chi-square table with 20 d.f., we see that the probability of a value less than 25.0 is 0.80. There is a 20% probability that the engineer's ration of lamps will last for a year. □

5.12. THE LOGNORMAL DISTRIBUTION

The lognormal distribution is another useful probability model for the lifetimes of electronic devices. Let Y denote the lifetime of a device. The model assumes that $\ln(Y)$ is normally distributed with mean μ and variance σ^2. The probability density function for Y is

$$f(y) = \frac{1}{y} \frac{1}{\sigma\sqrt{2\pi}} \exp\left[-\frac{[\ln(y) - \mu]^2}{2\sigma^2}\right],$$

where μ is the expected ln(life), and σ^2 is the variance of the ln(life). The median of the lognormal distribution is

$$\tilde{y} = \exp(\mu).$$

The expectation is

$$E(Y) = \exp\left(\mu + \frac{\sigma^2}{2}\right).$$

Example 5.12.1. The lifetime of an electronic device (in hundreds of hours) has a lognormal distribution with $\mu = 6.0$ and $\sigma = 0.800$. What percentage of the devices fails before 10,000 hours?

$$\ln(100) = 4.605 \quad \text{and} \quad z = \frac{4.605 - 6.00}{0.800} = -1.744.$$

From the normal tables, $F(-1.744) = 4\%$; hence, only 4% of the devices fail before 10,000 hours. The median lifetime is $100e^{6.00} = 40,300$ hours; the average lifetime is $100e^{6.4} = 60,200$ hours. □

5.13. THE CUMULATIVE NORMAL DISTRIBUTION CURVE AND NORMAL SCORES

The graph of the cumulative normal distribution is shown in figure 5.13.1. It plots the cumulative probability along the vertical axis and $X(-\infty, +\infty)$ along the horizontal axis. This is the familiar ogive curve, or lazy S. We see it again in chapters seven and eleven.

Normal probability paper changes the scale on the vertical axis and converts the ogive into a straight line. We can turn this around and argue that if we plot the cumulative distribution curve for some distribution on normal probability paper and the plot is *not* a straight line, then the distribution is not normal.

A similar result can be obtained by using normal scores. Table 5.13.1 lists a random sample of 20 observations from a standard normal distribution and their normal scores.

The first observation, 0.59, is the eighteenth member of the sample, in order of size; only two observations exceed it. It is called the eighteenth-order statistic for the sample. The average value of the eighteenth-order statistic from a random sample of 20 observations from a standard normal population is 1.13. That is the normal score of the first observation in the sample. Similarly, the fourth observation, -1.15, is the third-order statistic; its normal score is -1.13. The plot of the normal scores (Y) against the observations (X) should be a straight line. It is shown as figure 5.13.2.

THE NORMAL DISTRIBUTION

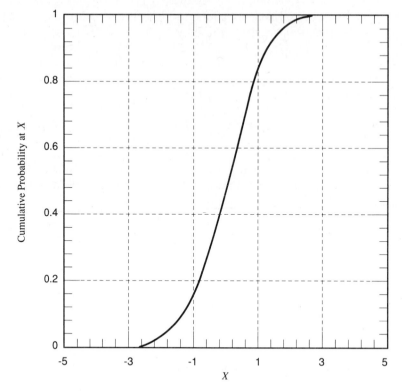

Figure 5.13.1. Cumulative normal distribution function.

On the other hand, table 5.13.2 shows a sample of 20 observations from an exponential population with $\theta = 100$ and their normal scores. One would not expect the order statistics for an exponential distribution to behave like order statistics from a normal population, and they do not. The plot of normal scores against the observations is shown in figure 5.13.3. It is clearly not a straight line!

Table 5.13.1. Twenty Normal Observations ($\mu = 0.0$, $\sigma = 1.0$)

0.59	−0.33	−0.11	−1.15	−1.20	−0.16	−0.36	−0.30
−0.32	1.46	−1.99	0.52	0.21	0.13	−0.52	0.18
−0.97	0.96	−0.80	0.07				

Normal Scores for the Sample

1.13	−0.31	0.19	−1.13	−1.40	0.06	−0.45	−0.06
−0.19	1.87	−1.87	0.92	0.74	0.45	−0.59	0.59
−0.92	1.40	−0.74	0.31				

THE CUMULATIVE NORMAL DISTRIBUTION CURVE AND NORMAL SCORES 85

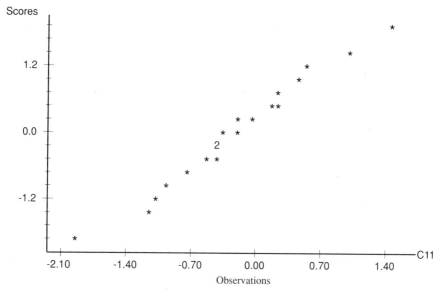

Figure 5.13.2. Normal score plot for a normal sample.

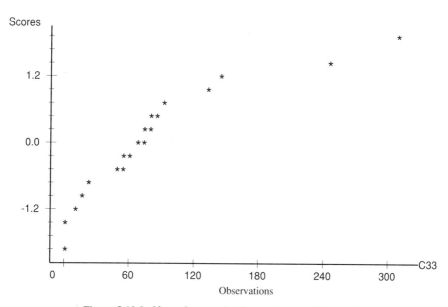

Figure 5.13.3. Normal score plot for an exponential sample.

Table 5.13.2. Twenty Exponential Observations ($\theta = 100$)

24	130	2	86	17	60	69	80	314	93
248	76	56	86	1	53	146	10	49	90

Normal Scores

−0.74	0.92	−1.40	0.45	−0.92	−0.19	−0.06	0.19	1.87	0.74
1.40	0.06	−0.31	0.31	−1.87	−0.45	1.13	−1.13	−0.59	0.59

EXERCISES

5.1. The amount of product made per hour by a chemical plant is normally distributed with $\mu = 72.0$ kg and $\sigma = 2.5$ kg. Let Y denote the amount. Find the probabilities of the following events: (a) $Y > 75.0$, (b) $Y < 74.2$, (c) $Y < 69.5$, (d) $75.8 < Y < 77.1$, and (e) $68.1 < Y < 69.3$.

5.2. A machine makes metal rods. The lengths are normally distributed with $\mu = 19.8$ cm and $\sigma = 5$ mm. The specifications call for the length of a rod to be in the interval $19.5 < x < 20.5$. What percentage of the rods meets specifications?

5.3. A manufacturer makes rods and sockets, both of which have circular cross sections. The diameters of the rods are normally distributed with $\mu = 1.0$ cm and $\sigma = 2$ mm; the diameters of the sockets are normal with $\mu = 1.05$ cm and $\sigma = 1.5$ mm. An engineer picks a rod and a socket at random. What is the probability that the rod will go into the socket?

5.4. An oil company has a contract to sell grease in 500-g containers. The quantities of grease that the filling machine puts in the containers are normally distributed with the mean at the designated value and $\sigma = 25$ g. To what value should the filling machine be set if the company does not wish to have more than 2% of the containers rejected as below specified weight?

5.5. The containers in exercise 5.4 are metal cans. Their empty weights are normally distributed with $\mu = 90$ g and $\sigma = 8$ g. The inspector weighs the cans of grease and expects them to weigh at least 590 g. To what nominal weight should the filling machine be set so that only 2% of the cans of grease fail inspection?

5.6. A device is designed to have an average lifetime of 1200 hours. Ten of those devices are put on test and their lifetimes are recorded. Suppose that the lifetimes are normally distributed with $\mu = 1200$ and $\sigma = 400$.

EXERCISES 87

Let Y denote the average lifetime of the devices that were tested. Find the probabilities that (a) $Y > 1300$, (b) $Y < 1150$, and (c) $1320 < Y < 1400$.

What are the first and third quartiles of the average lifetime? How many devices would you have to test if you wanted $P(Y > 1100) = 0.90$?

5.7. Repeat exercise 5.6 assuming that the lifetimes have exponential distributions with $\theta = 1200$.

5.8. The life of a device measured in hours has a lognormal distribution with $\mu = 8.700$ and $\sigma = 0.4$. What percentage of the devices operates longer than 10,000 hours? What are the median lifetime and the average lifetime?

5.9. The lifetimes of locomotive controls (in thousands of miles) have a lognormal distribution with $\mu = 5.65$ and $\sigma = 0.6$. What is the median lifetime? What percentage of the controls lasts more than 500,000 miles?

5.10. Calculate the skewness, g_3, and the kurtosis, g_4, for the exponential distribution. *Hint*: You should use the following formulas (see exercise 4.9):

$$m_3 = m'_3 - 3m'_2\mu + 2\mu^3$$
$$m_4 = m'_4 - 4m'_3\mu + 6m'_2\mu^2 - 3\mu^4$$

5.11. Calculate the skewness, g_3, and the kurtosis, g_4, for the gamma distribution. Show that as parameter α increases, g_3 tends to zero and value of the kurtosis converges to 3. This verifies that as the number of degrees of freedom increases, the shape of the chi-squared curve becomes closer to the normal curve. It also confirms that we are justified in using the central limit theorem to claim that, as n increases, the average of n exponential variables tends to behave like a normal variable.

5.12. The moment-generating function of a binomial variable is

$$M(t) = (q + pe^t)^n.$$

Calculate the third moment about the mean: $m_3 = npq(q - p)$.

5.13. Derive the moment-generating function for the normal distribution, equation (5.5.1).

CHAPTER SIX

Some Standard Statistical Procedures

6.1. INTRODUCTION

Chapters three, four, and five can be regarded as a short course in probability theory, designed to be the foundation for the statistical applications that are the main purpose of this book. In this chapter, we turn our attention to applications by considering three basic experiments, or statistical investigations. In each experiment, we select an appropriate mathematical model: the experiments are chosen to illustrate the binomial, exponential, and normal distributions. These models involve unknown parameters that have to be estimated from the data: p in the binomial, θ in the exponential, μ and σ in the normal. We consider both point and interval estimates for the parameters, and make our first venture into testing hypotheses.

In each experiment, we take a sample of n observations. The data are a set of n numbers that can be regarded as the realizations of n independent random variables, each of which comes from the same population, or distribution. They can be written as $x_1, x_2, x_3, \ldots, x_n$. Alternatively, we can write them in a column and denote the column by the vector $\mathbf{x}$. A geometer would regard that vector as a point in an n-dimensional sample space.

6.2. THE THREE EXPERIMENTS

6.2.1. Experiment One

We have a process for producing electronic components. Before they leave the plant, an inspection is made. A sample is taken from each batch of components. Each component is tested and classified as either defective or nondefective. We wish to estimate from the data the percentage of defectives being produced.

6.2.2. Experiment Two

A sample of n light bulbs is taken from a production line. They are put on test and their lifetimes are recorded. We wish to find the average lifetime of the lamps being produced on the line that day.

6.2.3. Experiment Three

We have a process for producing resistors with a nominal resistance of 90 ohms. Unfortunately, the resistance of the resistors varies from item to item. We wish to estimate the average resistance of resistors being produced by the process that day.

6.3. THE FIRST EXPERIMENT

Suppose that the batch that was being inspected had 5000 components and that the sample size was $n = 100$. The data vector is a sequence of 100 zeros or ones; x_i denotes the state of the ith component tested. It takes the value one if the component is defective and zero if the component is nondefective. An appropriate model for X_i is that of a Bernoulli trial, with probability of success p that is the same for all n trials. Let X be the total number of defectives in the sample. Then X has a binomial distribution with parameters $n = 100$ and p, which is unknown. (It can be argued that since we are not talking about sampling with replacement, a hypergeometric model would be correct, but, with the batch size 50 times the sample size, the binomial model is adequate.) We have to estimate p from the data.

A reasonable estimate is

$$\hat{p} = \frac{X}{n}. \qquad (6.3.1)$$

We recall from chapter three that $E(X) = np$, so that $E(\hat{p}) = np/n = p$. On the average, the estimate that we have chosen is correct.

This does not say that the estimate will be exactly equal to p for every sample. It does say that if we follow this procedure, day in, day out, on the average, we will get the correct value. An estimator, or recipe for obtaining an estimate, that has the property that it gives, on the average, the correct value is said to be an unbiased estimator.

6.4. THE SECOND EXPERIMENT

Let x_i be the lifetime of the ith bulb. An appropriate model for x_i is the exponential model with density function

$$f(x) = \left(\frac{1}{\theta}\right) e^{-x/\theta} \qquad x \geq 0. \qquad (6.4.1)$$

We want to estimate the parameter θ from the data. A reasonable estimate is

$$\hat{\theta} = \frac{\sum x_i}{n} = \bar{x}. \qquad (6.4.2)$$

6.5. THE THIRD EXPERIMENT

Let X_i denote the resistance of the ith item. Let us assume that the resistances are normally distributed; $X_i \sim N(\mu, \sigma^2)$. Now we have two parameters, μ and σ^2, to be estimated from the data. Since $E(X) = \mu$, it is reasonable to estimate it by

$$\hat{\mu} = \bar{x}. \qquad (6.5.1)$$

We discuss estimating the variance from the data later in this chapter.

6.6. THE BEHAVIOR OF THE SAMPLE AVERAGE

In each of the three experiments, we have estimated a parameter, p, θ, or μ, by the sample average: the average number of defectives per component, the average lifetime of the light bulbs, the average resistance of a large batch of resistors.

In the last chapter, we discussed the probability distribution of $\bar{x}$ at some length. We recall that in the normal case, $\bar{x}$ itself has a normal distribution, with $E(\bar{x}) = E(X)$ and $V(\bar{x}) = V(X)/n$, and so

$$z = \frac{\bar{x} - \mu}{\sigma/\sqrt{n}}$$

is a standard normal variable.

In the other two experiments, we can use the central limit theorem and act as if $\bar{x}$ were normally distributed with expectation p and variance pq/n in the first experiment and with expectation θ and variance θ^2/n in the second experiment.

6.7. CONFIDENCE INTERVALS FOR THE NORMAL DISTRIBUTION

The estimate, $\bar{x}$, of $E(X)$ is a point estimate in the sense that it is a single point, or value. The simple statement that our estimate is, say, 77.9 gives no indication of how precise that estimate might be. If we were reasonably sure that the true value of the parameter would lie within 0.1 of our estimate, we

CONFIDENCE INTERVALS FOR THE NORMAL DISTRIBUTION

might feel that we had a good estimator. On the other hand, if we thought that the true value was quite likely to be anywhere in the interval $\bar{x} \pm 10.0$, we should be less comfortable.

The two estimates, 77.9 ± 0.1 and 77.9 ± 10.0, are examples of interval estimates. In this section, we develop a class of interval estimates known as confidence intervals. They were introduced by J. Neyman and E. S. Pearson. We start by obtaining a 95% confidence interval for the mean of a normal distribution.

Let Z be a standard normal variable. The following statement is true with probability 0.95:

$$-1.96 \leq z \leq +1.96.$$

Hence,

$$-1.96 \leq \frac{\bar{x} - \mu}{\sigma/\sqrt{n}} \leq +1.96,$$

and multiplying by -1,

$$-1.96 \leq \frac{\mu - \bar{x}}{\sigma/\sqrt{n}} \leq +1.96.$$

Then

$$-\frac{1.96\sigma}{\sqrt{n}} \leq (\mu - \bar{x}) \leq +\frac{1.96\sigma}{\sqrt{n}}$$

and

$$\bar{x} - \frac{1.96\sigma}{\sqrt{n}} \leq \mu \leq \bar{x} + \frac{1.96\sigma}{\sqrt{n}}. \qquad (6.7.1)$$

Equation (6.7.1) is a 95% confidence interval for μ. We have 95% confidence in the interval because it was deduced from a statement that is correct with probability 0.95. If, day after day, we were to take samples of size n from this normal population and compute the interval given by equation (6.7.1), in the long run, 95% of the intervals would indeed contain μ. In 2.5% of the cases, μ would be outside the interval on the high side; in the remaining 2.5%, μ would be outside the interval on the low side.

Example 6.7.1. A sample of four rods is taken from a process that produces rods about 30.0 cm in length. The average length of the four rods in the sample is 29.575 cm. Assuming that the lengths are normally distributed with standard deviation 0.500 cm, obtain 90%, 95%, and 99% confidence intervals for the process mean.

We have $\sigma/\sqrt{n} = 0.500/2 = 0.250$. The 95% confidence interval is given by equation (6.7.1):

$$29.575 - 1.96(0.250) \leq \mu \leq 29.575 + 1.96(0.250)$$
$$29.085 \leq \mu \leq 30.065 . \qquad (6.7.2)$$

For the 90% interval, we substitute 1.645 for 1.96 in equation (6.7.1):

$$29.164 \leq \mu \leq 29.986 . \qquad (6.7.3)$$

For 99% confidence, we substitute 2.575:

$$28.931 \leq \mu \leq 30.219 . \qquad (6.7.4)$$

The length of the 95% interval is 0.98 cm. The 90% interval is shorter, 0.822 cm. On the other hand, if we are not content with an interval in which we have only 95% confidence and call for 99% confidence, the length of the interval increases to 1.288 cm. The cost of increased confidence is a longer interval; a shorter interval can be obtained with a decrease in confidence. Shorter intervals can also be obtained by taking larger samples. □

Example 6.7.2. In example 6.7.1, how large a sample should the engineer take to have a 95% confidence interval no longer than 0.500 cm, i.e., $\bar{x} \pm 0.250$ cm?

We wish to have

$$1.96 \left(\frac{0.500}{\sqrt{n}} \right) \leq 0.250$$

so that

$$\sqrt{n} \geq 1.96 \left(\frac{0.500}{0.250} \right) = 3.92 ,$$
$$n \geq 15.36 .$$

A sample of at least 16 observations is needed. □

The interval defined by equation (6.7.1) is a two-sided interval. It sets both upper and lower bounds to the value of $E(X)$. In some problems, one-sided intervals are more appropriate. We can start with this statement: With probability 0.95,

$$z \geq -1.645 .$$

Then

$$\frac{\mu - \bar{x}}{\sigma / \sqrt{n}} \geq -1.645$$

CONFIDENCE INTERVALS FOR THE NORMAL DISTRIBUTION 93

and

$$\mu \geq \bar{x} - \frac{1.645\sigma}{\sqrt{n}}. \qquad (6.7.5)$$

Alternatively, we can start with this statement: With probability 0.95,

$$z \leq +1.645,$$

whence

$$\mu \leq \bar{x} + \frac{1.645\sigma}{\sqrt{n}}. \qquad (6.7.6)$$

Intervals like those of equation (6.7.5) would be used if the engineer wants to be confident that μ exceeds some lower bound and is not concerned that the value of μ might be too high. A purchaser of acid might specify that the vendor deliver acid whose strength is at least 60%; if the vendor wants to send him 65% strength at the same price, so be it. Intervals like those of equation (6.7.6) are used if the engineer wants assurance that μ is not too high. One example might be water quality; X could be the percentage of some undesirable contaminant. We wish to be sure that the amount of contaminant present does not exceed a specified level. If it does, we take some action. We are not concerned about taking action if the water is purer than required.

The choice between using a one-sided or a two-sided interval depends on the application. Sometimes it is obvious that the situation calls for a one-sided interval. A good working rule is to use two-sided intervals unless the situation clearly calls for a one-sided interval. In case of doubt, use two-sided intervals. We say more about this when we discuss hypothesis testing in the next chapter.

Example 6.7.3. Ten measurements are made of the percentage of butenes, an impurity, in a crude oil shipment. The data are

2.73, 2.84, 2.92, 2.57, 2.98, 2.68, 2.59, 2.82, 2.74, 2.58.

Find an upper 95% confidence limit for the amount of butenes present.

For this data, we have $\bar{x} = 2.745$ and $\sigma = 0.15$. The value of z that corresponds to a probability of 5% in *one* tail is $z = 1.645$.

The upper confidence interval is given by

$$\mu < \bar{x} + 1.645\left(\frac{0.15}{\sqrt{10}}\right),$$

i.e.,

$$\mu < 2.83. \qquad \square$$

6.8. CONFIDENCE INTERVALS FOR THE BINOMIAL DISTRIBUTION

As early as 1718, De Moivre showed that for a fair coin, the binomial distribution approached the normal distribution in the limit as n became large. This was the first version of the central limit theorem. Laplace generalized De Moivre's results to values of p other than $p = 0.5$ nearly 100 years later. They showed that for large enough values of n, the distribution of X, the number of successes in n trials, is well approximated by a normal distribution. We may assume that X is normally distributed with $E(X) = np$ and $V(X) = npq$. How large is "large enough"? A useful rule of thumb is to choose n large enough for $np > 5$, and $nq > 5$. In these formulas, we substitute prior estimates of p and q. If $p = 0.5$, the normal approximation will be adequate for a sample of only ten observations. As p moves farther away from 0.5, the required sample size increases.

This means that we can obtain a 95% confidence interval for p by writing

$$z = \frac{np - x}{\sqrt{npq}},$$

and, following the development in section 6.7, this leads to the interval

$$\frac{x}{n} - 1.96\sqrt{pq/n} \leq p \leq \frac{x}{n} + 1.96\sqrt{pq/n}. \tag{6.8.1}$$

There is an obvious problem. The ends of the interval involve pq, and we do not know p. The best solution to this difficulty is to substitute our point estimates for the unknown quantities: $\hat{p} = x/n$ for p, and $\hat{q} = 1 - x/n$ for q. Then equation (6.8.1) becomes

$$\frac{x}{n} - 1.96\sqrt{\hat{p}\hat{q}/n} \leq p \leq \frac{x}{n} + 1.96\sqrt{\hat{p}\hat{q}/n}. \tag{6.8.2}$$

Example 6.8.1. Suppose that the sample of 100 electronic components in the experiment in section 6.3 has 20 defectives and 80 nondefectives. Find a 95% confidence interval for the fraction of defectives in the whole batch. $\hat{p} = 20/100 = 0.20$; $\hat{q} = 0.80$. Following equation (6.8.2),

$$0.20 - 1.96\sqrt{(0.2)(0.8)/100} \leq p \leq 0.20 + 1.96(0.04)$$

$$0.12 \leq p \leq 0.28. \tag{6.8.3}$$

If the sample size had been 1000 with 600 defectives, the confidence interval would have been

$$0.60 - 1.96\sqrt{0.24/1000} \leq p \leq 0.60 + 1.96\sqrt{0.24/1000},$$

$$0.57 \leq p \leq 0.63. \tag{6.8.4}$$

CONFIDENCE INTERVALS FOR THE BINOMIAL DISTRIBUTION

There are two other ways of handling the difficulty with the ends of the interval in equation (6.8.1). One is to take the upper and lower inequalities as quadratics and solve them for p. That is a more lengthy procedure, and the degree of improvement in the estimate does not really justify the effort.

The end points of the interval can be written as

$$p = \hat{p} \pm 1.96\sqrt{pq/n}$$

or

$$p - \hat{p} = \pm 1.96\sqrt{pq/n}.$$

Squaring both sides,

$$p^2 - 2p\hat{p} + \hat{p}^2 = \frac{3.84p(p - \hat{p})}{n}.$$

When the terms are rearranged, this equation becomes the following quadratic in p:

$$(n + 3.84)p^2 - (2n\hat{p} + 3.84)p + n\hat{p}^2 = 0.$$

In this example, we substitute 100 for n and 0.20 for $\hat{p}$, so that the equation becomes

$$103.84p^2 - 43.84p + 4 = 0,$$

whence

$$p = 0.133 \quad \text{or} \quad p = 0.289.$$

The interval has approximately the same length as the interval that we obtained earlier in equation (6.8.3), but it is moved to the right. It is centered on 0.21, not on $\hat{p} = 0.20$; this is a point against the method.

□

The other approach, which is useful when p is closer to 0.5, is to note that the maximum value that pq can take is 0.25, when $p = 0.5$, and to substitute this value for pq in equation (6.8.1). This procedure is clearly conservative. It is reasonable for quite a broad range of values. When $p = 0.4$ or 0.6, $pq = 0.24$. Even with $p = 0.3$ or 0.7, we have $pq = 0.21$ and $pq = 0.458$, so that this procedure gives a confidence interval that is only 9% longer than that from equation (6.8.1).

For example 6.8.1, where p is only 0.20, this estimate would give the interval

Table 6.8.1. Presidential Election Results

Year	Candidate	Percentage of Popular Vote
1960	Kennedy	50.09
1964	Johnson	61.34
1968	Nixon	43.56*
1972	Nixon	60.91
1976	Carter	50.57
1980	Reagan	50.99*
1984	Reagan	59.17

* The opposition vote was split between two candidates.

$$0.10 \leq p \geq 0.30,$$

which is longer than the other two intervals.

Finally, this leads to a quick method of obtaining approximate confidence intervals. In equation (6.8.1), replace 1.96 by 2.0 and pq by $\frac{1}{4}$. The equation then simplifies to

$$\frac{x}{n} - \frac{1}{\sqrt{n}} \leq p \leq \frac{x}{n} + \frac{1}{\sqrt{n}}. \tag{6.8.5}$$

One hopes that this approximation will not be relevant to industrial situations in which p represents the percent defectives! It is a useful approximation for political polling, since it is rare for one candidate in a two-person race to receive more than 70% of the votes. The percentages of the votes obtained by the winning candidates in the presidential races from 1960–1984 are shown in table 6.8.1.

Another example is found in baseball. Since 1910, no pennant winner in the National League, and only four in the American League, has won as many as 70% of the games that they played in the season, and few of the bottom teams have failed to win at least 30%.

6.9. CONFIDENCE INTERVALS FOR THE EXPONENTIAL DISTRIBUTION

The binomial distribution is not the only distribution whose mean approaches normality for large n. The central limit theorem, which was introduced in chapter five, says that for most densities, the distribution of $\bar{x}$ approaches the normal distribution as n increases. Applying the central limit theorem to the exponential distribution, which is the model in the third experiment, section 6.5, we recall that $V(X) = \theta^2$. The formula for the 95%

interval is

$$\bar{x} - \frac{1.96\theta}{\sqrt{n}} \le \theta \le \bar{x} + \frac{1.96\theta}{\sqrt{n}}. \tag{6.9.1}$$

We substitute the point estimate $\bar{x}$ for θ in each of the end points, and the confidence interval becomes

$$\bar{x}\left(1 - \frac{1.96}{\sqrt{n}}\right) \le \theta \le \bar{x}\left(1 + \frac{1.96}{\sqrt{n}}\right). \tag{6.9.2}$$

Example 6.9.1. Suppose that the average of $n = 100$ observations is $\bar{x} = 252$. The 95% confidence interval is

$$252(0.804) \le \theta \le 252(1.196)$$

$$202.6 \le \theta \le 301.4.$$

A preferable method for small samples appears in section 6.13. □

6.10. ESTIMATING THE VARIANCE

We recall that $V(X) = E(X - \mu)^2$. If we know μ, an obvious estimator for the variance is the average of the square of the deviations of the observations, i.e.,

$$\hat{\sigma}^2 = \frac{\sum (x_i - \mu)^2}{n}. \tag{6.10.1}$$

If we do not know μ, we can substitute the point estimate, $\bar{x}$. We must, however, change the denominator to $n - 1$. This change is justified in the next few paragraphs.

From the definition of the variance, we deduce that

$$\begin{aligned} V(X) = \sigma^2 &= E[(X - \mu)^2] \\ &= E[X^2 - 2\mu x + \mu^2] \\ &= E(X^2) - 2\mu E(X) + \mu^2 \\ &= E(X^2) - 2\mu^2 + \mu^2 \\ &= E(X^2) - \mu^2, \end{aligned}$$

whence

$$E(X^2) = \mu^2 + \sigma^2 = [E(X)]^2 + V(X) \tag{6.10.2}$$

and

$$E(\bar{x}^2) = [E(\bar{x})]^2 + V(\bar{x}) = \mu^2 + \frac{\sigma^2}{n}. \tag{6.10.3}$$

Let

$$\begin{aligned} S = \sum (x_i - \bar{x})^2 &= \sum (x_i^2 - 2x_i\bar{x} + \bar{x}^2) \\ &= \sum x_i^2 - 2\bar{x}\sum x_i + n\bar{x}^2 \\ &= \sum x_i^2 - n\bar{x}^2. \end{aligned}$$

Then

$$\begin{aligned} E(S) &= n(\mu^2 + \sigma^2) - n\left(\mu^2 + \frac{\sigma^2}{n}\right) \\ &= (n-1)\sigma^2, \end{aligned} \tag{6.10.4}$$

and so we estimate $V(X)$ by

$$s^2 = \frac{S}{n-1}. \tag{6.10.5}$$

S is called the sum of squares of deviations (of the observations from their average). Although S has n terms, they are not independent because the deviations add up to zero. The first $n-1$ of them may take any values, but the last deviation is dictated by the restriction that the sum is zero. For that reason, the estimate of the variance is said to have $n-1$ degrees of freedom. We estimate the standard deviation of X by the square root of the estimated variance. It should be noted that the derivation of s^2 in equation (6.10.5) does not depend on the data being normally distributed.

Example 6.10.1. A sample of five observations is 1, 8, 6, 4, and 6. Estimate $V(X)$.

For this sample, $\bar{x} = 5$, and it is easy to calculate the deviations by subtracting $\bar{x}$ from the observed values. The deviations are $-4, +3, +1, -1$, and $+1$, and

$$S = 16 + 9 + 1 + 1 + 1 = 28,$$

which leads to an estimate

$$s^2 = \frac{S}{4} = 7.0.$$

ESTIMATING THE VARIANCE

That calculation worked smoothly because n is small and x is an integer. If x is not a "convenient" number, the squares of the deviations will not be so easy to calculate, and the method used before may introduce roundoff error. In practice, it is preferable to compute S by the formula

$$S = \sum x_i^2 - \frac{\left(\sum x_i\right)^2}{n}. \qquad (6.10.6)$$

In this example,

$$S = 1 + 64 + 36 + 16 + 36 - \frac{(25)^2}{5} = 28.$$

□

Example 6.10.2. Estimate $V(X)$ from the following sample of six observations:

17.021, 17.028, 17.026, 17.024, 17.026, 17.022 .

The estimate can be obtained by following the procedures of the previous example. It is easier to subtract 17.020 from each observation first; recall from equation (3.10.3) that $V(X - b) = V(X)$. This is only a shift of the origin and does not change the spread of the data, which now become

0.001, 0.007, 0.006, 0.004, 0.006, 0.002 .

Multiplying by 1000 changes the data to

1, 7, 6, 4, 6, 2 ,

which is an easier set of numbers to handle.

Formally, we have made the transformation

$$Z = 1000(X - 17.020) \text{ with } V(Z) = 10^6 V(X).$$

We estimate $V(Z)$ by

$$5\sigma_z^2 = 1 + 49 + 36 + 16 + 36 + 4 - \frac{(26)^2}{6} = 30.33$$

$$\sigma_z^2 = 6.07.$$

The estimate of $V(X)$ is 6.07×10^{-6}.

□

6.11. STUDENT'S t STATISTIC

In section 6.7, we derived confidence intervals for the mean of a normal distribution when the variance is known. The intervals were obtained from the standard normal variable

$$z = \frac{\bar{x} - \mu}{\sigma/\sqrt{n}}.$$

Usually, the standard deviation is unknown and has to be estimated from the data. We substitute s for σ in the equation for z. The resulting statistic is denoted by

$$t = \frac{\bar{x} - \mu}{s/\sqrt{n}}. \tag{6.11.1}$$

This statistic does not have a standard normal distribution because the known variance has been replaced by its estimate, which is itself a random variable. The properties of the t statistic were first investigated by W. S. Gosset, who was one of the first industrial statisticians. He worked as a chemist for the Guinness Brewing Company. Since Guinness would not allow him to publish his work, it appeared under the pen name of "Student." Thereafter, Gosset was known to statisticians as Student and his statistic was called Student's t.

The graph of the density function of t looks much like the standard normal curve, except that it is more spread out. To obtain confidence intervals, we repeat the earlier derivation starting with t instead of z, and replacing 1.96 by t^*, the corresponding tabulated value of t in Table II (see Appendix). The value of t^* depends not only on the level of confidence desired, but also on the number of d.f. in the estimate of the variance. With infinite d.f., the density of t is the same as the standard normal.

Example 6.11.1. In the third experiment mentioned at the beginning of this chapter, a sample of ten resistors was taken from the production line. The resistances of the items in ohms were

114.2, 91.9, 107.5, 89.1, 87.2, 87.6, 95.8, 98.4, 94.6, 85.4.

Obtain a 95% confidence interval for the average resistance of the items being produced.

A dotplot of the data appears as figure 6.11.1 and a boxplot as figure 6.11.2.

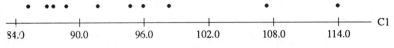

Figure 6.11.1. Dotplot for example 6.11.1.

CONFIDENCE INTERVALS FOR THE VARIANCE OF A NORMAL POPULATION

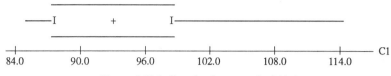

Figure 6.11.2. Boxplot for example 6.11.1.

Since there is no indication that a one-sided interval is needed, it is appropriate to use a two-sided interval.

The estimates of the mean and standard deviation are

$$\bar{x} = 95.17 \quad \text{and} \quad s = 9.36,$$

respectively. The value of t^* for a 95% interval based on $10 - 1 = 9$ d.f. is read from the table as 2.262. The confidence interval is

$$95.17 - \frac{(2.262)(9.36)}{\sqrt{10}} \leq \mu \leq 95.17 + \frac{(2.262)(9.36)}{\sqrt{10}}$$

$$88.47 \leq \mu \leq 101.87.$$

Nowadays, one rarely has to carry out these tedious calculations by hand. The printout for obtaining this interval using Minitab is given in the following. The data were entered in column one of Minitab's work sheet. The command TINT 95 C1 called for a 95% t interval for the data in C1.

SET in cl
  114.2 91.9 107.5 89.1 87.2 87.6 95.8 98.4 94.6 85.4

TINT 95 cl
  N = 10   MEAN =   95.170   ST.DEV. =   9.36

A 95.00 PERCENT C.I. FOR MU IS (88.4697, 101.8703)

□

6.12. CONFIDENCE INTERVALS FOR THE VARIANCE OF A NORMAL POPULATION

We showed in chapter five that the sum of the squares of independent standard normal variables has a chi-square distribution. Hence,

$$W = \frac{\sum (x_i - \mu)^2}{\sigma^2}$$

has a chi-square distribution with n d.f.

It can be shown that S/σ^2 also has a chi-square distribution, but with $n-1$ d.f. This property can be used to obtain confidence intervals for the variance as follows.

Example 6.12.1. Obtain a two-sided 95% confidence interval for the variance of the resistances in example 6.11.1.

In this example, $S = 789.141$ and there are $n - 1 = 9$ d.f. The distribution of chi-square is not symmetric. We read from Table III (see Appendix) in the row for 9 d.f. that the 2.5% value for chi-square is 2.700 and the 97.5% value is 19.023, i.e., with probability 0.025 chi-square will be less than 2.700. Then, with probability 0.95, we have

$$2.700 \leq \frac{S}{\sigma^2} \leq 19.023 .$$

Inverting this inequality leads to the desired confidence interval

$$\frac{S}{19.023} \leq \sigma^2 \leq \frac{S}{2.700} .$$

Substituting 789.141 for S gives

$$41.48 \leq \sigma^2 \leq 292.27$$

and

$$6.44 \leq \sigma \leq 17.10 .$$

Our point estimate of the variance is $s^2 = 789.141/9 = 87.68$. The confidence interval is not symmetric about this value. One should also note that the interval is quite wide. It stretches from 47% of the point estimate to 333%. Admittedly, there are only ten data points, but this does indicate that we need considerably more data if we want to get a really reliable estimate of variance. □

6.13. CHI-SQUARE AND EXPONENTIAL VARIABLES

We pointed out in sections 4.10 and 5.11 that the exponential and chi-square distributions are both special cases of the gamma family of distributions. We show in this section how the chi-square tables can be used to obtain confidence intervals for the parameter of an exponential distribution.

The method depends on the fact that if we have a random sample of n exponential variables, $Y = 2 \Sigma x_i/\theta = 2n\bar{x}/\theta$ will have a chi-square distribution with $2n$ d.f.

Example 6.13.1. Suppose that only $n = 10$ lamps had been tested in the second experiment, and that $\bar{x} = 252$, as before. Obtain a two-sided 95% confidence interval for θ.

In this example, $2 \Sigma x_i = 5040$. We read values of chi-square with 20 d.f. and obtain, with probability 0.95,

$$9.591 \leq \frac{5040}{\theta} \leq 34.170,$$

whence

$$147.5 \leq \theta \leq 525.5.$$

□

The method that we used in section 6.9 gives a different inverval,

$$95.8 \leq \theta \leq 408.2.$$

For small n, the method of this section is preferable because it is based on the exact distribution of the statistic and not an approximation to it. For larger values of n, the chi-square distribution approaches normality, and the two methods give almost the same answers.

This method can also be used to obtain one-sided confidence intervals (upper and lower bounds).

Example 6.13.1 (cont.). To obtain one-sided intervals, we argue that with probability 0.95,

$$\frac{5040}{\theta} < 31.4,$$

whence

$$\theta > 160.5,$$

or that

$$\frac{5040}{\theta} > 10.85,$$

whence

$$\theta < 464.5.$$

□

6.14. ANOTHER METHOD OF ESTIMATING THE STANDARD DEVIATION

In setting up quality control charts, it has been customary to use a different method of estimating the standard deviation based on the sample range. The range is the difference between the maximum and minimum observations in the sample. The advantage of this method is that the range is much easier to calculate than s^2; modern advances in computing have made that advantage less important. The estimate, s^2, that was obtained in section 6.10 is more efficient than the estimate that is derived from the range. However, for the small samples that are normally used in x-bar charts ($n = 4$ or 5), the differences in efficiency are trivial, and little is lost in using the simpler range method. The validity of the range method is very dependent on the assumption that the samples come from a normal population.

The range estimate is obtained by multiplying the sample range by an appropriate constant. For example, it can be shown that for samples of four observations from a normal distribution, the expected value of R is 2.06σ. The corresponding estimate of σ is $R/2.06$.

The multiplier 2.06 is denoted in quality control work by d_2. Table 6.14.1 gives values of d_2 for some values of n. The standard deviation is then estimated by

$$\hat{\sigma} = \frac{R}{d_2}. \qquad (6.14.1)$$

The range of the first five observations in example 6.11.1 is $R = 114.2 - 87.2 = 27.0$. The value of d_2 for $n = 5$ is 2.33. The estimate of the standard deviation from that subsample by the range method is

$$\tilde{\sigma} = \frac{27.0}{2.33} = 11.6;$$

the estimate from the other subsample, the last five observations, is

$$\frac{98.4 - 85.4}{2.33} = 5.58.$$

We cannot say which of these estimates or the previous estimate is closer to the "true" value for the population that provided this sample. They are each estimates of the unknown standard deviation. It is obvious that the highest resistance, 114.7, made the first estimate larger than the second. The

Table 6.14.1. Expected Values of the Range, d_2

n	2	3	4	5	6	7
d_2	1.13	1.69	2.06	2.33	2.53	2.70

optimist should resist the temptation to calculate several estimates and choose the smallest one!

6.15. UNBIASED ESTIMATORS

Equation (5.8.3) shows that $\bar{x}$ is an unbiased estimate of $E(X)$, which means that if we use $\bar{x}$ to estimate $E(X)$, we will, on the average, have the correct answer. Before giving a formal definition of the property of unbiasedness, we define the term statistic.

Definition. Let **x** be a data vector. A statistic, $t(\mathbf{x})$, is a function of the data, i.e., a number calculated from the data that does not involve any unknown parameters; $\bar{x}$ is a statistic. □

Definition. A statistic, $t(\mathbf{x})$, is said to be an unbiased estimate, or estimator, of a parameter, say, ϕ, if

$$E[t(\mathbf{x})] = \phi . \tag{6.15.1}$$

□

Unbiasedness is an attractive property for an estimator to possess, but it is not adequate just to have an unbiased estimate. In the first experiment, section 6.3, $E(X_1) = p$. This says that if we throw away all the data except the observation in the first component, and say that $p = 1$ if that component is defective and $p = 0$ if the component is okay, we will have an unbiased estimate: unbiased but also unwise! How might one choose between estimators? Which is the best "recipe" for estimating ϕ? One course of action is to use unbiased estimators wherever possible, and, when faced with two unbiased estimators, pick the one with the smaller variance. It is a good principle.

In many problems, it is possible to find an unbiased estimator of ϕ that we know to have a smaller variance than any possible competitor. Such an estimator is called the MVUE, minimum-variance unbiased estimator. There is an interesting mathematical theory behind the search for MVU estimators, but to investigate it would take us beyond the scope of this book. It suffices at this stage to say that under most circumstances, the sample mean, $\bar{x}$, is the MVU estimator of $E(X)$.

Definition. An estimator of ϕ that is a linear combination of the observations is called a linear estimator. It can be written as

$$\mathbf{a}'\mathbf{x} = \sum a_i x_i . \tag{6.15.2}$$

□

Linear estimators constitute an important class of estimators. They are used extensively when we fit equations by the method of least squares. The sample mean is a linear estimator of $E(X)$. The unbiased linear estimator of ϕ that has the smallest variance is sometimes called the BLUE, best linear unbiased estimator. This idea is best illustrated by a proof of a result that common sense would regard as obvious.

Example 6.15.1. Show that the sample mean is the best linear unbiased estimator of $E(X)$.

We are considering estimators of the form

$$W(\mathbf{x}) = \sum a_i x_i .$$

Since W is an unbiased estimator, $E(W) = E(X)$; hence, by equation (5.7.4),

$$\sum a_i E(X) = E(X),$$

and

$$\sum a_i = 1. \tag{6.15.3}$$

By equation (5.7.5), $V(W) = \sum a_i^2 V(X)$, and so we must minimize $\sum a_i^2$ subject to the restriction of equation (6.15.3). Direct application of the method of Lagrange multipliers shows that this occurs when all the coefficients, a_i, are equal, in which case, $a_i = 1/n$ for all i. □

6.16. MOMENT ESTIMATORS

In the previous sections, we have discussed estimates of parameters. The estimators that we chose were reasonable, but we gave little indication of why they were chosen in particular. Are there perhaps better estimators available? In this section and the next, we discuss two standard procedures for finding estimators: the method of moments and the method of maximum likelihood.

The method of moments goes back a long way in the history of statistics. It consists of equating sample moments to the expected values of the corresponding population moments and solving for the unknown parameters. That was the rationale for the estimates that were used in the three initial experiments.

In the binomial experiment, we equated the number of successes X to $E(X) = np$ and solved for the unknown p. In the normal and exponential examples, we equated the first sample moment $\bar{x}$ to $E(X)$ to estimate μ and θ.

MOMENT ESTIMATORS   107

Example 6.16.1. A quality control engineer samples items from a production line until five defective items are found. The fifth defective was the eightieth item inspected. Estimate the percentage of defectives in the process.

The number of items, X, inspected has a negative binomial distribution (section 3.16) with $r = 5$. Then $E(X) = r/p$, and so

$$80 = \frac{5}{p} \quad \text{and} \quad p = \frac{5}{80} = 0.0625.$$

If there are two parameters to be estimated, we equate the first two sample moments to their expectations and solve the simultaneous equations.

□

Example 6.16.2. Estimate the variance of the resistances in example 6.11.1.

We denote the estimate of the variance obtained by the method moments by $\tilde{\sigma}^2$.

The first two sample moments are

$$\bar{x} = 95.17 \quad \text{and} \quad m_2' = \frac{\sum x_i^2}{10} = 9136.24,$$

and so we write

$$\hat{\mu} = \bar{x} = 95.17 \quad \text{and} \quad \bar{x}^2 + \tilde{\sigma}^2 = 9136.24,$$

whence

$$\tilde{\sigma}^2 = 78.91.$$

This estimate is S/n rather than $S/(n-1)$; it is biased on the low side.

□

Example 6.16.3. A sample of n observations is taken from a uniform distribution on the interval $(0, \theta)$. Estimate the unknown parameter θ.

The probability density function is

$$f(x) = 1/\theta \quad 0 \leq x \leq \theta$$
$$= 0 \quad x > \theta.$$

$E(X) = \theta/2$, and so the estimate of θ is $\tilde{\theta} = 2\bar{x}$.

□

This can sometimes give an unreasonable answer. Suppose that we have five observations: 1, 2, 3, 4, and 10. Then $\bar{x} = 4$ and $\tilde{\theta} = 8$, which is not compatible with the largest observation, 10.

6.17. MAXIMUM-LIKELIHOOD ESTIMATES

Another standard method obtaining estimates of parameters is the method of maximum likelihood, which was developed by R. A. Fisher. The idea is that we look at the data and pick the value of the parameter(s) that, in retrospect, maximizes the likelihood of that particular sample.

Suppose that we have a continuous variable and only one unknown parameter, denoted by ϕ. The density function can be written as $f(x; \phi)$ to emphasize that it is a function of both x and ϕ.

Definition. Consider a sample of observations $\mathbf{x} = (x_1, x_2, \ldots, x_n)$. The *likelihood* of the sample is defined as the product

$$L(\mathbf{x}; \phi) = f(x_1; \phi) f(x_2; \phi) \cdots f(x_n; \phi) . \qquad \square$$

In the case of a discrete distribution, we have a probability mass function $P(x)$, which we can write as $P(x; \phi)$, and

$$L(\mathbf{x}; \phi) = P(x_1; \phi) P(x_2; \phi) \cdots P(x_n; \phi) .$$

The maximum-likelihood estimate of ϕ is the value that maximizes $L(\mathbf{x}; \phi)$. This value is usually obtained by differentiation. It is often more convenient to maximize the natural logarithm, $\ln(L)$ rather than L itself.

Example 6.17.1. Find the maximum-likelihood estimate of the parameter in an exponential distribution.

We have

$$f(x) = \left(\frac{1}{\theta}\right) \exp\left(\frac{-x}{\theta}\right)$$

and

$$L(\mathbf{x}; \theta) = \left(\frac{1}{\theta}\right)^n \exp\left(-\frac{n\bar{x}}{\theta}\right)$$

$$\ln(L) = -n \ln(\theta) - \frac{n\bar{x}}{\theta} ,$$

whence, differentiating,

$$\frac{d[\ln(L)]}{d\theta} = -\frac{n}{\theta} + \frac{n\bar{x}}{\theta^2}$$

and

$$\hat{\theta} = \bar{x} .$$

MAXIMUM-LIKELIHOOD ESTIMATES   109

Example 6.17.2. The normal distribution. The likelihood function is

$$L = \frac{1}{(\sigma\sqrt{2\pi})^n} \exp\left[-\frac{\sum (x_i - \mu)^2}{2\sigma^2}\right].$$

Then

$$\ln(L) = -\left(\frac{n}{2}\right)\ln(2\pi) - n\ln(\sigma) - \frac{\sum (x_i - \mu)^2}{2\sigma^2}.$$

Differentiating, we have

$$\frac{\ln(L)}{d\mu} = -\frac{\sum (x_i - \mu)}{2\sigma^2} = 0,$$

and

$$\frac{\ln(L)}{d\sigma} = -\frac{n}{\sigma} - \frac{\sum (x_i - \mu)^2}{\sigma^3} = 0.$$

Solving the simultaneous equations gives the estimates

$$\hat{\mu} = \bar{x} \quad \text{and} \quad \sigma^2 = \frac{S}{n}.$$

Notice that we would have obtained the same estimates if we had written $v = \sigma^2$ and differentiated with respect to v rather than σ.

Usually, the two methods, moments and maximum likelihood, agree on the same estimator. There are exceptions. One is the gamma distribution, which appears as an exercise. Another is the uniform distribution. □

Example 6.17.3. Find the maximum-likelihood estimator of θ in the uniform distribution. See example 6.16.3.

The likelihood function is

$$L(\mathbf{x}; \theta) = \left(\frac{1}{\theta}\right)^n.$$

Differentiating L does not find a maximum; it finds a minimum when $\theta = +\infty$. The maximum is obtained by taking θ to be as small as we can, consistent with the data, i.e., by setting θ equal to the largest observation. In the data set that follows example 6.16.3, the maximum-likelihood estimate is 10. Since the largest observation is usually less than θ and never exceeds it, this estimate is biased on the low side. □

EXERCISES

6.1. The standard deviation of a measurement of the octane number of a gasoline is 0.3. How many independent measurements must be taken for an engineer to know the octane number to within 0.1 number with 90% confidence?

6.2. A machine produces ball bearings; the diameters are normally distributed with $\sigma = 0.22$ mm. A sample of 20 bearings has an average diameter of 12.61 mm. Find a two-sided 95% confidence interval for the average diameter of the bearings being produced.

6.3. A sample of 25 devices is tested for speed. Their speeds in MHz are

10.90	10.30	10.50	10.60	11.50	10.10	13.80	14.50
14.60	14.10	14.10	14.80	10.10	15.20	14.10	14.70
10.50	10.30	10.30	11.50	11.20	10.40	10.50	10.30
10.30 .							

Make a dotplot of the data. Obtain a two-sided 95% confidence interval for the mean speed.

6.4. The mean grain sizes in a sample of six thin films of Al–1%Si on Si measured in μm are

0.563   0.934   0.673   0.863   1.102   1.061 .

Obtain a one-sided 95% confidence interval that gives an upper bound on the mean grain size.

6.5. Sixteen wafers are tested for sheet resistance. The resistances are

7.5	7.4	7.3	7.6	7.5	7.7	7.6	6.9
7.5	7.3	7.4	7.3	7.4	7.5	7.2	7.4 .

Make a dotplot of the data. Obtain a two-sided 95% confidence interval for the sheet resistance.

6.6. Repeat exercise 6.3, omitting the eighth wafer.

EXERCISES   111

6.7. A sample of 50 parts from a large batch contains six defectives. Obtain a two-sided 95% confidence interval for the percent defective in the batch.

6.8. A television newscaster reports the result of a national poll in which 1000 people split 556 in favor of a question and 444 against. He says that the figure 55.6% in favor has a margin of error ±3%. How is that figure calculated?

6.9. Twenty devices are placed on life test. The average lifetime of the devices tested is 1832 hours. The lifetimes are assumed to be exponentially distributed. Obtain a two-sided 95% confidence interval for the average lifetime of a device using:
(a) the central limit theorem,
(b) the chi-square distribution.

6.10. Obtain a two-sided 90% confidence interval for the variance of the sheet resistances from the sample reported in exercise 6.5.

6.11. A sample of 47 items from a large batch contains 18 defectives. Use the normal approximation to obtain a two-sided 90% confidence interval for the percent defective in the batch.

6.12. A quality control inspector takes four devices off the production line at the end of each shift and tests them for speed. The speeds for five such samples in MHz are (samples are rows)

16.0	13.2	17.2	10.6
17.3	12.5	16.2	15.5
12.2	15.8	14.5	13.0
18.0	17.0	14.0	22.0
13.5	19.4	14.4	17.3.

Obtain estimates of the standard deviation, σ, (a) using the sample ranges averaged over the five samples, and (b) using the more usual method of computing $s^2 = \Sigma (y - \bar{y})^2/3$ for each sample and taking the square root of the average of these estimates of σ^2.

6.13. If you have a software package that draws random samples from a normal distribution, verify that the value of $d_2 = 2.06$ by taking several samples of size 4, computing their ranges, and showing that the average range is close to 2.06. You can do this with Minitab in the following way:

```
random 500 c1-c4;
normal 0 1.
rmax c1-c4 c5
rmin c1-c4 c6
let c7 = c5-c6
describe c7
```

6.14. An engineer tests devices until ten defectives are found. The tenth defective is the 267th device tested. Estimate the fraction defective.

6.15. A sample of 100 observations from a gamma distribution has $\Sigma\, y = 492$ and $\Sigma\, y^2 = 1024$. Use the method of moments to obtain estimates of parameters α and β.

6.16. The beta distribution was introduced in exercise 4.12. A random sample of 20 observations has $\Sigma\, y = 16.0$ and $\Sigma\, y^2 = 13.00$. Use the method of moments to estimate parameters α and β.

6.17. A variable, X, has a uniform distribution on the interval $(-\theta, +\theta)$.

$$f(x) = \frac{1}{2\theta} \quad -\theta < x < +\theta,$$
$$= 0 \quad \text{elsewhere}.$$

A sample of six observations is $-3, +5, -2, -9, +10,$ and $+8$. Use the method of moments to obtain an estimate of θ. The moments of this distribution were calculated in exercise 4.15.

6.18. Use the method of moments to find estimates of α and β from a sample of n independent observations from a gamma distribution.

CHAPTER SEVEN

Hypothesis Testing

7.1. INTRODUCTION

Scientists make decisions on the basis of data. They postulate (set up) hypotheses, collect data, and, after analyzing the data, accept or reject the hypotheses. This a fundamental part of the scientific method. In quality control work, engineers test hypotheses and make statistical decisions without actually using those expressions.

An engineer looks at a quality control chart on which the averages of the daily samples are plotted. If one of the points falls outside the control lines, he decides to take action as if the process were out of control and starts looking for an assignable cause for the bad point. When he plots a point on the chart, he is testing the hypothesis that the process mean is at its usual value. If the point falls within the control lines, the engineer accepts the hypothesis and decides that the process is in control. If the point falls outside that acceptance region, the engineer rejects the hypothesis and takes the appropriate action.

A quality control inspector takes a random sample of items from a lot. If there are too many defectives, he rejects the lot; otherwise, he accepts it. The engineer is testing the hypothesis that the percent defectives in the lot is at some predetermined acceptable value. The number of defectives is the test statistic; if it falls in the acceptance region, the engineer accepts the hypothesis; if it is too high and falls outside the acceptance region, the engineer rejects the hypothesis. Testing for attributes, using the binomial distribution, is discussed further in chapters 11 and 12 under the heading of acceptance sampling.

In these two examples, the hypothesis being tested involved a statistical model. In the control chart, the engineer looked at a random variable that was taken to be (normally) distributed with a certain expectation, μ. The hypothesis was that μ was equal to some particular value, μ_0. In the second example, the hypothesis was made about the parameter, p, of a binomial variable.

Hypothesis testing is also used in comparative experiments. A chemical engineer may have a choice of two alkaline reagents in a process. A few runs using sodium hydroxide and a few using potassium hydroxide are made and their averages are compared. The engineer may conclude that one reagent is clearly better than the other or else that there is not an important difference between them so far as the quality of the product is concerned. For the latter, the choice might be made on cost or other grounds. The engineer is testing the hypothesis that the means for the two reagents are equal against the alternative that they are not.

There is a strong connection between confidence intervals and hypothesis testing. The chemical engineer in the laboratory could set up a confidence interval for the difference between the two means. If the interval included zero, the engineer could accept the hypothesis that the means were equal. The data are compatible with that hypothesis. If the interval did not include zero, the hypothesis of equality would be rejected. The confidence interval and the hypothesis test are mathematically the same. In this case, the engineer should calculate the interval and report that as well. The information on the size of the difference between the means is important to future decision making.

In the previous examples, the emphasis was on what are called decision rules. The quality control engineer and the inspector both need rules to apply to take appropriate actions. The inspector does not want a confidence interval for the fraction defective. He wants a rule for saying yes or no. This is the classical hypothesis-testing situation.

7.2. AN EXAMPLE

In the next eight sections, we develop the theory of testing hypotheses about the mean of a normal distribution in the framework of an example. Suppose that we purchase 200-ohm resistors from a supplier. One morning, a large batch is delivered. We wish to determine whether they meet the specification of 200 ohms. We do not expect that each resistor will measure exactly 200 ohms; our experience with this supplier has been that their resistors have a standard deviation of 8 ohms (a number chosen to make the arithmetic simple). We can live with that amount of variability. Our problem is to decide whether the average resistance of the items in the batch is 200 ohms.

Our probability model for the process is that the resistance, X, of an item is a normal random variable with expectation μ and standard deviation 8. We wish to test the hypothesis

$$H_0: \quad \mu = 200.$$

We decide (arbitrarily for the present) to take a random sample of $n = 16$ resistors, measure the resistance of each, and calculate the average resistance, $\bar{x}$. This will be our test statistic. We know that it is an unbiased estimator of μ and that its standard deviation is $8/\sqrt{n} = 2$.

7.3. THE DECISIONS

There are two decisions that can be made on the basis of the test data. We can accept the hypothesis or we can reject it. Rejecting the hypothesis means deciding that the average resistance is not 200 ohms, and, presumably, refusing to buy that batch. If we accept the hypothesis, we decide that the batch of resistors meets our specification, and we buy it. How do we decide? If $\bar{x}$ is close to the hypothesized value of 200, we argue that the sample is compatible with the hypothesis. If $\bar{x}$ is not close to 200, we argue that the hypothesis is false and reject it.

How do we define close? Look first at two extreme possibilities. We might decide to accept the hypothesis if $\bar{x}$ fell within the range $150 < \bar{x} < 250$, and reject if $\bar{x}$ fell outside. The interval $150 < \bar{x} < 250$ would be the acceptance region. The two intervals $\bar{x} \geq 250$ and $\bar{x} \leq 150$ would constitute the rejection (or critical) region. Neither we nor the supplier would doubt that the batches we rejected were indeed substandard, but, with acceptance limits so far apart, many bad batches would get through. On the other hand, if the acceptance interval were $199.8 < \bar{x} < 200.2$, we should, since $P(\bar{x} \geq 200.2) = P(\bar{x} \leq 198.8) = P(z \geq 0.10) = 0.46$, reject 92% of the batches for which μ was actually equal to 200. Those limits are unreasonably restrictive.

We can make two errors in the decision process. The type I error is to reject the hypothesis when it is true. The type II error is to accept the hypothesis when it is false. The probabilities of making these errors are conventionally denoted by alpha and beta, respectively. The choice of a suitable acceptance region involves balancing these risks in some way. Both risks can be reduced by increasing the sample size at the price of increased inspection costs. For a given sample size, one risk can only be reduced at the expense of increasing the other risk.

We reject the hypothesis if $\bar{x}$ is "far" from 200. There is strong evidence that the hypothesis is false. When the acceptance limits are wide, we are strongly persuaded by data with $\bar{x}$ outside those limits that the hypothesis is false. On the other hand, accepting the hypothesis does not really mean agreeing that the hypothesis is correct. Accepting the hypothesis means that there is *not* strong evidence to reject it. Hypothesis testing is like a judicial process. We start by assuming that the prisoner is innocent. The prisoner is declared to be guilty only if there is a preponderance of evidence that causes the jury to reject the hypothesis of innocence. Otherwise, the hypothesis of innocence is accepted; the prisoner may actually be guilty, but the evidence is not strong enough for conviction. Scottish criminal law used to allow three possible verdicts: guilty, not guilty, and the middle ground—not proven.

7.4. NULL AND ALTERNATE HYPOTHESES

We actually consider two hypotheses. The hypothesis that we have mentioned, H_0, is called the null hypothesis. The other hypothesis, denoted by H_a, or H_1, is called the alternate hypothesis. When we reject H_0, we accept H_a; when we accept H_0, we reject H_a. In the example of the resistors, the alternate hypothesis is

$$H_a: \quad \mu \neq 200.$$

It is the two-sided alternative. The null hypothesis is to be rejected if the average resistance is either much less than 200 or much greater than 200. Note that the null hypothesis gets the benefit of the doubt—the cases in which the old Scottish juries would have returned a verdict of not proven.

7.5. ACCEPTANCE AND REJECTION REGIONS

The usual procedure in designing a test to choose the test statistic and the desired level of alpha, and then to calculate the acceptance region. We mean by the acceptance region the set of values of the test statistic that result in our accepting the hypothesis. The other values of the statistic constitute the rejection region, or the critical region. Values of the test statistic that are in the critical region are said to be (statistically) significant. The traditional choices of level for alpha are 5% and 1%. An engineer testing with $\alpha = 5\%$ is arguing that a value of $\bar{x}$, which is so far from the hypothesized value that it could only have happened less than one time in twenty, is strong evidence that the null hypothesis is false.

To calculate the acceptance region with $\alpha = 5\%$ for the resistor example, we argue that if H_0 is true, then x has a normal distribution with mean 200.0 and standard deviation 2.0. The rejection region is chosen so that $\bar{x}$ will fall inside it only 5% of the time. The acceptance region is, therefore, to contain $\bar{x}$ 95% of the time, and so it is defined by

$$200.0 - 1.96(2) < \bar{x} < 200.0 + 1.96(2)$$

$$196.1 < \bar{x} < 203.91.$$

Now we have a decision rule, or test procedure: take 16 resistors at random from a batch, test them, and average the 16 readings. If $\bar{x} \leq 196$ or if $\bar{x} \geq 204$, reject the batch. If $196 < \bar{x} < 204$, accept the batch. In this example, rejecting the batch may mean several possible courses of action, from scrapping the whole batch to testing 100% of them.

The acceptance region that we have just obtained looks like a confidence interval; there is a very strong connection. In this example, a two-sided 95%

confidence interval for μ is

$$\bar{x} - 3.92 < \mu < \bar{x} + 3.92.$$

The interval contains the hypothesized value, $\mu = 200$, if

$$\bar{x} - 3.92 < 200$$

and

$$\bar{x} + 3.92 > 200,$$

i.e.,

$$196.1 < \bar{x} < 203.9,$$

which is identical with the acceptance region.

7.6. ONE-SIDED TESTS

Suppose that a company that runs a large fleet of cars contracts to purchase gasoline in large quantities from a major oil company. A tank truck of gasoline arrives at the buyer's garage. Several samples of the gasoline are drawn and their octane numbers are measured. (In the real world, this would be very unlikely because few buyers have such test facilities.) The buyer wants to have gasoline with an octane number of 90.0 or higher. The buyer does not care if the oil company supplies higher-octane gasoline than specified, but does want protection against gasoline with an ON less than 90.0. The truckload will only be accepted if $\bar{x}$ exceeds 90.0 by some "safety factor," so the buyer will use a decision rule such as "buy the load if, and only if, $\bar{x} > 90.1$, or some other figure."

In terms of hypothesis testing, the buyer would take for hypotheses

$$H_0: \quad \mu = 90.0 \text{ (or } \leq 90.0\text{)} \quad \text{and} \quad H_a: \quad \mu > 90.0.$$

When the null hypothesis is rejected, the buyer agrees to buy that load of gasoline. The rejection region that the buyer chooses is the upper tail of the distribution of $\bar{x}$, i.e., the set of values defined by $\bar{x} > c$, where c is a value calculated for this test. This is a one-sided test, and the buyer will buy the batch only if the data convince him that the gasoline is not below specification.

The vendor may look at the problem differently. The vendor would prefer a situation in which the customer had to buy the batch of gasoline unless it was established that the batch was below specification, i.e., unless $\bar{x}$

was significantly *less* than 90.0. The vendor would prefer the hypotheses

$$H_0: \quad \mu = 90.0 \text{ (or } \geq 90.0) \quad \text{and} \quad H_a: \quad \mu < 90.0.$$

They are looking at the situation from two different sides; each party is looking after its own interests.

This example illustrates two points. We can have either one-sided or two-sided alternatives to the null hypothesis. In the same situation, different parties may choose different alternatives. The choice between one and two sides rests with the person designing the test. The essential question is this: Do we want to reject H_0 only if $\bar{x}$ is too high, or only if $\bar{x}$ is too low, or if $\bar{x}$ is either too high or too low? The correct choice is usually clear from the context of the problem. A good rule of thumb is to use a two-sided alternative unless a convincing case can be made for either of the single-sided alternatives.

If in the example of the resistors, the alternative had been $H_a: \mu > 200.0$ with $\alpha = 5\%$, the critical region would have been

$$\bar{x} > 200 + 1.645(2), \qquad \bar{x} > 203.3.$$

Notice that the upper limit in the one-tailed test is lower than the limit in the two-tailed test. A value $\bar{x} = 203.5$ would call for accepting H_0 in the two-tailed case and for rejecting it in the other. By choosing the critical region appropriate to the one-sided alternative, we have improved the power of the test to spot departures from specification on the high side. The cost that we pay is that we can no longer spot departures on the low side, but in a particular situation, we may not care about that.

7.7. THE CHOICE OF ALPHA

The choice of alpha depends on the consequences of making a type I (or a type II) error. This is illustrated in the following example. Suppose that we are considering replacing our present process for making a chemical product by another process. The current process converts 60% of the raw material to finished product; we would like to increase that figure. The chemists in the laboratory come up with a possible new process. We make some runs with their new process and calculate the average yield. We wish to test

$$H_0: \quad \mu = 60.0 \text{ against } H_a: \quad \mu > 60.0.$$

If the average yield with the new process is significantly greater than 60.0, we reject the null hypothesis. This means that we decide to carry out further investigation into the new process at the pilot plant level. The cost of a type I error is a few weeks' work on the pilot plant. The cost of a type II error is the loss of an exciting new process that could make the company a lot of money. At this stage we can afford to take a fairly large value of alpha (and a correspondingly small value of beta).

At the next stage, the consequences of an incorrect decision are different. The hypotheses tested after the pilot plant runs are the same, but now the consequence of rejecting the null hypothesis is that the company will choose the new process and build a new plant at a cost of several million dollars. Now we had better be sure that we do not erect a costly monument to our misjudgment. We want alpha to be small, a very conservative test. As for beta, that is a matter of opinion; a small beta would be nice, too. With luck, we will be allowed enough time and research budget to achieve that also.

The last example in this section concerns drug testing. Suppose that a pharmaceutical company proposes a new drug that allegedly alleviates the symptoms of some illness. The usual procedure is to select two groups of patients. One group is given a course of pills made from the new drug. The others are given the placebo—pills that do not contain any drug (although the recipients are unaware of this). The responses of the patients are noted and a test statistic is calculated. It might be the difference in the numbers of attacks per patient in the two groups. The null hypothesis is that the new drug is of no use—the two means are equal. Unless the scientists think that there is some danger that the new drug will actually make the patients worse, the test has a one-tailed alternative. The new drug will be approved only if the patients who receive it have significantly fewer attacks per person than the control group. Rejecting the null hypothesis means approving the new drug for use. That is a very serious step, which calls for a low value of alpha.

7.8. THE z-TEST

There are two important advantages in the choice of $\bar{x}$ as the test statistic in the problems of the previous sections. It is a good estimator of μ. Indeed, by most standards, it is the best estimator. This brings us back to the subject that was only briefly touched upon in section 6.15: $\bar{x}$ is the minimum-variance unbiased estimator of the parameter that is being investigated. The second virtue is that we know how the statistic behaves. It is normally distributed; that property was used to obtain the acceptance limits. We can rephrase the development of the test in the following way.

We know that $\bar{x}$ is the statistic that we would use to estimate μ, and that $\bar{x}$ takes values (at least in theory) anywhere on the real line from $-\infty$ to $+\infty$. We can regard $\bar{x}$ as a summary, or reduction, of the set of n data points into a single value. The mathematician sees the sample as a point in n-dimensional space. Thus, $\bar{x}$ is a mapping of that point on to the real line. The real line is then divided into two regions, acceptance and rejection; there is the test. We could just as well use any equivalent mapping. We do not have to use $\bar{x}$. We can use any function of $\bar{x}$ that has a one-to-one correspondence. So we may as well focus on a function of $\bar{x}$ that we can handle.

This is the rationale for the z-test. Instead of focusing on $\bar{x}$ directly, we focus on

$$z = \frac{\bar{x} - \mu_0}{\sigma/\sqrt{n}},$$

where the null hypothesis is H_0: $\mu = \mu_0$. When the null hypothesis is true, z has a standard normal distribution. Thus, if, for example, we wish to have a two-sided test with $\alpha = 5\%$, we accept the null hypothesis if $|z| < 1.96$.

The z-test is no different from the test using $\bar{x}$ itself as the statistic. It has the advantages, at least pedagogically, of being more general in its statement and of pointing out more clearly the connection between tests and confidence intervals.

7.9. THE TYPE II ERROR

The one-sided test that was developed earlier for resistors tests $\mu = 200$ against the alternative, $\mu > 200$. With $\sigma = 8$ and $n = 16$, the rejection region is $\bar{x} > 203.3$. What if the null hypothesis is false, and the mean resistance is not 200.0, but has some other value? What, for example, is the type II error when $\mu = 205.0$?

In this case, $\bar{x}$ is normally distributed with mean 205.0 and standard deviation 2.0, and the corresponding z statistic is $(\bar{x} - 205)/2$. Measuring in units of the standard deviation of $\bar{x}$, we have $203.3 - 205.0 = -1.7 = -0.855\sigma(\bar{x})$. We write $F_z(z)$ for the cumulative distribution of the standard normal distribution. Then

$$\beta = F_z(-0.855) = 0.20.$$

The alpha risk is 5%; the beta risk is 20%.

Similar calculations show that when $\mu = 204.0$, $\beta = 36\%$, and when $\mu = 203.0$, $\beta = 56\%$. Thus, the value of beta is seen to depend on the value of μ.

Beta is the risk of failing to reject the null hypothesis when we should do so. The power of the test is defined as $1 - \beta$. It measures the ability of the test to discriminate between μ and the value μ_0 in the null hypothesis. A plot with μ along the horizontal axis and the power along the vertical axis is called the Operating Characteristic (OC curve) for the test. The OC curve for this example is shown in figure 7.9.1.

The OC curve is the plot of

$$1 - F_z\left(\frac{c - \mu}{\sigma/\sqrt{n}}\right) = F_z\left(\frac{\mu - c}{\sigma/\sqrt{n}}\right),$$

THE TYPE II ERROR

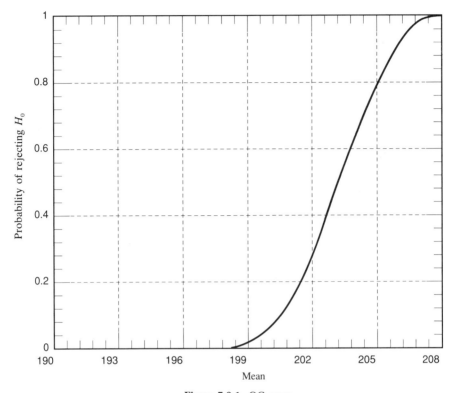

Figure 7.9.1. OC curve.

where the rejection region is $\bar{x} > c$. It is the cumulative normal curve, or ogive. When it is plotted on normal probability paper, the graph is a straight line. It is then determined by only two points, which leads to the next consideration.

Suppose that the engineer is prepared to take a 5% alpha risk of accepting a batch with resistance as low as 200.0 ohms. Suppose also that the engineer wants to have a 90% power of rejecting the null hypothesis if the resistance is as high as 203.0 ohms. How should the test be designed? The engineer will choose a one-sided test with the critical region defined by $\bar{x} > c$. Two numbers have to be specified: c and the sample size, n.

When the null hypothesis is true, $\bar{x}$ is normally distributed about a mean 200.0 with standard deviation $8/\sqrt{n}$. To satisfy the requirement that $\alpha = 5\%$, we have

$$c = 200.0 + (1.645)\left(\frac{8}{\sqrt{n}}\right). \tag{7.9.1}$$

If $\mu = 203.0$, $\bar{x}$ is normally distributed with mean 203.0 and the same standard deviation. The power requirement leads to

$$c = 203.0 - (1.28)\left(\frac{8}{\sqrt{n}}\right). \tag{7.9.2}$$

Solving equations (7.9.1) and (7.9.2), we get

$$n = 61 \quad \text{and} \quad c = 201.7.$$

The derivation of a general formula is left as an exercise for the reader. Let z_α and z_β denote the values of z that correspond to our choices of alpha and beta (in the example, they were 1.645 and 1.28, respectively), and let the difference between the means in the two hypotheses be δ. Then the sample size required is given by

$$n = \frac{(z_\alpha + z_\beta)\delta}{\sigma}. \tag{7.9.3}$$

7.10. THE *t*-TEST

If we do not know the variance of the hypothesized normal population, we have to estimate it from the data. We did this earlier in section 6.11 with confidence intervals. The normal test statistic z is replaced by Student's t:

$$t = \frac{\bar{x} - \mu_0}{s/\sqrt{n}}. \tag{7.10.1}$$

The t statistic has the same number of degrees of freedom as the estimate, s^2, of the variance. In this case, the estimate is obtained from a sample of n observations, and so there are $n-1$ d.f. The value of t obtained from the sample is compared to t^*, which is the tabulated value for $n-1$ d.f. and the desired value of alpha. If the observed value falls in the critical region, the null hypothesis is rejected.

Example 7.10.1. In example 6.11.1, a sample of ten resistors was taken and their resistances were measured. The average resistance was $\bar{x} = 95.17$ with estimated standard deviation $s = 9.36$. Suppose that we wish to test

$$H_0: \ \mu = 90.0 \text{ against } H_a: \ \mu \neq 90.0.$$

The test statistic is

$$t = \frac{95.17 - 90.0}{9.36/\sqrt{10}} = 1.746.$$

If $\alpha = 5\%$ with the two-sided alternative, there is to be probability 2.5% in each tail, and the appropriate tabulated value for t with nine d.f. is

$$t^* = 2.262.$$

The calculated value 1.746 is less than 2.262, and so the null hypothesis is accepted. □

The engineer could just report the bald fact that the t value was not significant, and, therefore, one cannot conclude that $\mu \neq 90.0$. That is not good enough. The 95% confidence interval obtained earlier is

$$88.47 \leq \mu \leq 101.87.$$

There is much more information here. The interval gives us a range of values of μ; it contains the hypothesized value 90.0. Thus, the data are compatible with the null hypothesis, but only barely. As the sample size increases, t^* decreases. We can see from the confidence interval that a larger sample with a correspondingly lower value of t^* might have led to rejection of the null hypothesis. We can also go back to the tables and note that had we taken alpha to be 10%, we should have had $t^* = 1.833$, which is much closer to the calculated value of 1.746. Clearly, in some problems, we might accept the null hypothesis at $\alpha = 5\%$ and reject it if we chose $\alpha = 10\%$.

The Minitab printout for the t-test is as follows:

```
MTB > set in c11
MTB > 114.2  91.9  107.5  89.1  87.2  87.6  95.8  98.4  94.6  85.4
MTB > end
MTB > ttest 90 c11
TEST OF MU = 90.000 VS MU N.E. 90.000
        N      MEAN    STDEV   SE MEAN      T    P VALUE
C11    10     95.170   9.364    2.961     1.75    0.11
```

The P value is the smallest value of alpha at which the observed value of t, 1.75, would be significant. It is the area under the density curve for the t distribution between 1.75 and infinity and can now be easily calculated by modern computers. The observed difference between $\bar{x}$ and μ would have been declared significant if, for some peculiar reason, the engineer had chosen to assign alpha a value of 0.11 or higher. Calculating and printing the P value clearly avoids having to look up t^* in the table.

7.11. TESTS ABOUT THE VARIANCE

In section 6.12, we used the chi-square distribution to construct confidence intervals for the variance of a normal population. We can construct tests in a similar way. Examples 6.11.1 and 6.12.1 both deal with the sample of ten resistors. The sum of squares of deviations is $S = 789.141$, and we calculated a 95% confidence interval for the variance

$$41.48 \leq \sigma^2 \leq 292.27.$$

The point estimate is $s^2 = 87.68$.

Suppose that we wish to test the hypothesis

$$H_0: \sigma^2 = 60.00$$

against H_a: $\sigma^2 > 60.00$.

Under H_0, the statistic $S/60 = 13.153$ has a chi-square distribution with nine d.f. For a test with $\alpha = 0.05$, therefore, we compare this value to the 95% value, 16.92, from the tables.

Since $13.153 < 16.92$, we accept the null hypothesis. Even though the point estimate of 87.88 is nearly 50% higher than the hypothesized value, we cannot say, with alpha set at 5%, that the hypothesis is false. The reason is that there was not enough data, only nine d.f.

We can rewrite the test statistic, S/σ^2, as $(n-1)s^2/\sigma^2$. Suppose that we had the same estimate, 87.68, based on 31 observations. Then the test statistic would have been $30(87.68)/60 = 43.84$, which is greater than the tabulated value of 43.77 with 30 d.f. In that case, the null hypothesis would have been rejected.

7.12. THE EXPONENTIAL DISTRIBUTION

Chi-square statistics can also be used in tests about exponential distributions. In example 6.13.1, we obtained a 95% confidence interval for the parameter of an exponential distribution. There were ten lamps in the sample, and their average lifetime was 252 hours.

To test the hypothesis H_0: $\theta = 160$ against H_a: $\theta > 160$, we recall that $2 \Sigma x_i/\theta$ has a chi-square distribution with $2n$ d.f.

If H_0 is correct, the statistic

$$\frac{2 \Sigma x_i}{\theta} = \frac{20(252)}{160} = 31.5$$

has a chi-square distribution with 20 d.f. With $\alpha = 5\%$, we compare the calculated value 31.5 to the tabulated 95% value, 31.4, and reject the null hypothesis.

7.13. TESTS FOR THE PERCENT DEFECTIVE

In example 6.8.1, we considered a sample of 100 components, of which 20 were defective. We obtained a confidence interval for the percent defective in the population by using a normal approximation to the binomial distribution. The same approximation can be used for testing hypotheses about the percentage, p. This is illustrated in example 7.13.1.

CHI-SQUARE AND GOODNESS OF FIT   125

Example 7.13.1. A sample of 100 components has 20 defectives. Test at $\alpha = 0.05$ the null hypothesis: H_0: $p = 0.25$ against the alternative H_1: $p \neq 0.25$.

We argue that under the null hypothesis, the number of defectives, X, has a normal distribution with expectation $np = 25$ and variance $npq = 18.75$. The test statistic is

$$z = \frac{x - np}{\sqrt{npq}} = \frac{-5}{4.33} = 1.155,$$

and the evidence is not strong enough for us to reject the null hypothesis. It would not be strong enough for the one-tailed test with the alternative H_1: $p < 0.25$ either. □

7.14. CHI-SQUARE AND GOODNESS OF FIT

We have been using the chi-square distribution as the distribution of the sum of squares of standard normal variables. It has led us to tests about the variance of a normal distribution and the expectation of an exponential variable. However, the history of chi-square goes back earlier. In 1900, Karl Pearson introduced chi-square tests in the context of binomial and multinomial distributions. In this section, we discuss these goodness-of-fit tests, beginning with another look at the binomial test of the previous section.

Example 7.13.1 (cont.). We consider a model with two "bins": defective and nondefective. The observed numbers in the bins are 20 and 80, respectively; the expected numbers are 25 and 75, respectively. The deviations in the bin numbers are thus $20 - 25 = -5$ and $80 - 75 = +5$, respectively (notice that the deviations sum to zero). This is shown in table 7.14.1.

The chi-square test statistic is

$$U = \sum \frac{(O_i - E_i)^2}{E_i} = \sum \frac{\text{dev}^2}{E}.$$

In this example,

$$U = \frac{25}{25} + \frac{25}{75} = 1.33.$$

Table 7.14.1

	Defective	Nondefective
Observed (O)	20	80
Expected (E)	25	75
Deviation (dev) = $O - E$	−5	+5

We compare the value of U to the tabulated value, 3.84, for chi-square with one d.f., and since $1.33 < 3.84$, accept the null hypothesis that $p = 0.25$. There is only one d.f. because the two deviations are forced to sum to zero. In the example, $U = 1.33 = (1.155)^2$ and the tabulated value of chi-square is $3.84 = (1.96)^2$. Comparing 1.33 to 3.84 is equivalent to comparing 1.155 to 1.96; the z-test and the chi-square test are mathematically the same. This equivalence is always true for the binomial test; the derivation appears as an exercise.

It is essential that the test be applied using the actual numbers in the bins and *not* the percentages. Had there been 300 components with 60 defectives, the percent defective would have been the same, but the value of U would have been tripled.

$$U = \frac{225}{75} + \frac{225}{225} = 4.00,$$

which exceeds the critical value, 3.84. □

7.15. MULTINOMIAL GOODNESS OF FIT

This test can be extended to the multinomial distribution with several bins. The extension is illustrated by two more examples.

Example 7.15.1. Suppose that an electronic component undergoes a two-stage testing procedure before being put in service. Out of 160 components tested, 18 failed at the first stage and were rejected, 20 failed at the second stage, and the others passed. It had been expected that 15% would fail at the first stage and 20% at the second, with 65% passing both stages. We wish to test the hypothesis H_0: $p_1 = 0.15$, $p_2 = 0.20$, and $p_3 = 0.65$ against the alternative that those are not the correct values.

	Stage 1	Stage 2	Pass
O	18	20	122
E	24	32	104
$O - E$	-6	-12	$+18$

$$U = \frac{36}{24} + \frac{144}{32} + \frac{324}{104} = 9.12.$$

We note again that the sum of the deviations is zero. There are, therefore, two d.f., and we compare U to the tabulated value of chi-square, 5.99. Since $U > 5.99$, we reject the null hypothesis. □

Example 7.15.2. In table 3.2, we listed the frequencies of the digits 1 through 9 in a sample of 320 digits. The frequencies were

	0	1	2	3	4	5	6	7	8	9
frequency	40	36	29	33	29	34	32	30	31	26

A quick glance might suggest that there are too many zeros and not enough nines. Is the data compatible with the assumption that each digit occurs 10% of the time in the parent population?

For each digit, the expected number of occurrences is 32. The deviations are, therefore,

$$8 \quad 4 \quad -3 \quad 1 \quad -3 \quad 2 \quad 0 \quad -2 \quad -1 \quad -6$$

and $U = 144/32 = 4.5$. That is considerably less than the tabulated value of chi-square with nine d.f. and $\alpha = 5\%$, which is 16.92. □

EXERCISES

7.1. A sample of ten devices was tested for speed. The frequencies of the devices in MHz were

10.80 10.90 7.60 9.90 11.00 8.90 8.20 11.30 8.50 12.00

Test at $\alpha = 5\%$ the hypothesis $\mu = 11.25$ vs. the alternative $\mu \neq 11.25$.

7.2. A research group is trying to produce silicon-on-sapphire films with a thickness of 0.2 μm. A sample of nine of their films gives the following thicknesses:

0.20 0.11 0.19 0.17 0.26 0.17 0.27 0.25 0.17

Test the hypothesis that the actual average thickness by their method is indeed 2.0 μm against the alternative that it is not.

7.3. Five measurements of the octane number of a gasoline are

80.0 79.8 79.6 80.2 80.2

The standard deviation of an octane determination is 0.3. Test at $\alpha = 5\%$ the hypothesis that the octane number is 79.7 against the alternative that it exceeds 79.7.

7.4. In exercise 7.3, how many measurements should be made if β is to be 10% when the octane number is 80.0?

7.5. In exercise 6.2, test the hypothesis H_0: $\mu = 12.5$ against the alternative $\mu \neq 12.5$.

7.6. In exercise 6.3, test the hypothesis H_0: $\mu = 13.00$ against the alternative $\mu > 13.0$ at $\alpha = 10\%$.

7.7. A sample of seven items from a normal distribution has a sample variance $s^2 = 6.04$. Test, at $\alpha = 5\%$, the hypothesis $\sigma^2 = 4.0$ against the alternative hypothesis $\sigma^2 > 4.0$. Is an apparent 50% increase in the variance significant if it is based on only seven observations?

7.8. We wish to test the variance of a normal distribution. Following exercise 7.7, how many observations should we take in order for a 50% increase in the sample variance, s^2, over the hypothesized variance to be significant?

7.9. A sample of 18 items are put on test. The average lifetime is 2563 hours. Assuming that the lifetimes are exponentially distributed, test the hypothesis H_0: $\theta = 2400$ against the alternative H_a: $\theta > 2400$, using $\alpha = 5\%$. What is the value of β if $\theta = 3000$?

7.10. In a sample of 80 freshmen engineering students, 32 were exempt from the first semester of calculus. In previous years, the fraction of students exempt was $p = 0.31$. In this an unusually good year?

7.11. A poll is conducted on the eve of a closely contested congressional race. Out of 1300 voters polled, 703 are Democrats. Let p denote the fraction of Democrats in the electorate. Test H_0: $p = 0.50$ against the alternative H_a: $p > 0.50$.

7.12. The company gardener plants 120 flowers in a large bed outside the main entrance to the factory. When they bloom, 81 are white, 20 are red, and the rest are particolored. Is this sample compatible with the claims of the nursery that white, particolored, and red flowers ought to occur in the ratios $9:6:1$?

7.13. A company buys components from two vendors. A sample of 50 components from vendor A has a mean time between failures (MTBF) = 402 hours; a sample of 50 from vendor B has MTBF = 451 hours. Test the hypothesis that the average MTBF is the same for both vendors against the alternative that it is not.

CHAPTER EIGHT

Comparative Experiments

8.1. INTRODUCTION

In chapters six and seven, we discussed some common statistical procedures in which the emphasis was on drawing conclusions about a population from a single sample. We considered, for example, estimating and testing hypotheses about the mean, μ, and variance, σ^2, of a normal distribution from a sample of n observations. We turn now to comparative experiments in which we compare two populations and ask whether they are the same or whether they differ. Is the yield of our plant the same whether we use hydrochloric or nitric acid as reagent? Is the yield of wafers the same with two different etching regimes? Do we get less variability in our product when we lower the operating temperature?

We begin by comparing the means of two normal populations, first, when their variances are known, and, later, when the variances have to be estimated from the data. This is followed by Wilcoxon's nonparametric test, which does not assume normality. Then we discuss comparing the variances of two normal populations and comparing the means of two exponential populations, which are similar problems. The chapter ends with a discussion of comparing two binomial populations and the use of chi-square for analyzing 2×2 tables.

8.2. COMPARING TWO NORMAL MEANS WITH KNOWN VARIANCE

Suppose that we have two random samples. The first, $x_1, x_2, \ldots, x_m$, comes from a normal population with unknown mean μ_1 and known variance σ_1^2. The second sample, consisting of $y_1, y_2, \ldots, y_n$, comes from a normal population with unknown mean μ_2 and known variance σ_2^2. We wish either to estimate the difference, $\mu_1 - \mu_2$, or to test the hypothesis H_0: $\mu_1 = \mu_2$.

We estimate μ_1 by $\bar{x}$. Since the first sample has m independent observations, $\bar{x}$ is normally distributed with expectation μ_1 and variance σ_1^2/m.

Similarly, $\bar{y} \sim N(\mu_2, \sigma_2^2/n)$. Their difference, $d = \bar{x} - \bar{y}$, is an unbiased estimate of $\mu_1 - \mu_2$ with variance $\sigma_1^2/m + \sigma_2^2/n$.

If, in equation (6.7.1), we replace $\bar{x}$ by $\bar{x} - \bar{y}$ and $\sigma/\sqrt{n}$ by $\sqrt{\sigma_1^2/m + \sigma_2^2/n}$, we obtain the following two-sided 95% confidence interval for $\mu_1 - \mu_2$:

$$\bar{x} - \bar{y} - 1.96\sqrt{\sigma_1^2/m + \sigma_2^2/n} < \mu_1 - \mu_2 < \bar{x} - \bar{y} + 1.96\sqrt{\sigma_1^2/m + \sigma_2^2/n}. \tag{8.2.1}$$

Similarly, a z-test for the hypothesis H_0: $\mu_1 - \mu_2 = 0$ is obtained by setting

$$z = \frac{\bar{x} - \bar{y}}{\sqrt{\sigma_1^2/m + \sigma_2^2/n}}. \tag{8.2.2}$$

If $\sigma_1^2 = \sigma_2^2 = \sigma^2$, equations (8.2.1) and (8.2.2) take the simpler forms

$$\bar{x} - \bar{y} - 1.96\sigma\sqrt{1/m + 1/n} < \mu_1 - \mu_2 < \bar{x} - \bar{y} + 1.96\sigma\sqrt{1/m + 1/n} \tag{8.2.3}$$

and

$$z = \frac{\bar{x} - \bar{y}}{\sigma\sqrt{1/m + 1/n}}. \tag{8.2.4}$$

Example 8.2.1. A chemical engineer has a process in which either of two reagents, A or B, can be used. Twelve runs with each of the reagents are made and the following yields are observed:

$A(x)$: 71.1 68.3 74.8 72.1 71.2 70.4
         73.6 66.3 72.7 74.1 70.1 68.5
$B(y)$: 73.3 70.9 74.6 72.1 72.8 74.2
         74.7 69.2 75.5 75.8 70.0 72.1.

A dotplot of both samples on the same scale (see Figure 8.2.1) suggests that the yields with B might be higher. The two sample means are $\bar{x} = 71.1$ and $\bar{y} = 72.93$.

If we assume that the variance of A yields is 6.0, and the variance for B is 4.0, a 95% confidence interval for the difference between the population means is

$$71.10 - 72.93 - 1.96\sqrt{6/12 + 4/12} < \mu_1 - \mu_2 < -1.83 + 1.79$$
$$-3.62 < \mu_1 - \mu_2 < -0.04.$$

The z statistic for the hypothesis that $\mu_1 = \mu_2$ is

$$z = \frac{1.83}{0.913} = 2.00,$$

which is barely significant. □

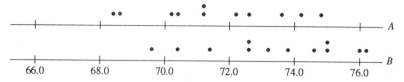

Figure 8.2.1. Dotplot for example 8.2.1.

8.3. UNKNOWN VARIANCES

The situation when the variances are unknown has to be separated into two cases. We deal first with the case in which it is reasonable to assume that the two variances are equal to σ^2.
Let

$$S_1 = \sum (x_i - \bar{x})^2 \quad \text{and} \quad S_2 = \sum (y_i - \bar{y})^2.$$

Then S_1/σ^2 has a chi-square distribution with $m - 1$ d.f., and S_2/σ^2 has a chi-square distribution with $n - 1$ d.f.

$$E(S_1) = (m-1)\sigma^2 \quad \text{and} \quad E(S_2) = (n-1)\sigma^2,$$

and so

$$\begin{aligned} s^2 &= \frac{S_1 + S_2}{m + n - 2} \\ &= \frac{(m-1)s_1^2 + (n-1)s_2^2}{m + n - 2} \end{aligned} \quad (8.3.1)$$

is an unbiased estimate of σ^2 with $m + n - 2$ d.f. By substituting s for σ in equation (8.2.4), the test statistic becomes

$$t = \frac{\bar{x} - \bar{y}}{s\sqrt{1/m + 1/n}}. \quad (8.3.2)$$

This statistic has Student's t distribution with $m + n - 2$ d.f.

In a similar way, the confidence interval is obtained by substituting s for σ and t^* for 1.96 in equation (8.2.3), where t^* is the tabulated value for t with $m + n - 2$ d.f.

$$\bar{x} - \bar{y} - t^*s\sqrt{1/m + 1/n} < \mu_1 - \mu_2 < \bar{x} - \bar{y} - t^*s\sqrt{1/m + 1/n}. \quad (8.3.3)$$

We illustrate this by returning to example 8.2.1.

Example 8.2.1 (cont.). If we do not assume that the variances are known, we calculate estimates from the two samples: $s_1^2 = 6.513$ and $s_2^2 = 4.592$. The pooled estimate of the common variance is $s^2 = 5.541$.

The 5% value for t with 22 d.f. is $t^* = 2.074$, and so the two-sided 95% confidence interval for the difference is calculated by equation (8.3.3):

$$-1.83 - 2.074\sqrt{5.541/6} < \mu_1 - \mu_2 < -1.83 + 1.99$$
$$-3.82 < \mu_1 - \mu_2 < 0.16.$$

This interval contains zero, and so the data are compatible with the hypothesis H_0: $\mu_1 = \mu_2$.

The t-test statistic is

$$t = -\frac{1.83}{0.917} = -1.90,$$

which is less, in absolute value, than the tabulated value for alpha = 5%, 2.074, and so we do not reject the hypothesis of equality. □

8.4. UNEQUAL VARIANCES

When it is not reasonable to assume that the two variances are equal, the situation becomes difficult mathematically and is known to statisticians as the Behrens–Fisher problem. We can certainly estimate the two variances by

$$s_1^2 = \frac{S_1}{m-1} \quad \text{and} \quad s_2^2 = \frac{S_2}{n-1},$$

and we can then write the test statistic as

$$t' = \frac{\bar{x} - \bar{y}}{\sqrt{s_1^2/m + s_2^2/n}}. \qquad (8.4.1)$$

The mathematical problem is that this statistic does not have Student's distribution and the actual distribution cannot be derived in a tractable form. There does exist an approximation; before the days of computers, this approximation was awkward to use, but now it is routinely available in many software packages.

If m and n are greater than about 20, you are not going to get into serious trouble by acting as if equation (8.4.1) has a normal distribution, in which case a 95% confidence interval is given by

$$\bar{x} - \bar{y} - 1.96\sqrt{s_1^2/m + s_2^2/n} < \mu_1 - \mu_2 < \bar{x} - \bar{y} + 1.96\sqrt{s_1^2/m + s_2^2/n}.$$
$$(8.4.2)$$

Example 8.2.1 (cont.). For this data set, the Minitab program gives the value of t' as

$$t' = -1.91,$$

which should be compared to the tabulated value of t with approximately 22.4 d.f. The value of t' in this case is not much different from the t value of the previous section. □

8.5. THE PAIRED t-TEST

We treated the data in example 8.2.1 as if the 24 observations were all independent of each other. Suppose that the engineer was aware that there were two sources of variation in the process: random noise and also variability between batches of raw material. The engineer, therefore, took 12 batches of raw material and divided each into two halves; one-half received reagent A and the other received reagent B. Then the difference between the two observations made on the same batch was a measure of $\mu_1 - \mu_2$, free of the batch effect, which had canceled out.

This is called a paired experiment, and the appropriate analysis is called the paired t-test. Other expressions are also used. In semiconductor work, split-lot experiments are run in which a "lot," perhaps 24, of wafers is split into two halves for one step of the processing; for that stage and that stage alone, there are 12 wafers receiving treatment A and 12 receiving treatment B; the other sources of variation have canceled out.

In the earlier treatment of the reagent data, we had a simple model:

$$x_i = \mu_1 + e_{1i} \quad \text{and} \quad y_i = \mu_2 + e_{2i},$$

where μ_1 and μ_2 are the unknown constants whose difference we are trying to measure, and e_{1i}, e_{2i} are independent random errors, which we assume to have the same variance σ^2. Now we add the batch term and write

$$x_i = \mu_1 + \beta_i + e_{1i} \quad \text{and} \quad y_i = \mu_2 + \beta_i + e_{2i},$$

where β_i denotes the contribution from being in the ith batch. The variance of a single observation has two components, σ^2 and σ_b^2, the variance between batches; $V(x_i) = V(y_i) = \sigma_b^2 + \sigma^2$. The difference, $x_i - y_i$, is

$$d_i = \mu_1 - \mu_2 + (e_{1i} - e_{2i}).$$

The batch component has vanished and $V(d_i) = 2\sigma^2$. The test procedure consists, therefore, of subtracting y_i from x_i in each pair and carrying out a one-sample t-test on the differences.

To illustrate this, we take another look at the data of example 8.2.1.

Example 8.2.1 (cont.). The data have been rewritten to show the pairs and the differences:

Pair	1	2	3	4	5	6
A	71.1	68.3	74.8	72.1	71.2	70.4
B	73.3	70.9	74.6	72.1	72.8	74.2
d	−2.2	−2.6	0.2	0.0	−1.6	−3.8

Pair	7	8	9	10	11	12
A	73.6	66.3	72.7	74.1	70.1	68.5
B	74.7	69.2	75.5	75.8	70.0	72.1
d	−1.1	−2.9	−2.8	−1.7	0.1	−3.6

The one-sample t-test on the differences gives

$$t = 4.53 \quad \text{with 11 d.f.},$$

which is clearly significant. Notice that the number of degrees of freedom has dropped to 11 (12 − 1). Fewer degrees of freedom means that a higher value of t^* must be overcome to establish significance, but if the variance between batches is considerable, the increase in precision from eliminating it will more than compensate for that increase in the critical value of t.

□

8.6. WILCOXON'S TWO-SAMPLE TEST

During World War II, an industrial statistician, Frank Wilcoxon, developed a nonparametric test, i.e., a test that does not depend on assuming that the populations are normally distributed. At about the same time, H. Mann and D. R. Whitney, working at Ohio State University, developed a similar test that bears their names and is mathematically equivalent; the Mann–Whitney version is more popular among social scientists. Theoretically, the two-sample t-test is the "best" test when the conditions of normality and equal variance are met. However, the superiority of the t-test over Wilcoxon's test is not great even then. When there are departures from normality, Wilcoxon's test can be as good or better than the t-test.

The first step in this procedure is to arrange all $m + n$ observations in increasing order of size. From now on, we work with the ranks of the observations. The smallest observation has rank one; the largest has rank $m + n$, and so on. If there are ties, we give each of the tied observations the average of their ranks. Formally, we are testing the equality of the population medians rather than their means. If the medians are the same, the average rank of the X variables should be about the same as the Y average. If the X median is appreciably lower than the Y median, the X observations will tend to be clustered to the left and the Y observations on the right.

The test statistic is the sum of the ranks of either the X or the Y observations; they add up to $(m + n)(m + n + 1)/2$. One chooses whichever is more convenient, usually the sum that has the fewer observations to be added.

Consider the simple case where $m = n = 3$. The theory of the test can be derived by thinking of a box with three red balls and three white balls. We remove the balls from the box one at a time and note the order of the colors. There are 20 possible arrangements of six balls, three red and three white, and so there are 20 possible orderings (ignoring ties) of X and Y. They are

XXXYYY	XXYXYY	XXYYXY	XXYYYX	XYXXYY
XYXYXY	XYXYYX	XYYXXY	XYYXYX	XYYYXX
YXXXYY	YXXYXY	YXXYYX	YXYXXY	YXYXYX
YXYYXX	YYXXXY	YYXXYX	YYXYXX	YYYXXX

The sums of the X ranks for the 20 sequences are $W =$

6   7   8   9   8   9   10   10   11   12   9   10   11   11   12
13   12   13   14   15 .

The lowest value is 6, the highest 15; each of those scores has probability 1/20. We can now formulate a simple test for the hypothesis that the two populations have the same median, namely, reject the hypothesis if, and only if, $W = 6$ or $W = 20$, i.e., if X has the three lowest observations or if X has the three highest observations. For this test, $\alpha = 1/20 + 1/20 = 0.05$.

Tables of the critical values of W are available for various values of m and n. In the Mann–Whitney test, each X observation is given a score equal to the number of Y observations that exceeds it, and the test statistic is the sum of those scores.

When m and n are ten or more, we can act as if the sum of the X ranks is approximately normal with $E(W) = m(m + n + 1)/2$ and $V(W) = mn(m + n + 1)/2$; the sum of the Y ranks has $E(W) = n(m + n + 1)/2$ and the same variance. Note that we are not making any normality assumptions about the actual data. We are merely applying that approximation to the orderings of m red balls and n white balls as they come out of the box!

This is illustrated by going back to the original data of example 8.2.1 and ignoring the remarks about pairing that were made in section 8.5. We suppose only that we have 12 observations on A and 12 on B.

Example 8.2.1 (cont.). The 24 observations in ascending order are

1	2	3	4	5	6	7	8
66.3	68.3	68.5	69.2	70.0	70.1	70.4	70.9
X	X	X	Y	Y	X	X	Y

9	10	11	12	13	14	15	16
71.1	71.2	(72.1	72.1	72.1)	72.7	72.8	73.3
X	X	Y	Y	X	X	Y	Y

17	18	19	20	21	22	23	24
73.6	74.1	74.2	74.6	74.7	74.8	75.5	75.8
X	X	Y	Y	Y	X	Y	Y

Observations 11, 12, and 13 in order are all equal. They each receive the same rank, $(11 + 12 + 13)/3 = 12$.

The sum of the X ranks is

$$1 + 2 + 3 + 6 + 7 + 9 + 10 + 12 + 14 + 17 + 18 + 22 = 121.$$

(It is a good idea to check your calculations by adding the Y ranks, too; that total is 179; the two should add to $(24 \times 25)/2 = 300$.) Under the hypothesis of equal medians, W is normally distributed with $\mu = 150$ and $\sigma^2 = 300$. The test statistic is

$$z = \frac{121 - 150}{\sqrt{300}} = -1.67. \qquad \square$$

8.7. THE DUCKWORTH TEST

In the first issue of *Technometrics*, John Tukey (1959) presented a simple test that he had constructed to meet "Duckworth's specifications." Duckworth wanted a quick-and-easy test for comparing the medians of two populations that did not need any tables and could be used for a wide range of m and n. Tukey produced a test that is, in his words, "not only quick but *compact*." This test is not as powerful as some of the more elaborate procedures, but it is easy to use and the table of critical values can be carried in your head.

The test only works if the largest and smallest values of the observations come from different populations. It is preferable that the two samples have approximately the same number of observations, or at least that neither m/n or n/m should exceed 2. Suppose that the smallest observation is from the X population and the largest from the Y population. The test statistic, D, is the sum of the two overlaps, the number of X observations that are smaller than the smallest Y plus the number of Y observations that are larger than the largest X. If either $3 + 4n/3 \leq m \leq 2n$, or vice versa, we subtract 1 from D.

Under these circumstances, the table of critical values consists of the three numbers 7, 10, and 13. If $D \geq 7$, we reject the hypothesis of equal medians at $\alpha = 5\%$; 10 is the 1% value; 13 is the 0.1% value.

Example 8.2.1 (cont.). Using the Duckworth test for the data of example 8.2.1, we note that there are three X observations below all the Y's and one Y above all the X's. The total is $D = 4$, which is not significant at $\alpha = 5\%$.

□

8.8. COMPARING VARIANCES

In section 8.4, we obtained estimates s_1^2 and s_2^2 of the variances. How do we test the hypothesis $\sigma_1^2 = \sigma_2^2$? The test statistic is the variance ratio

$$F = \frac{s_1^2}{s_2^2}. \qquad (8.8.1)$$

If F is close to unity, we accept the hypothesis that the variances are equal. If F is large or small, we conclude that the variances are different.

To derive the distribution mathematically, we start with the ratio

$$F = \frac{s_1^2/\sigma_1^2}{s_2^2/\sigma_2^2}.$$

The numerator is $(S_1/\sigma_1^2)/(m-1)$; the denominator is $(S_2/\sigma_2^2)/(n-1)$, and so F is the ratio of two independent chi-square variables, each divided by its degrees of freedom. In the present situation, we hypothesize that the two variances are equal, and F reduces to the simple variance ratio s_1^2/s_2^2 of equation (8.8.1).

The distribution of F depends both on the number of d.f. in the numerator and the number of d.f. in the denominator, which we can denote by ϕ_1 and ϕ_2, respectively. The F statistic will sometimes be denoted by $F(\phi_1, \phi_2)$. Table IV (see Appendix) lists critical values of the F statistic. Because there are two parameters involved, a complete table would occupy many pages; some books economize by letting s_1^2 be the *larger* of the two estimates and s_2^2 the *smaller*, which implies that only values of F greater than unity need to be given. The use of the table is illustrated in this example.

Example 8.8.1. Suppose that two samples have the following estimates of the variance

$$s_x^2 = 80.0, \text{ 10 d.f.}; \quad \text{and} \quad s_y^2 = 40.0, \text{ 15 d.f.}$$

We compute the variance ratio with the larger estimate in the numerator:

$$F(10, 15) = \frac{80}{40} = 2.0.$$

The table gives the value of F that corresponds to the given percentage in the upper tail. Since we have already taken care to put the larger estimate in the numerator, we want to compare F with the value that would be exceeded by chance 5% of the time (not $\alpha/2$). The appropriate value is found in the table under $\alpha = 0.05$ in the tenth column and the fifteenth row:

$$F^*(10, 15) = 2.54.$$

It is perhaps discouraging to see that even though the one estimated variance is twice the other, that difference is not significant. In fact, good estimates of variance need plenty of data, and estimates based only on a few d.f. can be a long way off. □

8.9. CONFIDENCE INTERVALS FOR THE VARIANCE RATIO

We saw in the previous section that the statistic that actually has the F distribution is

$$F = \frac{s_1^2/\sigma_1^2}{s_2^2/\sigma_2^2} = \left(\frac{\sigma_2^2}{\sigma_1^2}\right)\left(\frac{s_1^2}{s_2^2}\right).$$

We can, therefore, construct a $(1-\alpha)\%$ confidence interval by writing

$$F_{1-\alpha/2}(\phi_1, \phi_2) < F < F_{\alpha/2}(\phi_1, \phi_2);$$

then, dividing through by s_1^2/s_2^2, we get

$$\left(\frac{s_2^2}{s_1^2}\right) F_{1-\alpha/2}(\phi_1, \phi_2) < \frac{\sigma_2^2}{\sigma_1^2} < \left(\frac{s_2^2}{s_1^2}\right) F_{\alpha/2}(\phi_1, \phi_2).$$

Unfortunately, this is upside down, and so we have to switch the subscripts throughout. The confidence interval that we are seeking is thus

$$\left(\frac{s_1^2}{s_2^2}\right) F_{1-\alpha/2}(\phi_2, \phi_1) < \frac{\sigma_1^2}{\sigma_2^2} < \left(\frac{s_1^2}{s_2^2}\right) F_{\alpha/2}(\phi_2, \phi_1). \quad (8.9.1)$$

One more difficulty remains. For a 95% interval, the value of F that appears on the right-hand side of equation (8.9.1) is found in the table under $\alpha = 0.025$. The value on the left side is not in the tables. It can be obtained by using the following equation:

$$F_{1-\alpha/2}(\phi_2, \phi_1) = \frac{1}{F_{\alpha/2}(\phi_1, \phi_2)}. \quad (8.9.2)$$

Example 8.8.1 (cont.). We can now obtain a 95% confidence interval for the variance ratio:

$$(2.5)\left(\frac{1}{3.06}\right) < \frac{\sigma_1^2}{\sigma_2^2} < (2.5)(3.52)$$

$$0.82 < \frac{\sigma_1^2}{\sigma_2^2} < 8.80 \, . \qquad \square$$

8.10. COMPARING EXPONENTIAL DISTRIBUTIONS

Suppose that we have samples from two exponential distributions. The first sample, $x_1, x_2, \ldots, x_m$, comes from an exponential population with parameter θ_1; The second, $y_1, y_2, \ldots, y_n$, comes from a population with expectation θ_2. We estimate θ_1 by $\bar{x}$ and θ_2 by $\bar{y}$. We can then compare θ_1 and θ_2 in two ways. First, we can act as if $\bar{x}$ and $\bar{y}$ are both approximately normally distributed with variances $\bar{x}^2$ and $\bar{y}^2$, respectively, and then follow the procedure of section 8.2. There is an alternative way that has the virtue of not being an approximation.

We recall that $2 \Sigma x/\theta_1$ and $2 \Sigma y/\theta_2$ have chi-square distributions with $2m$ and $2n$ degrees of freedom, respectively. If we divide each of them by their d.f. and take the ratio, we have an F statistic:

$$F = \frac{\bar{x}}{\bar{y}} \frac{\theta_2}{\theta_1}$$

has the $F(2m, 2n)$ distribution. We can, therefore, test the null hypothesis H_0: $\theta_1 = \theta_2$ by comparing the test statistic

$$F = \frac{\bar{x}}{\bar{y}} \qquad (8.10.1)$$

to the tabulated values of $F(2m, 2n)$.

A $(1 - \alpha)\%$ confidence interval for θ_1/θ_2 can be obtained by the methods of the previous section as

$$\frac{\bar{x}/\bar{y}}{F_{\alpha/2}(2m, 2n)} < \frac{\theta_1}{\theta_2} < (\bar{x}/\bar{y}) F_{\alpha/2}(2n, 2m) \, . \qquad (8.10.2)$$

Example 8.10.1. Twenty components from supplier A are put on test; their average lifetime is 178 hours; 30 components from supplier B have an average lifetime of 362 hours.

Method 1. We assume that

$$V(\bar{x}) = \frac{(178)^2}{20} = 1584.2 \quad \text{and} \quad V(\bar{y}) = \frac{(362)^2}{30} = 4368.1.$$

$$z = \frac{178 - 362}{\sqrt{V(\bar{x}) + V(\bar{y})}} = 2.38,$$

and at $\alpha = 5\%$, we would decide that the reliabilities differed.

A 95% confidence interval for the difference between average lifetimes is

$$(178 - 362) - 1.96\sqrt{1584.2 + 4368.1} < \theta_1 - \theta_2 < -184 + 151.2$$

$$-335 < \theta_1 - \theta_2 < -33.$$

Method 2

$$F = \frac{362}{178} = 2.03,$$

to be compared to $F_{0.05}(60, 40) = 1.64$.

The confidence interval is

$$\left(\frac{178}{362}\right)\left(\frac{1}{F_{0.025}(40, 60)}\right) < \frac{\theta_1}{\theta_2} < \left(\frac{178}{362}\right)[F_{0.025}(60, 40)]$$

$$0.28 < \frac{\theta_1}{\theta_2} < 0.88. \qquad \square$$

8.11. COMPARING BINOMIAL POPULATIONS

A manufacturer receives supplies of a certain part from two suppliers and wishes to compare them. A random sample of 200 parts from supplier A contains 30 defectives; a random sample of 100 parts from supplier B contains six defectives. Are the two suppliers producing at the same rate of defectives?

We can act as if the fraction defectives in the two samples are normally distributed and construct a test and a confidence interval along the lines of section 8.2. Suppose that the first sample contained x defectives out of m parts and that the second contained y defectives out of n parts; the estimates of the binomial parameters are

$$\hat{p}_1 = \frac{x}{m} \quad \text{and} \quad \hat{p}_2 = \frac{y}{n}.$$

We recall that the variance of the number of defectives in a sample of size n is npq. Under the null hypothesis that $p_1 = p_2 = p$, the common value p is estimated by pooling the two samples:

$$\hat{p} = \frac{x+y}{m+n}.$$

In constructing the z statistic, we take $V(x/m) = \hat{p}\hat{q}/m$ and $V(y/n) = \hat{p}\hat{q}/n$, whence

$$z = \frac{x/m - y/n}{\sqrt{\hat{p}\hat{q}(1/m + 1/n)}}. \tag{8.11.1}$$

For the confidence interval, we no longer have to assume that $p_1 = p_2$. We take the best estimates of the variances:

$$V\left(\frac{x}{m}\right) = \frac{\hat{p}_1\hat{q}_1}{m} \quad \text{and} \quad V(y) = \frac{\hat{p}_2\hat{q}_2}{n}.$$

A two-sided 95% interval is given by

$$\frac{x}{m} - \frac{y}{n} - 1.96\sqrt{(V(x/m) + V(y/n))} < p_1 - p_2$$

$$< \frac{x}{m} - \frac{y}{n} - 1.96\sqrt{(V(x/m) + V(y/n))}.$$

Example 8.11.1. For the data at the beginning of this section, $x = 30$, $m = 200$, $y = 6$, $n = 100$; $\hat{p}_1 = 0.15$ and $\hat{p}_2 = 0.06$. Pooling the samples, $\hat{p} = 36/300 = 0.12$.

$$z = \frac{0.15 - 0.06}{\sqrt{(0.12)(0.88)(1/200 + 1/100)}} = 2.26.$$

For the confidence interval, $\hat{p}_1\hat{q}_1/m = 0.0006375$, $\hat{p}_2\hat{q}_2/n = 0.000564$, and the interval is

$$(0.15 - 0.06) - 1.96\sqrt{0.0012} < p_1 - p_2 < 0.09 + 0.07$$

$$0.02 < p_1 - p_2 < 0.16. \quad \square$$

8.12. CHI-SQUARE AND 2 × 2 TABLES

A commonly used alternative to the z-test of the previous section is to set up the data in a 2 × 2 table (a table with two rows and two columns) and make a chi-square test. For this data set, the table would be

	Supplier	
	A	B
nondef.	170	94
def.	30	6

We can add the marginal totals and see immediately that if a part were to be chosen at random from the combined sample, the probability that it would be a nondefective is $264/300 = 0.88$, and the probability that it came from supplier A is $200/300$.

	A	B	
nondef.	170	94	264
def.	30	6	36
	200	100	$N = 300$

If the probability of a nondefective were independent of supplier, the expected number of nondefectives from A would be $300(0.88)(2/3) = 176$. Thus, we have for that cell

$$O = 170, \quad E = 176, \quad O - E = -6.$$

Similar calculations show that the expected values in the other cells are 88, 12, and 24, with deviations $+6$, -6, $+6$, respectively. The deviations all have the same absolute value; there is only one degree of freedom.

The value of the chi-square statistic is

$$U = \frac{36}{176} + \frac{36}{88} + \frac{36}{12} + \frac{36}{24} = 5.11 = z^2.$$

In the general case, we can denote the cell frequencies by a, b, c, and d:

	a	b	a + b
	c	d	c + d
	a + c	b + d	$N = a + b + c + d$

The deviation in each of the four cells is $\pm(ad - bc)/N$. The chi-square statistic may be calculated by the formula:

$$U = \frac{N(ad - bc)^2}{(a+b)(c+d)(a+c)(b+d)}.$$

EXERCISES

8.1. Twenty wafers were sent to a vendor for a boron transplant. Twenty received the standard deposition. After the treatment, all 40 wafers were processed together and the probe yields were measured. The yields were

Control

54.7 57.6 31.2 44.1 59.7 63.2 34.3 46.5 52.1 48.7
31.4 46.5 54.4 62.3 45.7 55.3 66.9 48.9 55.4 43.4

EXERCISES

Boron

| 54.5 | 34.3 | 54.8 | 82.4 | 52.7 | 51.6 | 57.8 | 49.8 | 52.5 | 74.2 |
| 56.1 | 52.3 | 84.2 | 55.6 | 48.3 | 49.7 | 75.4 | 49.5 | 84.1 | 61.3 |

Was the boron implantation effective?

8.2. Devices from two manufacturers were said to have the same nominal speed (frequency of operation). Twenty-two devices from manufacturer A and 25 from manufacturer B were tested at room temperatures. Their observed frequencies in MHz were

A: 13.6 13.5 9.8 13.7 12.3 14.2 11.0 9.3 14.2
16.0 15.7 10.0 13.7 13.8 13.9 13.7 13.6 16.8 13.2
14.1 13.1 13.7

B: 13.3 12.7 13.0 13.1 14.2 12.5 16.7 17.5 17.4
16.9 16.9 17.7 12.4 18.0 16.9 17.4 12.9 12.6 12.6
13.8 13.9 12.7 12.9 12.6 12.6

Do the devices from the two manufacturers have the same average speed? Make dotplots of both samples on the same scale. Use three methods for this exercise: the t-test, Wilcoxon's test, and Duckworth's test.

8.3. During wafer preparation, a layer of gold, 3000 Angstroms thick, is sometimes sputter-deposited onto the back of the wafer. In a comparative test of two vendors, the following results were obtained:
For vendor A, 32 out of 45 units tested were void-free. For vendor B, 39 units out of 76 were void-free. Test the hypothesis that the percentages of void-free units for the two vendors are equal.

8.4. The mean times between failures (MTBF) of an electronic component have an exponential distribution. Eighty failures were investigated and the mean time to failure was 282 hours. Denote the MTBF by θ. The specification calls for θ to be 325 hours or longer. Test the hypothesis H_0: $\theta = 325$ against the alternative H_1: $\theta < 325$.

8.5. An oil company is considering introducing a new additive in its gasoline that it hopes will increase the mileage per gallon. The engineers in the research group take ten cars and run them on the company's regular brand and another ten cars that they run on the regular gasoline with additive. A summary of their results is:

Control average mpg = 31.2, $s = 1.8$
Additive average mpg = 32.5, $s = 2.4$

Is the new additive effective in increasing the mpg?
Is there greater variability in mpg with the new additive?

CHAPTER NINE

Quality Control Charts

9.1. INTRODUCTION

A great deal of the industrial production in this country comes from production line work—a method of production that calls for carrying out a repetitive process again and again and again. It would be nice to think that once a production line has been set up, all we have to do is to put in raw material at one end, and out of the other will come a steady succession of objects, called the product, that are identical with one another and each conforming exactly to the standards that we had in mind. Unfortunately, life is not like that. No matter how hard we try, we cannot construct a process that will continue forever to produce absolute clones. Components in the production machinery wear out, and there is variability in the raw material. Whether we like it or not, each production process has built into it a certain amount of natural variability, which we can call noise. Perhaps, if we try hard enough, we can eliminate almost all the noise. Some of the noise we cannot eliminate, although we can reduce it.

In making a cake on a production line in a bakery, we cannot persuade the hens to lay eggs that are absolutely uniform in size or the cows to produce cream that does not vary in the percentage of fat from day to day. We can construct ovens that maintain a desired temperature within 5 degrees, or within a tenth of a degree. The latter cost more, both in initial capital investment and in maintenance. If the cake recipe is robust to temperature changes, which is statistical jargon for saying that a deviation of a few degrees in the oven temperature will not seriously affect the palatability of the product, it will not be cost efficient to invest in the highest-costing ovens. Indeed, if you are selling cake mixes to be made into cakes in the buyers' own kitchens, you would do well to have a mix that is robust to the fluctuations in ovens from house to house.

We are going to assume in the next few chapters that we have a production process, out of which comes a product. It is convenient to say that we are producing widgets. We also assume that we have some measure of the quality of each widget. These measures may be discrete, as in chapter three. Then each widget is classified as either acceptable or not acceptable, defective or nondefective, conforming or nonconforming. In the jargon of quality control, that situation is described as sampling and classification by attributes; it is the subject of the next two chapters. In this chapter, we discuss quality control charts for continuous variables. The measure of quality is something like the weight or length of the widget.

We can get into arguments when we try to measure characteristics, such as palatability, that are difficult to quantify. If we either accept each cake or reject it, we clearly have the attribute case. On the other hand, we could grade each cake on a scale of 1 through 4. Grade 1 cakes we sell to the best restaurants at the premium price that they deserve, grade 2 goes to the grocery stores, grade 3 is sold at a discount to college residence halls, and grade 4 is inedible. That is a more complicated situation than we want to discuss at this time.

9.2. QUALITY CONTROL CHARTS

Statistical quality control charts were introduced about 60 years ago by Dr. Walter Shewhart, a scientist at Bell Labs, the research arm of American Telephone and Telegraph Company. In 1931, he wrote a book entitled *Economic Control of Quality of Manufactured Product*.

What happened in those days in too many shops, and still does, is that people set up the machinery and start production. They continue to produce widgets as long as the items look satisfactory and there is not a preponderance of complaints from the clients. The clients in this sense may be the actual end users or they may be the people in another plant some miles away who are using the widgets in the construction of some more complex pieces of equipment. Production continues until it dawns on someone that they are now producing unacceptable widgets. At that stage, there is a stir in the shop. Perhaps they make some ad hoc adjustments to the machinery, "tweaking" this knob or that, which seems to improve things. Perhaps they introduce inspection of the outgoing product so that they can, at least, cut down on the number of unacceptable widgets that leave the premises. It is often rather haphazard.

What Shewhart did was to introduce a simple graphic procedure for monitoring the output of the process that enables the manager to tell whether the process is in a state of control. If the process has gone out of control, the manager gets a signal to take action. The basic idea is that we take a sample of four, or five, widgets off the line at regular intervals and measure their lengths. We then prepare two charts. We plot the average length of each sample on one chart and the range of each sample on the other chart. These charts are then used to tell us whether or not the process is in control.

Shewhart's ideas depend upon two statistical concepts. The first we have mentioned earlier: a certain amount of random noise occurs naturally in any production process. The other is that in a random process, there is a certain amount of regularity. With a normal distribution of noise, only one time in twenty will a variable differ from its mean by more than two standard deviations.

We can now give a definition of control in this context. A process is said to be in (statistical) control if it is performing within the limits of its capability. We may find, when we look closely, that the capability of the process is not what we would wish, that the product being manufactured and the customer's specifications are some way apart. The average length of our widgets may not be what the customer wants, or the variability may be too great. How do we make adjustments to get the process to meet specifications?. That is the topic of process improvement (off-line experimentation), which is to be discussed in the final chapters. For the present, we are concerned with the stability of our process. Is it moving along steadily?

9.3. THE x-BAR CHART

At regular intervals, such as every hour, or once a shift, we take a random sample of n widgets from the production line. Commonly used values of n are 4 or 5. Some engineers call each sample a rational subgroup. We plot the sample averages, usually denoted by $\bar{x}$, along the vertical axis, with time, or sample number, along the horizontal axis. The piece of paper on which the sample means are plotted has three horizontal lines. The center line corresponds to the process mean. The upper control line is at a distance $3\sigma/\sqrt{n}$ above the center line, or at our estimate of that distance. The lower line is $3\sigma/\sqrt{n}$ below it. In the United Kingdom, it is more usual to place the outside lines at a distance $3.09\sigma/\sqrt{n}$ above and below the center line.

As long as the plotted points fall between the two outside lines, we say that the process is under control. If a point lies either above the upper line or below the lower line, we say that the process is out of control. We stop production and, in the jargon, we seek an assignable cause. This cause, when found, may turn out to be something serious that calls for corrective action such as replacing a part of the machinery or an operator who has not been properly trained. It may only have been a passing phenomenon that is easily explained and is not expected to recur. It could have happened by chance, but only about one time in 400.

From the point of view of hypothesis testing, we have a null hypothesis that if the process is under control, the random variable, $\bar{x}$, is normally distributed with expectation μ and variance σ^2/n. With each sample, we carry out a test. The area outside the outer control lines is the critical area, or rejection region. Under the null hypothesis of control, the probability that a point falls outside those limits is only one in 400. If the probability distribution of $\bar{x}$ is really $N(\mu, \sigma^2/n)$, the probability is 0.0025 that we will mistakenly stop production and that there will be no assignable cause to find. That figure is the alpha risk.

We now have a simple graphical procedure that shows how the process is behaving, and a simple action, or decision, rule. Its simplicity is one of its great virtues. The chart can be kept on the shop floor. One of the workers is assigned to make the entries. Everyone, workers as well as management, can see how their process is doing. Everyone is involved. Everyone can focus on the problems of quality.

9.4. SETTING THE CONTROL LINES

When we speak of $\bar{x}$ having expectation μ and variance σ^2/n, we are talking about the mean and variance of the process as it actually is and not about the specifications. We have to estimate those two parameters from a pilot set of data taken from actual production. A pilot set of about 20 rational subgroups usually suffices to get started. The average of the individual $\bar{x}$ is denoted by $\bar{\bar{x}}$, and called x-bar-bar. It provides our estimate of μ, and fixes the center line of the chart.

The standard deviation has traditionally been estimated from the ranges. The range, R_i, of each subgroup is recorded. Their grand average is denoted by $\bar{R}$. It can be divided by d_2, as in section 6.14, to obtain an estimate of the standard deviation. We recall that $d_2 = 2.06$ for $n = 4$ and 2.33 for $n = 5$. Then we can multiply that estimate of the standard deviation by $3/\sqrt{n}$.

In practice, the procedure is a little different. The two steps are combined into one. We define another multiplier,

$$A_2 = \frac{3}{d_2\sqrt{n}}.$$

The upper and lower control lines are then set at

$$\bar{\bar{x}} \pm A_2 R.$$

Values of d_2 and A_2 are given in table 9.4.1.

Table 9.4.1. Factors for the x-Bar Chart

n	2	3	4	5	6	7
d_2	1.128	1.693	2.059	2.326	2.534	2.704
A_2	1.880	1.023	0.729	0.577	0.483	0.419
Eff.	1.000	0.992	0.975	0.955	0.933	0.911

Alternatively, one could obtain an unbiased estimate of the variance from the squared deviations in the usual way. For each sample, one computes

$$s^2 = \frac{\sum (x_i - \bar{x})^2}{n-1}.$$

The square root, s, is not an unbiased estimate of the standard deviation. The unbiased estimate is s/c_4, where c_4 is a multiplier whose values are given later in table 9.12.1. The asymptotic value of $1/c_4$ for larger n is $1 + (4n-4)^{-1}$. Charts based on $\bar{s}$ are the topic of section 9.12.

The range estimate for σ has very high efficiency by comparison to s/c_4. Those efficiencies are also given in table 9.12.1.

The primary advantage of using the range to estimate σ is its ease of calculation. It can be argued that this advantage is less important today than it was 25 years ago; every modern engineer carries a hand calculator that can compute s^2 and s/c_4 in an instant. The simplicity should not, however, be dismissed. This is a diagnostic tool. The chart is meant to give everybody a picture of what is happening; we want everybody to be able to relate to it. Not only can the worker on the floor easily calculate the range of a sample of four observations, but the concept of range as a measure of variability is easier to understand than the concept of standard deviation.

9.5. THE R CHART

The x-bar chart keeps track of changes in the process mean over time. A similar chart of the sample range keeps track of changes in the variance. The sample range, R_i, is plotted along the vertical axis. Again, there are three control lines. The center line is set at R, the average from the pilot set. The upper and lower lines are obtained by multiplying R by D_4 (upper line) or D_3 (lower). Values of D_3 and D_4 are shown in table 9.5.1.

The rationale for computing multipliers D_3 and D_4 is now shown for the case of $n = 4$. Denote the range by w. It can be shown that if the observations come from a *normal* population, $E(w) = 2.059\sigma = d_2\sigma$ and that $V(w) = 0.7758\sigma^2$. Substituting w/d_2 for σ in the latter expression gives

$$V(w) = 0.183w^2 \quad \text{and} \quad sd(w) = 0.428w,$$

Table 9.5.1. Multipliers for the Range Chart

n	2	3	4	5	6	7
D_3	0	0	0	0	0	0.076
D_4	3.267	2.574	2.282	2.114	2.004	1.924

AN EXAMPLE

where $sd(w)$ denotes the standard deviation of w. Following the same procedure as for the x-bar chart, we set the upper control line at

$$D_4 = w + 3sd(w) = w + 3(0.428)w = 2.284w.$$

The difference between the factor 2.284 and the actual value of $D_4 = 2.282$ is due to roundoff in the calculations. To compute D_3, we should take

$$D_3 = w - 3(0.428)w,$$

but this quantity is negative, and so we replace it by zero.

There is a tacit assumption here that $w \pm 3sd(w)$ are the correct limits and that w is normally distributed, which is not true. The distribution of w is certainly not symmetric, even when the observations are normal. However, this is the rationale that has traditionally been used.

It can be argued that the engineer should look at the R chart before looking at the x-bar chart. If the variability of the process is out of control, as manifested by the R chart, the process is certainly not under control, even though inflated ranges may make the control limits on the x-bar chart far enough apart to include all the points. The R chart is obviously sensitive to outliers. A single bad observation can make the range of that sample so large as to exceed the upper control line. It is possible that the search for an assignable cause will find a reason that could justify the engineer in throwing out that point and recalculating the estimate of σ without that sample. That should only be done if the engineer is very sure that it is justified. Remember that, whether you like it or not, it is an observation, and a process that produces outliers has problems.

9.6. AN EXAMPLE

The following pilot set of 20 subgroups of four observations each was used to set up x-bar and R charts for a process. The data are in table 9.6.1. The charts of the pilot set are in figures 9.6.1 and 9.6.2. The process is clearly rolling along steadily.

The grand average $\bar{x}$ for the 20 groups is 10.13; the average range is 1.82. For the x-bar chart, the upper and lower control lines are set at

$$10.13 + (0.729)(1.82) = 11.45$$

and

$$10.13 - (0.729)(1.82) = 8.80.$$

The only point that comes close to an outer control line is the seventh, which has a sample mean 11.30. Had that value been, instead, 11.50, we

Table 9.6.1. Twenty Samples

	1	2	3	4	5	6	7	8	9	10
	10.6	10.2	10.1	10.1	8.7	10.1	11.2	10.6	9.8	10.0
	10.1	11.6	9.8	9.5	11.6	9.8	11.5	9.6	7.7	8.4
	11.3	10.5	8.8	10.3	9.7	10.8	10.9	10.3	9.4	10.6
	9.1	10.5	9.3	10.6	9.3	8.9	11.6	9.9	9.9	8.8
$\bar{x}$	10.28	10.70	9.50	10.13	9.83	9.90	11.30	10.10	9.20	9.45
R	2.2	1.4	1.3	1.1	2.9	1.9	0.7	1.0	2.2	2.2
	11	12	13	14	15	16	17	18	19	20
	11.0	11.4	10.6	10.7	9.6	8.8	11.7	11.9	10.6	10.2
	8.5	9.0	10.0	10.2	11.5	8.7	11.4	9.5	11.3	8.7
	9.2	8.7	10.9	10.4	11.1	10.5	10.1	8.9	11.9	9.5
	9.4	10.9	9.1	11.1	9.5	10.5	10.2	11.7	10.2	10.0
$\bar{x}$	9.53	10.00	10.15	10.60	10.43	9.63	10.85	10.50	11.00	9.60
R	2.5	2.7	1.8	0.9	2.0	1.8	1.6	3.0	1.7	1.5

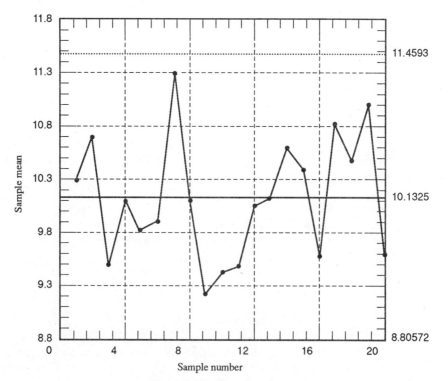

Figure 9.6.1. The x-bar chart.

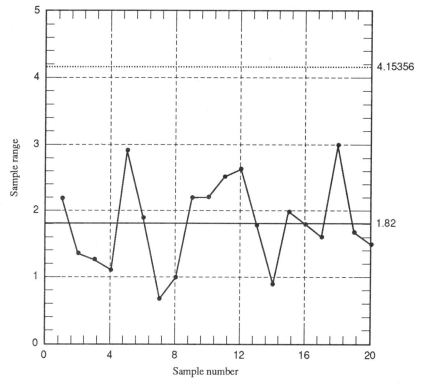

Figure 9.6.2. The range chart.

would have noted that it was above the upper control line and proceeded to look for an assignable cause. One outside point in a string of 20 points may not signify anything serious. It could occur by chance. It could, for example, be due to a new operator who was a bit heavy handed until he acquired the knack. Certainly, in this case, the (supposed) problem seems to have been temporary.

The upper limit for the range chart is taken at

$$(2.282)(1.82) = 4.15;$$

the lower limit is zero. All the points fall between the limits.

9.7. ANOTHER EXAMPLE

The process that is charted in figure 9.7.1 is, on the other hand, an example of a process that is manifestly out of control. There are numerous points outside the limits on both the high and low sides.

One might suspect the following scenario. The process has an unfortunate tendency for the mean to increase. As soon as this is noticed, the operator

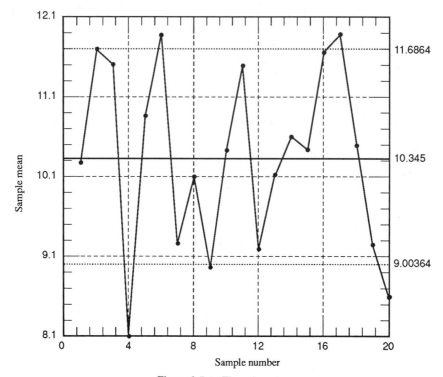

Figure 9.7.1. The x-bar chart.

tweaks one of the variables to bring the value of the mean back down, and usually overdoes it. Then another cycle starts. The process drifts upwards and is "adjusted." One piece of evidence to support this theory is that points above the upper line are immediately followed by points that are markedly lower.

What if the outside points had not occurred in such a systematic pattern, but came, seemingly, at random among the 20 points recorded? How might such a phenomenon be explained? It is important to recall how the estimate of the process variance was obtained. We obtained it from the ranges of the individual samples. It is a measure of the variability of four items made on the same machine on the same shift. It does not take into account variability between shifts. In some cases, the variance that we have measured will be relatively small—the variation between four items produced on a precision machine under virtually identical circumstances. But the variability between shifts could include differences between operators, who might adjust the machine in different ways when they start their work turn; it could also include differences between batches of raw material, and other external sources of variation.

Table 9.8.1. Values of $\bar{x}$ for 30 More Samples

9.25	10.13	10.58	9.60	10.30	9.65	10.25	9.88	10.35	10.30
11.08	10.70	11.28	11.23	10.60	10.90	10.88	10.75	10.63	11.18
11.18	11.23	10.95	11.00	11.50	11.00	11.23	10.65	10.93	10.95

9.8. DETECTING SHIFTS IN THE MEAN

Figure 9.8.1 is an x-bar chart for the same process that produced figures 9.6.1 and 9.6.2. Thirty more samples have been taken and added to the previous 20. Their averages are given in table 9.8.1.

After the thirtieth sample (the 20 in table 9.6.1 and the first ten in table 9.8.1), something happened to the process. There was an increase in μ of about one standard deviation. After a while, there was a value of $\bar{x}$ that exceeded the upper limit. That happened at the forty-fifth sample, 15 samples after the change had taken place. Is that an unreasonable delay in detecting a change? In the sections that follow, we discuss alternative stop rules and a criterion for assessing them: average run length.

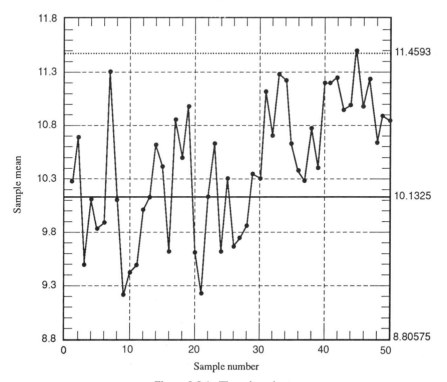

Figure 9.8.1. The x-bar chart.

9.9. ALTERNATIVE STOPPING RULES

The term *stopping rule* is used by statisticians working with stochastic (random) processes. We have been assuming that we are dealing with a random normal process with mean μ and standard deviation σ. A stopping rule is a rule for making the decision that the process no longer satisfies the original conditions. Our only stopping rule so far has been

Rule 1. One point outside the three sigma lines. Some engineers draw two more lines on their charts at a distance $2.0\sigma/\sqrt{n}$ above and below the center line, and call them inner control lines, or warning lines. These lines are used for the second rule. □

Rule 2. Two consecutive points, either both above the upper warning line or both below the lower warning line. □

There are also rules that are based on the simple coin tossing model. If our process is behaving as we expected, roughly half the points should be above the center line and half should be below it. Since they should occur at random, the situation is the same as a sequence of coin tosses with heads for points above the line and tails for points below. This leads to two classes of rules based upon runs. A run is a string of successive points, either all above the line or all below it.

In the data of table 9.6.1, there is a run of five points below the center line from sample 8 through sample 12. The R chart has a run of four above the center line from 9 through 12.

Rule 3. A run of seven, or more, points, either all above the center line or all below it. □

In figure 9.8.1, this rule would have been triggered after the thirty-fifth sample. The probability of a run of seven is $2(1/128) = 0.016$.

There are also rules based on the number of runs, too many or too few. One could have too many runs in the extreme case of two operators working alternate shifts if one operator consistently made the widgets too long and the other consistently made them too short—the sawtooth effect. Too few runs might indicate a slow cyclic trend in the process average.

9.10. AVERAGE RUN LENGTHS

In hypothesis testing, we spoke of the alpha and beta risks. In the context of control charts, alpha is the probability that we (wrongly) trigger a stop rule when the process is still in control; beta is the probability that if there had been a shift in the process average, we would fail to detect it. We choose the

AVERAGE RUN LENGTHS

test that has the smallest value of beta for a given alpha. With control charts, the emphasis changes slightly to a different criterion—the average run number. The word run is used in a different sense from the previous section. Here it refers to the number of samples that are taken before the rule is triggered. In figure 9.8.1, the process average changed between the thirtieth and thirty-first samples. The change was detected in the forty-fifth sample—a run of 15 samples. How long a run will it take, on the average, to detect a change in the mean? We consider rule 1 in some detail.

We assume that our estimates of the mean and the standard deviation are correct. As long as the process remains in control, the chance that a point will be outside the control lines is 2(0.00135). If the samples are independent, the number of samples until one exceeds the limits has a negative binomial distribution with $p = 0.0027$. Its expectation is $1/p = 370$. Using rule 1, the average run length when there is a zero shift in the mean is 370.

Suppose that the mean shifts from μ to $\mu + \sigma/\sqrt{n}$, while the standard deviation remains the same. Now the process is operating about a new center line that is only $2.0\sigma/\sqrt{n}$ below the upper line, but is $4\sigma/\sqrt{n}$ above the lower line. The chance that a point will now fall below the lower line is virtually zero, but the chance that a point will fall above the upper line is 0.02275 and the average run length (ARL) is 44. For a shift of $2\sigma/\sqrt{n}$, the ARL is reduced to 6.3, since the distance to the upper line is now reduced to one standard deviation of $\bar{x}$.

Table 9.10.1. Average Run Lengths to Detect a Shift $d(\sigma/\sqrt{n})$ in the Mean

d	Rule 1	Rule 2
0.2	310	799
0.4	200	349
0.5	161	238
0.6	122	165
0.7	93	117
0.8	72	84
0.9	56	62
1.0	44	46
1.1	35	35
1.2	28	27
1.3	22	22
1.4	18	17
1.5	15	14
1.6	12	12
1.7	10	10
1.8	9	8
1.9	7	7
2.0	6	6

The ARL for rule 2 can be calculated by setting $r = 0$ and $m = 2$ in formula (13.3.4). The formula is $(1 + p)/p^2$, where p is the probability of a point outside a warning line.

Table 9.10.1 contains average run lengths, using rule 1 and rule 2, for several values of the shift in the process mean.

9.11. s CHARTS

So far in this chapter, we have used the R chart for the control of variability. We mentioned in section 9.5 that charts can be based, not on the range, but on the sample standard deviation, s. With modern computing facilities, the calculations are easily made. One can also make charts based on the sample variance in a similar way.

For small samples, such as $n = 4$ or $n = 5$, the relative efficiency of the range estimate to s is close to unity, and little is lost by using the method that is simpler arithmetically. However, if we wish to detect small shifts in the mean within a reasonable average run length, we can use larger samples, in which case the relative efficiency of the range decreases. This is shown later in table 9.12.1, where values of c_4 are given.

9.12. SETTING THE CONTROL LIMITS FOR AN s CHART

In these charts, the sample standard deviations, s_i, are plotted along the vertical axis. The center line is set at the average of the sample deviations, $\bar{s} = \Sigma s_i/N$, from the N samples in the pilot set. The control lines are drawn, as usual, three standard deviations above and below the center line.

It can be shown that the variance of s is given by the formula

$$V(s) = (1 - c_4^2)\sigma^2 .$$

For $n = 4$, $c_4 = 0.9213$, and so the standard deviation of s is 0.389σ. We now substitute $\bar{s}/c_4$ for σ and set the upper control limit at

$$\text{UCL} = \bar{s} + 3\left(\frac{0.389}{0.9213}\right)\bar{s} = 2.266\bar{s} .$$

In practice, the engineer sets the limits at

$$\text{UCL} = B_4\bar{s} \quad \text{and} \quad \text{LCL} = B_3\bar{s} ,$$

where the values of B_3 and B_4 are obtained from tables. Some values of c_4, B_3, and B_4 are given in table 9.12.1.

Table 9.12.1. Constants for s Charts

n	c_4	$1/c_4$	B_3	B_4	Efficiency
2	0.7979	1.2533	0	3.267	1.000
3	0.8862	1.1284	0	2.568	0.992
4	0.9213	1.0854	0	2.266	0.975
5	0.9400	1.0638	0	2.098	0.955
6	0.9515	1.0510	0.030	1.970	0.933
7	0.9594	1.0423	0.118	1.882	0.911
8	0.9650	1.0363	0.185	1.185	0.890
9	0.9693	1.0317	0.239	1.761	0.869
10	0.9727	1.0281	0.284	1.716	0.850
11	0.9754	1.0252	0.321	1.679	0.831
12	0.9776	1.0229	0.354	1.646	0.814
13	0.9794	1.0210	0.382	1.618	0.797
14	0.9810	1.0194	0.406	1.594	0.781
15	0.9823	1.0180	0.428	1.572	0.766
16	0.9835	1.0168	0.448	1.552	0.751
17	0.9845	1.0157	0.466	1.534	0.738
18	0.9854	1.0148	0.482	1.518	0.725
19	0.9862	1.0140	0.497	1.503	0.712
20	0.9869	1.0133	0.510	1.490	0.700

9.13. ALTERNATIVE CONTROL LIMITS

In the previous section, the control limits were calculated as the center line plus or minus three standard deviations. As in the R chart, this procedure tacitly, and wrongly, assumes that the distribution of s is normal. Nevertheless, it is the standard procedure. We can, however, calculate better limits in this case because we know the exact probability distribution of s when the data are normal. We now develop these limits for the case of samples of size 4.

The reader will recall from section 6.12 the procedure for obtaining a confidence interval for the variance of a normal population. With $n = 4$, the upper and lower 0.5% values of chi-square, with 3 d.f., are 0.07172 and 12.8381, respectively. A two-sided 99% confidence interval for s is, therefore, derived from the statement

$$0.07172 < \frac{3s^2}{\sigma^2} < 12.8381.$$

Then, after taking square roots and simplifying,

$$0.1546\sigma < s < 2.068\sigma.$$

We now replace σ by $\bar{s}/c_4 = \bar{s}/0.9213$ and obtain

$$0.168\bar{s} < s < 2.245\bar{s} \ .$$

The control limits are, thus,

$$\text{UCL} = 2.245\bar{s} \quad \text{and} \quad \text{LCL} = 0.168\bar{s} \ .$$

Notice two things: the limits are tighter, closer to the center line than the usual limits, and the lower limit is positive.

9.14. NONNORMALITY

The derivation of the control charts depends on the assumption that the observations are normally distributed. How essential is this? It is not serious as long as the distribution is roughly normal. The robustness of control charts under nonnormality has been investigated by Schilling and Nelson (1976).

We saw in chapter five that the distribution of the sample mean $\bar{x}$ approaches normality as n increases. Even for sample sizes as small as $n = 4$ and $n = 5$, the x-bar chart is robust to moderate departures from the normality assumption.

The situation with the R and the usual s charts is different. Even if the data are normal, the distributions of R and s are *not* normal, and so the rule of thumb of drawing the control lines at plus or minus three standard deviations does not pass muster mathematically. Instead of having only one chance in 400 of a point outside the limits under normality, the risk of a false stoppage may increase twofold or more. One should not, however, overreact to this. Even if the actual alpha risk is a little cloudy, one in 200 rather than one in 400, the principle of the R chart is sound, and it still serves as a powerful (and simple) diagnostic tool.

9.15. PROCESS CAPABILITY

Nothing has been said yet in this chapter about specifications. Our emphasis in process control has been that the process should be rolling along steadily. A process can, in theory, be under control but be way off specification! Two criteria tie process control and specifications together. They are measures of process capability.

We consider the case in which the specification is two-sided and symmetric. The measured characteristic, such as the length of the item, has to fall in the interval

$$(\mu_0 - \delta, \mu_0 + \delta) \ ;$$

PROCESS CAPABILITY

μ_0 is called the target value of the characteristic and 2δ is the specification width. We denote the actual average value for our process by μ (as represented by the center line on the control chart); if the process is off target, $\mu \neq \mu_0$.

The process capability index, C_p, is defined as the ratio of the specification width to the process spread:

$$C_p = \frac{2\delta}{6\sigma}$$
$$= \frac{\text{USL} - \text{LSL}}{6\sigma},$$

where USL and LSL are the upper and lower specification limits, respectively. C_p measures the potential cability of the process; as σ decreases, C_p increases.

If the process is on target—if μ is actually equal to the target value μ_0—a process with $C_p = 1.0$ has $\delta = 3\sigma$ and is capable in the sense that 99.73% of the items will fall in the interval $(\mu - \delta, \mu + \delta)$. If, however, the process is not properly centered, the percentage of items within specifications will fall below 99.73%. If, for example, $\mu = \mu_0 + \sigma$, the specification limits will be

$$\text{LSL} = \mu - 4\sigma \quad \text{and} \quad \text{USL} = \mu + 2\sigma.$$

Virtually every item produced will be above the LSL, but 2.27% will fall above the USL, and only 97.73% of the items will be within specifications. If $\mu = \mu_0 + 2\sigma$, only 84.13% will meet specifications. We now introduce another measure of process capability that includes the failure to be on target.

We define two new indices

$$\text{CPU} = \frac{\text{USL} - \mu}{3\sigma} \quad \text{and} \quad \text{CPL} = \frac{\mu - \text{LSL}}{3\sigma}.$$

If $\mu = \mu_0 = (\text{USL} + \text{LSL})/2$, both these indices are equal to C_p.

The second criterion is

$$C_{pk} = \text{minimum (CPL, CPU)} = C_p(1 - k),$$

where

$$k = \frac{|\mu_0 - \mu|}{\delta}.$$

If the process is correctly centered, so that the process mean is indeed at the target value, $k = 0$ and $C_{pk} = C_p$.

A desirable aim for a process is to have $C_{pk} > 1.0$. It was reported a few years ago that the minimum acceptable value of C_{pk} in some Japanese industries is 1.33, which corresponds to a fraction 0.00006 (6 in 100,000) of nonconforming units. Nowadays, in the semiconductor industry, manufacturers are striving for processes with $C_p = 2.0$ and $C_{pk} = 1.5$, which represents about 3.4 defective parts per million. This can be achieved by reducing the standard deviation until USL − LSL = 12σ and reducing the deviation of the mean value from target to $|\mu - \mu_0| = 1.5\sigma$.

9.16. I CHARTS

Thus far we have discussed setting up x-bar charts when the sample size is greater than $n = 1$. There are two choices for estimating the standard deviation, either to use the sample range or the sample standard deviation. What can be done when the samples are individual observations, i.e., $n = 1$? Such a chart is called an I chart (individuals).

The standard deviation is estimated from the moving ranges of pairs of adjacent observations. Let x_i denote the value of the ith individual. Then

$$r_i = |x_{i+1} - x_i|$$

is the range of the ith pair. The sequence of the values of r_i is called the sequence of moving ranges. Denote its average by $\bar{r}$. The estimate of σ is given by $\bar{r}/d_2 = \bar{r}/1.128$. The control lines are drawn at a distance $\pm 3\bar{r}/1.128 = 2.66\bar{r}$ from the center line.

Example 9.16.1. Suppose that the sample averages in table 9.6.1 had been reported as single observations without any information about σ. We have a set of twenty individual data points:

	1	2	3	4	5	6	7	8	9	10
x	10.28	10.70	9.50	10.13	9.83	9.90	11.30	10.10	9.20	9.45

	11	12	13	14	15	16	17	18	19	20
x	9.53	10.00	10.15	10.60	10.43	9.63	10.85	10.50	11.00	9.60

The first two values of r_i are

$$r_1 = 10.70 - 10.28 = 0.42 \quad \text{and} \quad r_2 = 10.70 - 9.50 = 1.20 .$$

The average of the moving ranges is $\bar{r} = 0.629$, and the center line is at $x = 10.13$. Therefore, control limits are set at

$$\text{UCL} = 10.13 + 2.66(0.629) = 11.80, \quad \text{LCL} = 8.46 .$$

EXERCISES

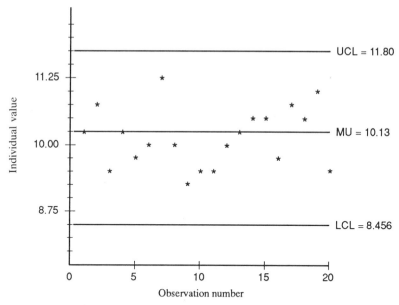

Figure 9.16.1. I chart for the averages in table 9.6.1.

It will be recalled that the control limits in the original x-bar chart were set at 11.45 and 8.80.

The Minitab I chart for this data is shown in Figure 9.16.1. □

EXERCISES

9.1. The following data set shows 30 samples, each with three observations. The first three samples are in the first row. Make an x-bar chart and an R chart for the data and mark any points that are out of control, using all three stopping rules.

22.83	25.39	23.41;	27.23	25.52	23.19;	22.22	23.15	22.75;
25.55	23.80	24.99;	26.22	22.10	24.32;	23.09	24.13	26.37;
18.69	25.51	22.86;	25.83	24.54	25.62;	20.01	22.77	23.14;
25.57	22.96	22.10;	28.77	23.72	28.55;	25.52	22.61	25.23;
25.24	23.10	23.57;	20.38	23.86	26.35;	25.48	21.58	22.76;
25.90	23.71	28.82;	19.82	23.29	23.53;	23.64	24.27	24.02;
23.59	23.11	23.50;	21.12	21.05	26.95;	23.48	25.84	21.42;
26.94	25.84	20.02;	23.76	21.61	21.83;	24.60	25.10	22.06;
24.91	25.70	25.73;	25.04	27.76	25.26;	22.70	25.31	22.76;
25.17	26.39	23.54;	26.78	23.32	26.88;	28.51	25.02	24.16.

9.2. Repeat exercise 9.1 assuming that you were given standard values $\mu = 24.0$ and $\sigma = 1.7$. Do your conclusions about points being out of control change?

9.3. This data set contains 100 observations taken from a production line in 25 samples of 4 each. The first two samples are in the first row. When the process was under control, the process average was 64.0 and the standard deviation was 1.5. Make an x-bar chart and an R chart for the data and mark any points that are out of control, using all three stopping rules.

64.0	64.0	64.0	63.6;	62.0	63.2	66.8	67.2;
64.4	67.6	60.8	66.8;	63.6	64.8	62.0	64.0;
64.4	63.6	62.8	65.2;	64.4	62.8	66.4	64.0;
62.0	65.6	62.8	65.6;	62.8	60.4	62.4	67.6;
63.2	61.2	64.8	64.4;	62.4	62.4	64.8	64.8;
66.0	65.2	66.4	62.0;	60.4	64.0	64.0	65.6;
65.6	61.6	65.0	67.8;	65.8	66.6	65.8	65.4;
66.6	65.4	65.8	65.0;	68.2	62.6	62.6	65.4;
63.8	65.0	64.6	67.0;	63.0	67.4	64.2	62.2;
63.0	65.0	66.6	63.8;	62.6	65.0	66.6	65.4;
66.2	69.4	65.4	64.2;	65.0	67.0	64.2	67.8;
65.0	66.6	63.2	64.0;	64.8	65.2	62.8	59.6;
65.2	63.2	64.4	63.2.				

9.4. Samples of three items are taken from the production line at the end of each shift for ten days (30 samples). The measurements of the items are given below with the first three samples in the first row.

17.3	17.3	16.8;	18.1	18.8	17.4;	16.9	17.9	17.2
19.4	19.5	17.7;	17.9	17.5	18.5;	16.3	18.2	16.4
16.5	18.0	16.8;	17.4	17.9	15.9;	15.5	16.4	18.4
16.5	16.1	15.9;	16.2	17.3	18.5;	18.6	18.0	17.8
20.0	18.6	17.6;	18.7	17.2	17.7;	18.7	17.6	18.1
18.0	15.6	18.0;	18.1	19.6	18.6;	20.4	17.4	20.2
19.6	18.0	19.5;	18.2	19.8	20.9;	17.8	18.8	16.8
17.5	17.3	18.1;	17.5	17.3	16.5;	20.5	17.7	16.0
17.6	17.4	17.8;	17.5	18.1	17.9;	18.2	18.0	18.7
18.6	19.4	18.2;	18.0	18.4	18.5;	17.4	18.7	18.7

9.5. This data set consists of 90 observations in three samples of three observations each. The first row contains the first three samples, and so on. Make an x-bar chart and an R chart of the data and note which samples are out of control.

40.01	38.97	38.55	38.80	39.95	39.95	40.36	40.94	39.77
40.47	40.46	39.68	39.41	40.43	40.60	38.98	40.20	39.30
39.98	37.10	41.01	40.32	39.47	40.18	39.56	40.66	39.33
39.67	38.92	39.52	38.48	39.79	39.57	40.68	39.07	41.14
40.35	40.14	39.20	39.90	39.54	40.84	41.56	39.76	40.86
38.54	40.33	41.53	41.38	40.29	39.92	39.94	40.99	39.97
41.13	40.41	41.16	40.19	39.25	40.92	40.87	40.15	40.14
40.49	39.38	40.86	40.53	40.79	40.69	39.73	40.57	39.55
40.95	40.38	41.15	41.83	40.44	40.80	41.16	41.02	40.17
40.93	40.58	41.31	40.84	40.65	41.07	40.53	40.80	41.34

9.6. Make an s chart for the data of exercise 9.1.

9.7. Make an s chart for the data in exercise 9.4.

9.8. Prove that s is a biased estimator of σ, i.e., that $E(s) \neq \sigma$.

9.9. Assuming that $E(s) = c_4 \sigma$, prove that $V(s) = (1 - c_4^2)\sigma^2$.

9.10. Make an I chart for the data in table 9.8.1.

CHAPTER TEN

Control Charts for Attributes

10.1. INTRODUCTION

In chapter nine, we discussed control charts in which the measurements were continuous random variables. Those are known as control charts for variables or control charts for measurements. In this chapter, we look at the situation in which no measurements are recorded. An item is examined and classified as either conforming, in which case it is accepted, or as nonconforming, in which case it is rejected. We can associate with each item tested a random variable x that takes the value zero if the item is acceptable and one if it is nonconforming. We are sampling, or testing, by attributes. Either the item has the attribute of acceptability or it does not. Another term used is go, no-go: either the item is acceptable (go) or it is nonconforming (no-go). In order to bring U.S. terminology more in line with international standards, the words nonconforming and nonconformity have lately replaced in some industries the older words defective and defect. Nonconforming units replaces the use of defective as a noun. Many engineers use the terms interchangeably.

One can picture an inspector drawing a random sample from a large batch of components and counting the number of nonconforming units. We recall from sections 3.12 and 6.3 that if the inspector takes a random sample of parts, the number of nonconforming units theoretically has a hypergeometric distribution. However, if the sample consists of only a small fraction of the batch, we can reasonably act as if the number of nonconforming units has a binomial distribution with parameters n and p, where n is the number of parts sampled, and p is the proportion of nonconforming units in the batch. In the first part of the chapter, we present control charts based upon the binomial distribution. Then we introduce another series of charts that use the Poisson distribution.

10.2. BINOMIAL CHARTS

We assume a process rolling along under control, producing parts of which a proportion p do not conform. An inspector draws a sample of size n_i, of

BINOMIAL CHARTS FOR A FIXED SAMPLE SIZE   165

which x_i are nonconforming. The fraction nonconforming for that sample is

$$p_i = \frac{x_i}{n_i}, \tag{10.2.1}$$

where p_i is an unbiased estimate of p. It is a random variable and its variance is pq/n. If n is large enough, we can act as if p_i is normally distributed, and think in terms of upper and lower control lines at a distance of three standard deviations from a center line as before.

How large should n be? A commonly used rule of thumb is to take a prior estimate of p and choose n so that $np > 5$. This means that if the fraction nonconforming is thought to be about 5%, a sample size of at least 100 is recommended. When we get the fraction of nonconforming units down to 0.005, we will need samples of 1000 items!

The value of p varies from application to application. Some semiconductor manufacturers are working in some stages in the region of one defective unit per million. On the other hand, an Associated Press report, printed in the *Austin American Statesman* on December 24, 1989, quoted an Air Force report that a certain defense contractor had, over a 12-month period, reduced its overall defect rate from 28% to 18.1%.

Two examples are enough to illustrate the general procedure. The first has a fixed sample size. In the second, the sample size varies.

10.3. BINOMIAL CHARTS FOR A FIXED SAMPLE SIZE

Figure 10.3.1 shows a control chart for an inspection scheme in which samples of $n = 150$ were taken from each batch. The process is in statistical control. The data are given in table 10.3.1. There are two equivalent methods of making a chart. One can either plot p_i along the y-axis, in which case it is called a p chart, or else one can plot x_i and obtain what is called an np chart. We will construct an np chart. The center line is set at x-bar, the average of all the x_i; $\bar{x}/n$ is our estimate of p. Then the variance of x_i is estimated by

$$V(x_i) = npq = \frac{\bar{x}(n - \bar{x})}{n}. \tag{10.3.1}$$

The upper and lower control lines are drawn at

$$\text{UCL} = \bar{x} + 3\sqrt{\bar{x}(n - \bar{x})/n} \tag{10.3.2}$$

Table 10.3.1. The Numbers Nonconforming in Samples with $n = 150$

6	4	9	3	4	6	4	8	
8	5	3	8	6	12	3	9	
1	10	8	4	5	5	4	5	4

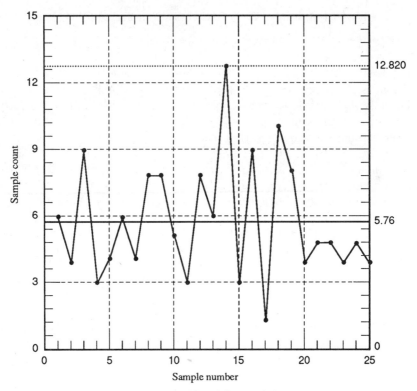

Figure 10.3.1. Binomial control chart.

and

$$\text{LCL} = \bar{x} - 3\sqrt{\bar{x}(n-\bar{x})/n}. \quad (10.3.3)$$

If the value given by equation (10.3.3) is negative, which is often the case, LCL is set at zero.

The average number of nonconforming units per sample is 5.76, which corresponds to approximately a 4% nonconformance rate. The center line is set at this value. The UCL is calculated by equation (10.3.2) as 12.8. Equation (10.3.3) gives a value of -1.30 and so the lower line is set at zero. All 25 points lie between the control lines, but there are two close calls. The fourteenth sample had 12 nonconforming units, and the seventeenth only one. In these two extreme cases, the percentage of nonconforming units in the samples was as large as 8% and as small as 0.7%.

The only difference between the np chart and the p chart is in the scale of the vertical axis. In the p chart, we plot the percentage nonconforming. In this example, the center line would be labeled 3.84% and the UCL 8.53%.

10.4. VARIABLE SAMPLE SIZES

Another strategy is to sample every item in the batch: 100% sampling. Sometimes 100% sampling is clearly out of the question, especially when testing is destructive. If the method of testing a fuse is to expose it to increased levels of current until it blows, the test destroys the item, and 100% testing would leave the engineer with no product. Serious questions have been raised about the practical efficiency of 100% testing. It would seem that this method ought to spot all the nonconforming units, but in practice it does not. Especially when the nonconforming rate is low and the procedure is tedious, inspectors tend to become inattentive. However, in some cases, 100% sampling can be virtually, or actually, automated. The test can merely be a matter of plugging a component into a test machine and seeing if the light goes on, or of having a piece of automatic testing equipment installed on-line.

Figure 10.4.1 is a p chart for a process to produce electrical parts, with batches of about 2000 items subjected to 100% inspection. The important difference between this example and the preceding one is that the sample

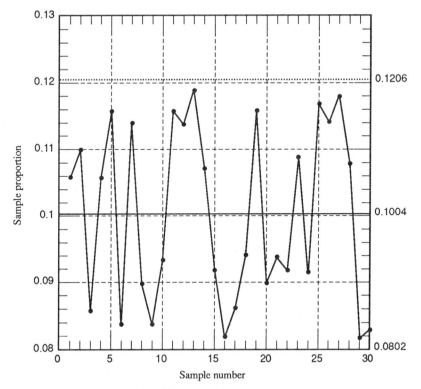

Figure 10.4.1. The p chart (unequal sample sizes).

168 CONTROL CHARTS FOR ATTRIBUTES

size is not constant. The batch size varied from day to day for various reasons. In this data set, the largest batch had 2488 parts and the smallest 1549. This variation makes the simpler *np* chart inappropriate. The percentages of nonconforming parts ranged from 8.2% to 11.9%. The data are given in table 10.4.1.

We can make a chart that ignores the variation in sample size. The center line is set at 0.1004, the average of the p_i (some engineers use $P = 100p$). The usual procedure is to compute the values for the outer control lines from equations (10.3.2) and (10.3.3), using the average batch size, 2015. This gives values 0.0802 and 0.1206. No points are out of control, although point 13 on the high side and points 16 and 29 on the low side come close.

Control lines based on the average batch size are adequate to get the general picture, but for the close cases, we must calculate the limits using the actual sizes. The outer lines are no longer horizontal. For each batch, we calculate the limits

$$\bar{p} \pm 3\sqrt{p_i q_i / n_i}. \qquad (10.4.1)$$

Again, the lower line is set at zero if this formula gives a negative value.

A chart prepared under these conditions is shown in figure 10.4.2. Point 13 is now out of control because it was the largest batch in the group. The lowest points, 16 and 29, are smaller batches.

Table 10.4.1. 100% Sampling of Batches of Components

	1	2	3	4	5	6	7	8
batch size	1705	1854	2319	2322	1796	2103	1950	1904
nonconf.	180	204	199	245	209	177	222	172
% nonconf.	10.6	11.0	8.6	10.6	11.6	8.4	11.4	9.0
	9	10	11	12	13	14	15	16
batch size	2171	2383	2153	1884	2488	2451	1549	1842
nonconf.	182	224	249	213	295	261	143	151
% nonconf.	8.4	9.4	11.6	11.3	11.9	10.7	9.2	8.2
	17	18	19	20	21	22	23	24
batch size	1755	1633	2086	2276	2047	1918	1774	2183
nonconf.	150	154	242	205	192	176	194	201
% nonconf.	8.6	9.4	11.6	9.0	9.4	9.2	10.9	9.2
	25	26	27	28	29	30		
batch size	1819	2231	1983	1584	1901	2374		
nonconf.	213	254	234	171	156	197		
% nonconf.	11.7	11.4	11.8	10.8	8.2	8.3		

Average batch size = 2015; $p = 0.1004$

INTERPRETING OUTLYING SAMPLES

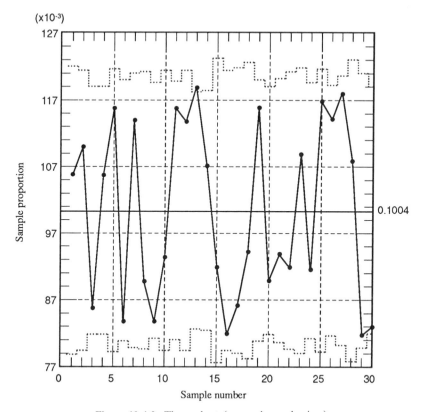

Figure 10.4.2. The p chart (unequal sample sizes).

10.5. INTERPRETING OUTLYING SAMPLES

In section 9.5, we made a passing reference to the question of handling outliers in the data set that is used to set up a control chart. The interest then was in R charts. An unreasonably large range can give the engineer an inflated estimate of σ. This will lead the engineer to set the control lines for the x-bar chart too far apart and to be lulled into a sense of security when all the averages fall comfortably between the unrealistically liberal limits. Nothing was said about samples in which R was too small, because with the usual sample sizes of $n = 4$ or 5, the lower control limit on the R chart is zero.

Similar problems arise with p and np charts. If a batch has an unusually high fraction of defective parts, the search for an assignable cause may lead to a reason, such as a malfunctioning piece of equipment that is replaced. That point can then be dropped from the data set and the control limits can be recalculated.

Table 10.5.1. Numbers of Defects in 25 Samples ($n = 150$)

17	19	24	17	15	22	19	26	20	21
13	17	17	23	16	19	22	2	3	4
24	18	20	21	31					

Points that fall below the LCL may present a different problem. There is a temptation to greet such points with applause. Look how well we can do when we really try hard! It is just as important to search for an assignable cause for a low point as for a high point. Unfortunately, it often happens that the low point is too good to be true. One of the standard reasons for low points is that they have been reported by an inexperienced inspector who fails to spot all the defects. This difficulty also occurs with c-charts.

The following data set illustrates the difficulty. The observations are shown in table 10.5.1. They are the numbers of defective items in 25 samples of $n = 150$ items each. The np chart is shown in figure 10.5.1.

The average value (center line) is $np = 18.0$. There are three points below the LCL—numbers 18, 19, and 20. The last sample has 31 defectives and lies above the LCL.

Suppose that the three points below the LCL are dropped from the data set and that the control lines are recalculated from the remaining 22 points. The new center line is at the new average, $np = 20.05$. The UCL is at $np = 32.55$. The last sample is now below the UCL and is under control.

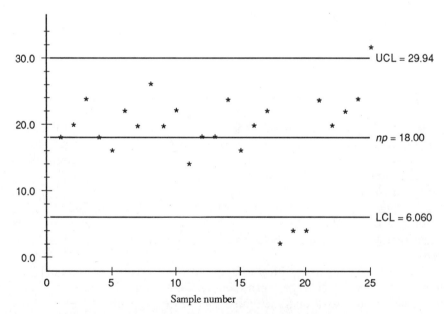

Figure 10.5.1. The np chart for table 10.5.1.

10.6. THE ARCSINE TRANSFORMATION

In the last section, doubts were cast about the importance of observations that fell below the LCL. That should not be interpreted as meaning that all points below the line are false. The LCL for the np chart is set at the larger of zero and $n\bar{p} - 3\sqrt{n\bar{p}(1-\bar{p})}$. It can be shown (and it appears as an exercise) that the LCL will be set at zero unless $np > 9(1-p)$. One can argue that when the LCL is zero, the engineer loses power to detect genuine improvements in quality. This is one of the reasons why the arcsine chart has been proposed as an alternative to the p chart when the sample size is constant. Another reason is that the arcsine transformation makes the plotted variable independent of p. It is also claimed that the transformed variable has a distribution that is closer to normal than does the binomial.

In applying the transformation, one calculates p, the percent nonconforming, in the sample and replaces it by

$$x = 2\arcsin(\sqrt{p}),$$

where $\arcsin(u)$ is the angle whose sine is u, expressed in radians. The variance of x is $1/n$, and the control limits are set at

$$\bar{x} \pm 3\,\text{sqrt}\left(\frac{1}{n}\right).$$

Nelson (1983a) suggests an alternative transformation:

$$x = \arcsin[\sqrt{x(n+1)}] + \arcsin[\sqrt{(x+1)/(n+1)}].$$

The difference between the two transformations is a matter of taste and is not important.

When the arcsine transformation, $x = 2\arcsin(\sqrt{p})$, is applied to the data in table 10.3.1, the transformed values are

```
0.4027   0.3281   0.4949   0.2838   0.3281   0.4027   0.3281
0.4661   0.4661   0.3672   0.2838   0.4661   0.4027   0.5735
0.2838   0.4949   0.1635   0.5223   0.4661   0.3281   0.3672
0.3672   0.3281   0.3672   0.3281 .
```

The center line is at 0.3844. The control limits are

$$\text{UCL} = 0.6294 \quad \text{and} \quad \text{LCL} = 0.1395.$$

There are no points outside the limits.

10.7. c-CHARTS

Suppose that we are looking at the output of a machine that makes yarn for knitting and weaving. At various places in a spool, the yarn will be weak enough to break. That is a nonconformity. If the number of nonconformities per reel has a Poisson distribution with $\mu = 5$, there are, on the average, five nonconformities per reel. If we take a sample of ten reels and count the total number of nonconformities, that total will have a Poisson distribution with $\mu = 50$.

Charts for the total number of nonconformities, based on the Poisson model, are called c-charts. The letter c is used instead of μ to denote the parameter of the distribution. The charts shown in figures 10.7.1 and 10.7.2 are taken from a process similar to that described before. The output of 50 days' work was inspected; ten reels were inspected in the course of each day, and the total number of nonconformities in the ten reels was recorded. The data are given in table 10.7.1.

Figure 10.7.1 is a c-chart made from the first 21 points. The average number of nonconformities per day during that period was $c = 56.95$. The center line of the chart was set at that level. Since the mean and variance of a Poisson variable are the same, the standard deviation of the process was taken as 7.55, and the control lines were drawn at

$$56.95 \pm 3(7.55) = 34.3, 79.6.$$

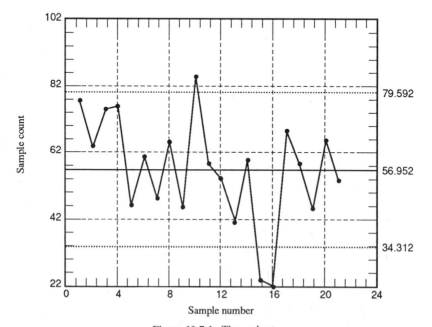

Figure 10.7.1. The c-chart.

C-CHARTS

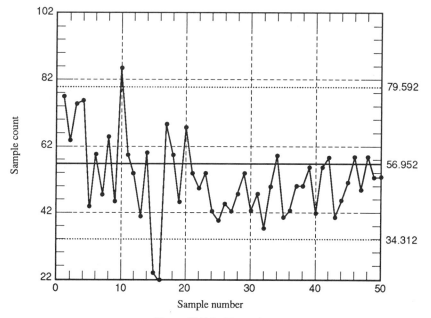

Figure 10.7.2. The c-chart.

There are three points outside the lines. Point 10 is too high. This occurred on a Friday, and the engineer noticed it immediately without the need for a chart. Investigation over the weekend found an assignable cause in a part of the machine. A temporary repair was made and production resumed on Monday. Points 15 and 16 are below the lower line. It would be nice to conclude that the process had suddenly pulled itself together and that those points might set a standard for future performance. Unfortunately, closer investigation revealed that the regular inspector was sick on those two days. Her duties were taken over temporarily by another employee who did not know exactly what to look for. As a result, he missed quite a few weak spots and reported too few nonconformities. Assignable cause—inadequately trained worker.

Throwing out those three points would have increased the average number of nonconformities to 59.2. That was not done because a new part

Table 10.7.1. Number of Nonconformities in Ten Reels of Yarn

77	64	75	76	45	61	49	65	45	85
59	54	41	60	24	22	69	59	45	67
54	49	54	43	40	46	43	49	55	43
49	38	51	59	41	43	50	50	56	42
56	59	41	46	51	59	49	59	53	53

arrived on the afternoon of the twenty-first day and was substituted next morning for the part that had been temporarily repaired. The replacement resulted in an immediate improvement. During the last 29 days, the average number of nonconformities per day dropped to 49.2. That represented a 10% reduction in the nonconforming rate, but it still left room for further improvement.

Figure 10.7.2 shows all 50 points plotted with the same control lines that were calculated from the first 21 points. The improvement in quality after the new part was installed is obvious. A chart of the last 29 points shows a process under control at the new level of c. It is interesting to note that rule 1 of the previous chapter—seek an assignable cause if there is a point outside the outer control lines—was not triggered; no doubt it would have been eventually. What justifies our concluding from the chart that there has been an improvement is the preponderance of points below the old center line, particularly the run of 13 consecutive points below the line.

10.8. DEMERITS

In constructing c-charts, we have assumed that one nonconformity is no better or no worse than another. That may not be so. Some defects, such as a scratch in the paint, are merely cosmetic; others are functional. To handle this situation, some companies use a system of demerits. A small defect may carry one demerit. A more serious defect may carry a penalty of ten demerits. This might be a more appropriate procedure in the evaluation of the overall performance of a defense contractor, to which reference was made in section 10.2, than computing an unweighted percentage figure. Charts for demerits are similar to c-charts.

Suppose that a process has defects with three demerit levels, one point, five points, and ten points. Let the numbers of occurrences of these defects be denoted by x_1, x_5, and x_{10}, respectively, and let them have Poisson distributions with parameters c_1, c_5, c_{10}, respectively. The total number of demerits is then

$$y = x_1 + 5x_5 + 10x_{10}.$$

Although Y is a linear combination of Poisson variables, it is not the simple sum of them, and so it does not, itself, have a Poisson distribution. We can, however, readily calculate its expectation and variance as

$$E(y) = c_1 + 5c_5 + 10c_{10} \quad \text{and} \quad V(Y) = c_1 + 25c_5 + 100c_{10}.$$

We then set the center line from the pilot data at

$$y = \bar{x}_1 + 5\bar{x}_5 + 10\bar{x}_{10} = y_0$$

EXERCISES

and the control lines at

$$y = y_0 \pm 3\sqrt{\bar{x}_1 + 25\bar{x}_5 + 100\bar{x}_{10}}.$$

EXERCISES

10.1. Twenty samples of 200 items each were taken from a production line. The number of defectives in the samples are given in the following. Make a p chart and a np chart for this data set.

4	7	2	8	8	2	13	5	8	7
3	4	4	7	8	11	7	8	7	5

Is the sample with 13 defectives out of control?

10.2. Twenty-five batches of devices were subjected to 100% inspection. The batch sizes and the number of defectives are given in the following. The average batch size is 2013. The average number of defectives is 73.6. Make a p chart of the data.

$n =$	1953	1446	1441	1954	1450	1554	1852	2038
$x =$	70	47	55	78	51	57	80	89

2450	2448	1753	2549	2248	2450	1552	2449	2552
85	83	64	90	75	66	54	120	88

1850	2154	1950	2340	1746	1949	2049	2149
64	90	80	82	49	65	84	74

Batch 16 is clearly a bad batch. Batch 2 has the fewest defectives, but it is next to the smallest batch. Is it out of control?

10.3. Make a (Poisson) c-chart for the last 29 observations in table 10.7.1, which were made after the new part was installed. Use $\mu = 49.2$. Is the process under control at the new level of defects?

10.4. Make a c-chart for this set of data assuming that the average number of defects is 9.5 and mark any points that are out of control:

13	10	8	14	18	12	12	9	15	10	10	11	8
9	9	17	12	14	13	19	15	8	11	10	14	

10.5. Make arcsine charts for the data in table 10.5, both with and without the three low points. Do your conclusions agree with those of section 10.5?

10.6. Five hundred devices are inspected and the number of defects in each device is recorded. Let x_i denote the number of devices that have n_i defects. A summary of the data is

$n =$	0	1	2	3	4	5 (i.e., >4)
$x =$	274	138	37	11	9	31

The average number of defects per device is 0.97. Show that the data are most unlikely to have come from a Poisson distribution, and so the defects cannot be regarded as occurring at random. This data set was constructed with 450 observations from a Poisson distribution with $\mu = 0.5$ and 50 from a Poisson distribution with $\mu = 5.0$. This corresponds to a process that produces "good" items with an average rate of one defect for every two items and is contaminated by about 10% "bad" items that have, on the average, five defects per item. When the devices are good, they are indeed good; when they are bad, they are dreadful.

CHAPTER ELEVEN

Acceptance Sampling I

11.1. INTRODUCTION

In developing the theory and practice of acceptance sampling, we introduce two people whose attitudes have traditionally been those of adversaries, the vendor and the buyer. The vendor manufactures widgets in a factory. When a batch of them is made (we henceforth use the jargon of acceptance sampling and use the word lot instead of batch), the lot is presented to the buyer, with the claim that it is of satisfactory quality. Before accepting the lot, the buyer calls for an inspection to determine whether the lot meets requirements. If the lot passes inspection, it is accepted. If the lot does not pass inspection, it is rejected. The reader will already have realized that we are going to talk about a special form of hypothesis testing. In practice, the role of buyer can be played by the actual purchaser or by the vendor's own quality control inspectors who stand between the production line and the factory gate.

The emphasis in this chapter is on single sample plans. This is the simplest case. The inspector takes a random sample of n items from a lot and rejects the lot if the sample contains more than c defective items. The numbers n and c are predetermined. The discussion is, for the most part, confined to sampling by attributes. Chapter twelve begins with multiple sampling plans and sequential sampling. It ends with a description of the use of one of the major sampling schemes, MIL-STD-105D, that is commonly used in defense work.

The field of acceptance sampling covers a very broad area and we have space for only a fraction of the topics that could be covered. The reader who wishes to pursue the matter further is referred to the books by Schilling (1982) and by Duncan (1986).

11.2. THE ROLE OF ACCEPTANCE SAMPLING

When a buyer is offered a lot, there are three choices:

(i) accept the lot as satisfactory without any inspection;

(ii) inspect every item (100% inspection);
(iii) use an acceptance sampling scheme in which only a fraction of the items are inspected.

There are several arguments against 100% inspection. Among them are the following:

(i) if testing is destructive, e.g., if testing consists of twisting an item until it breaks, the buyer will be left with no items;
(ii) testing introduces a chance of damaging good items;
(iii) the cost of testing every item may be excessive;
(iv) when the fraction of defective items is low, 100% inspection can become a tedious task, with the result that too many defects are missed.

A desirable goal would be a manufacturing process in which the first option—zero defects—zero sampling—is valid. That is the aim of manufacturers who seek first-class quality and first-class competitiveness in world markets. This is already happening in situations where a major manufacturer has built up such a strong, close relationship with a supplier that the expenditure of time and money on acceptance sampling is no longer warranted. In such a case, nonconforming units are so rare that the occurrence of one will justify a very serious investigation to identify the error and to eliminate it.

There is also a mathematical point. Whereas only a few years ago manufacturers talked about defective rates of 1% to 10%, we are now seeing in the microelectronics industry processes that have a defect rate as low as 500 or even 50 parts per million. If a manufacturer is only producing one nonconforming part in 10,000, there is only about a 1% chance that a sample of 100 parts will contain a defective part. If the nonconformity rate is one in 100,000, only about one such sample in 1000 will contain a defective.

The situation described in the last two paragraphs is a worthy aim. However, many processes are far from that ideal. For that majority, acceptance sampling remains a major feature of the production system. It is there to give protection both to the consumer and to the producer. Schilling (1984) argues that acceptance sampling has other uses than merely deciding the fate of individual lots. He argues that "when applied to a steady flow of product from a supplier it can be made a dynamic element supplementing the process control system." It can be the buyer's quality control chart for the vendor's production.

11.3. SAMPLING BY ATTRIBUTES

Each item in the sample is tested by an inspector who rates it as either defective or nondefective. The words conforming and nonconforming are often used instead of the older terms defective and nondefective. We stay with the older terminology, although we sometimes find it convenient to call items "good" or "bad."

The quality of a lot is measured by the percentage of defective items: the lower the percentage defective, the higher the quality. Deciding whether a lot should be accepted or rejected is called lot sentencing.

When a lot is rejected there are several choices:

(i) The entire lot can be scrapped, which may incur a large financial loss.
(ii) If the defects do not render them unusable, the items in the lot can be sold on an "as is" basis at a lower price. This can happen in the garment industry, where batches of shirts can be sold at discount outlets as "seconds."
(iii) 100% sampling can be made of the items in the condemned lot and the defective items replaced by good items; the lot is then said to have been rectified.

These choices are sometimes referred to collectively as lot disposition actions.

The standard inspection system for sampling by attributes that is used by the U.S. Department of Defense is MIL-STD-105D. It is discussed in chapter twelve.

11.4. SINGLE SAMPLING PLANS

We mentioned in section 11.1 that, in the simplest case, an inspector draws a random sample of n items from the lot and notes the number of defectives. He works according to a sampling plan that tells him how large the sample should be and gives an acceptance number, c. If the number of defectives found in the sample is c or fewer, the lot is accepted; if there are more than c defectives, the lot is rejected. Some use Ac (short for acceptance number) instead of c and Re (short for rejection number) for $c + 1$. The main question is how does one choose the values of n and c?

How large a sample size do we need? We will be taking samples that are only a small proportion of the lot, so that we can assume that the number of defectives has a binomial distribution. A small handful of items will not give enough power to discriminate between good quality and bad. Samples of about 100 or 200 will be the norm. It will not matter whether the lot size is 2000 or 5000, what will matter is the sample size.

A common mistake is always to test a fixed percentage, say, 10%, of a lot. This can lead to ineffective samples from small lots and unnecessarily large samples from big lots. Cynics have suggested that an unscrupulous vendor, who realized that the fraction defective in his manufacturing process had increased, might make it a point to present the substandard material in smaller lots. Then, since small samples have less power to reject null hypotheses than large samples, he could increase the chance that the poor material would pass inspection!

The quality level of a process is the percent defective when it is under control—the center line on the control chart. We denote it by p_0. The vendor wants a situation in which only a small portion of the product will be rejected when the process is under control. That is the vendor's risk, alpha. How small alpha should be depends upon the consequences of the rejection of a lot. If having too many lots rejected means loss of business to competitors who have better quality, or if it means large amounts of money spent on scrap or rework, the vendor is under pressure to work at a low value of alpha.

The buyer wants good product, with a low percentage of defectives. Zero defectives is a nice slogan and an important target, but it is not always practical. In some situations, one can obtain zero defectives by very slow and cautious workmanship at greatly increased labor cost. Perhaps it is in the buyer's interest to accept a small fraction of defectives to obtain a reduction in unit cost. This will depend on the consequences to him of receiving a defective item. If your office manager buys pens by the gross and the pen that you pick does not work, you can throw it away and take another without too much loss. On the other hand, if the lot consists of vital parts whose failure is a matter of life and death, that is a different situation. The buyer has learned that he can manage comfortably with a certain level of defectives. This is the target level.

It is convenient if this target level, which we denote as the acceptable quality level (AQL), coincides with the vendor's production level. On the other hand, the buyer also knows that the fraction defective in the lots received from even the most competent of vendors will vary; there is some upper bound on the fraction defective in a batch that can be accepted without extreme inconvenience or annoyance. This level is called the lot tolerance percent defective (LTPD). The buyer will want a sampling plan that gives a low risk, beta, of accepting so bad a lot.

11.5. SOME ELEMENTARY SINGLE SAMPLING PLANS

A single sampling plan is fixed by two numbers, n and c. Why should one choose one plan over another? The decision should be made on the basis of the relative alpha and beta risks for certain values of p. In this section, we look at four plans with small values of n. These plans are quite unrealistic. They are included only to give the reader some simple illustrative examples. In the next section, we look at some more realistic plans.

For purposes of illustration in this section, we make the following basic assumptions to make the arithmetic easy. The vendor has a process that,

SOME ELEMENTARY SINGLE SAMPLING PLANS

when it is under control, produces items that have $p = 10\%$ defectives. The buyer has learned to live with this rate of defectives and is prepared to accept lots that have a slightly higher percentage, but does not want to accept lots that have as much as 20% defectives. Thus, alpha, the vendor's risk, is the probability that a lot with $p = 0.10$ is rejected; beta, the buyer's risk, is the probability that a lot with $p = 0.20$ is accepted, an unfortunate situation for the buyer. We use the symbol $P_a(\pi)$ to denote the probability that a lot with percentage defectives equal to π will be accepted.

We can write

$$\alpha = 1 - P_a(0.1) \quad \text{and} \quad \beta = P_a(0.2).$$

We assume that the number of defectives, d, has a binomial distribution with the appropriate values of n and p; α is evaluated at $p = 0.1$; β is evaluated at $p = 0.2$.

Plan A. $n = 3$, $c = 0$

Draw a sample of three items and accept the lot only if all three are good.

If $p = 0.10$, $q = 0.90$, the probability that a sample will pass inspection is $(0.9)^3 = 0.729$.

$$\alpha = 1.0 - (0.9)^3 = 1 - 0.729 = 0.27$$
$$\beta = (0.80)^3 = 0.51.$$

Approximately a quarter of the vendor's "good" lots will be rejected. On the other hand, if the vendor presents lots that are 20% defective, about half of them will pass inspection and the buyer will have to accept them.

Plan B. $n = 5$, $c = 0$

$\alpha = 1.0 - (0.9)^5 = 0.41$, $\beta = (0.80)^5 = 0.33$. This plan is tougher on the vendor, but it gives the buyer better protection against bad lots.

Plan C. $n = 5$, $c = 1$

$$\alpha = 1.0 - P(d = 0) - P(d = 1)$$
$$= 1.0 - (0.9)^5 - 5(0.9)^4(0.1) = 0.082.$$
$$\beta = (0.8)^5 + 5(0.8)^4(0.2) = 0.738.$$

This plan gives much more protection to the vendor than the previous plans, but it is completely unacceptable to the buyer.

Plan D. $n = 10$, $c = 1$

$\alpha = 0.26$, $\beta = 0.38$. The two risks are more nearly equal, but they are clearly too big for practical purposes. We need to take larger samples.

11.6. SOME PRACTICAL PLANS

In this section, we consider three plans that are of more practical use. Two of them have $n = 100$; the other calls for $n = 200$. In comparing the plans, we assume that the vendor and the buyer have agreed on a target rate of $p = 2\%$ defectives.

Plan 1. $n = 100$, $c = 4$

Draw a sample of 100 items and accept the lot if there are four, or fewer, defectives.

To evaluate alpha, we need to compute, for $p = 0.02$ and $q = 0.98$,

$$\alpha = 1.0 - q^{100} - 100 q^{99} p - (100)\left(\frac{99}{2}\right) q^{98} p^2$$
$$- (100)(99)\left(\frac{98}{6}\right) q^{97} p^3 - (100)(99)(98)\left(\frac{97}{24}\right) q^{96} p^4.$$

This quantity is not easy to compute because of the number of decimal places required. During World War II, it was very difficult because electronic computers were not yet available. Statisticians turned to the Poisson approximation to the binomial, which is close enough for practical purposes and is, nowadays, easy to compute on a hand calculator. There are now good tables of cumulative probabilities available, from which we can read $\alpha = 5.1\%$.

To obtain the Poisson approximation, we equate the parameter, μ, of the distribution to np. In this example, $\mu = 2.0$, and

$$P(d \leq 4) = e^{-\mu}\left(1 + \mu + \frac{\mu^2}{2} + \frac{\mu^3}{6} + \frac{\mu^4}{24}\right)$$
$$= 7e^{-2} = 0.947,$$
$$\alpha = 5.3\%.$$

Table 11.6.1 gives both the binomial probabilities of accepting a batch for several values of p and the Poisson approximations.

Table 11.6.1. Values of P_a for Different Percentages Defective ($n = 100$, $c = 4$)

p	0.02	0.03	0.04	0.05	0.06	0.07	0.08
Binomial	0.949	0.818	0.629	0.436	0.277	0.163	0.090
Poisson	0.947	0.815	0.629	0.440	0.285	0.173	0.100

CHOOSING THE SAMPLE SIZE

Table 11.6.2. $n = 200$, $c = 7$

p	0.02	0.03	0.04	0.05	0.06	0.07	0.08
P_a	0.951	0.746	0.450	0.213	0.083	0.027	0.008

Plan 2. $n = 200$, $c = 7$

This plan also has α at approximately 5%, but it is more favorable to the buyer. The values of beta are less than those for plan 1. The sample size has been doubled. The values of P_a are given in table 11.6.2.

Plan 3. $n = 100$, $c = 7$

This plan is the most favorable of the three to the vendor. It gives him enormous protection with an alpha risk of only 0.1%; the buyer will accept 87% of those lots that are 5% defective. The sample size is the same as plan 1, but the acceptance number is the same as in plan 2. The probabilities are given in table 11.6.3.

11.7. WHAT QUALITY IS ACCEPTED? AQL AND LTPD

The buyer defines two particular levels of quality. The acceptable quality level (AQL) is the worst average level of quality that he is willing to accept. Often this is a level of percent defective that the buyer has learned to live with over the years, and both the buyer and the vendor may agree that such a percentage is reasonable for both sides.

There is also a lower level of quality (higher value of p) that the buyer finds particularly offensive (unacceptable), even in single lots. This is the lot tolerance percent defective (LTPD).

The buyer wants a plan with a low probability, beta, of accepting a lot at the LTPD. The vendor wants a plan that has a low probability, alpha, of rejecting a lot that is at the AQL. Example 11.8.1 shows how one might design a sampling plan with the AQL at 2% and the LTPD at 7%.

11.8. CHOOSING THE SAMPLE SIZE

In chapter seven, we showed how to calculate the sample size and the cutoff point for a one-sided test of the mean of a normal population. The same principle is used to construct a single sampling plan. Suppose that we wish to

Table 11.6.3. $n = 100$, $c = 7$

p	0.02	0.03	0.04	0.05	0.06	0.07	0.08
P_a	0.999	0.989	0.953	0.872	0.748	0.599	0.447

test the null hypothesis H_0: $p = p_0$ against the alternative H_a: $p = p_1$ at given levels of α and β. We can calculate n and c for the test (sampling plan) in the following way, using the normal approximations to the binomial distributions involved.

If $p = p_i$, the number of defectives is normally distributed with $\mu = np_i$ and $\sigma^2 = np_i q_i$. Then, if $p = p_0$,

$$c - np_0 = z_\alpha \sqrt{np_0 q_0},$$

and, if $p = p_1$,

$$np_1 - c = z_\beta \sqrt{np_1 q_1},$$

whence

$$\sqrt{n} = \frac{z_\alpha \sqrt{p_0 q_0} + z_\beta \sqrt{np_1 q_1}}{p_1 - p_0}.$$

Example 11.8.1. Suppose that the engineer wishes to design a sampling plan with risks $\alpha = 5\%$ when $p = 0.02$, and $\beta = 10\%$ when $p = 0.07$:

$$\sqrt{n} = \frac{1.645\sqrt{0.0196} + 1.28\sqrt{0.0651}}{0.05} = 11.138,$$

$$n = 124 \quad \text{and} \quad c = 5. \qquad \square$$

11.9. THE AVERAGE OUTGOING QUALITY, AOQ AND AOQL

Both the vendor and the buyer may also be interested in the average quality of the lots that pass inspection with a given plan. Two measures are used: the average outgoing quality (AOQ), and the average outgoing quality limit (AOQL).

The average outgoing quality is the quality that one gets by applying a sampling plan to a process in which the average percent defective is p and rectifying all the lots that fail inspection. In calculating the AOQ, we make the following assumptions about the inspection procedure.

When a lot is accepted, the defective items in the sample are replaced by good items and the rectified sample is returned to the lot. The remaining $N - n$ items in the lot may include some undetected defectives. Thus, on the average, each lot accepted by the buyer contains $(N - n)p$ defective items.

When a lot is rejected, it is rectified. This means that it is subjected to 100% inspection in which every defective is (we hope) detected and replaced by a good item. Thus, the rejected lots are replaced by lots that have zero defectives. Given the problems that exist with 100% inspection, this assumption may be too optimistic.

THE AVERAGE OUTGOING QUALITY, AOQ AND AOQL

A fraction, P_a, of the lots will pass inspection. A fraction, $1 - P_a$, will be rejected and rectified. When these are combined, the average number of defectives per lot is seen to be

$$P_a(N - n)p + (1 - P_a)(0) = P_a(n - n)p, \qquad (11.9.1)$$

and so the AOQ is

$$\frac{P_a(N - n)p}{N}.$$

If n/N is small, this quantity is well approximated by the formula

$$\text{AOQ} = P_a p. \qquad (11.9.2)$$

Average outgoing quality values for plans 1 and 2 of section 11.6 are shown in table 11.9.1. Note that under this scheme, bad lots come out better than good lots. In plan 1, with $p = 2\%$, few lots are rejected, and so few are rectified; the AOQ, 1.90%, is thus only fractionally lower than p. On the other hand, if the quality drops to 8% defectives, 91% of the lots will be rectified; virtually all the lots will be subjected to 100% inspection and the AOQ falls to less than 1%.

The average outgoing quality limit (AOQL) is the maximum of the AOQ for a given plan. The AOQL for plans 1 and 2 are underlined in table 11.9.1. For the first plan, the AOQL is 2.52% at $p = 0.04$, and for the second, it is 2.24% at $p = 0.03$. The AOQL represents the worst long-run average percent defective that the buyer should expect to endure when that plan is used. Note that we speak of "long-term average" and "expect to endure." We are saying that for plan 1, the worst fate that can befall the buyer, on the average, will occur in the case when the vendor produces 4% defectives. Then, with this sampling plan and rectification, the buyer will receive, on the average, 2.24% defectives. Note well "on the average"—

Table 11.9.1. AOQ for Plans 1 and 2

	Plan 1						
p	0.02	0.03	0.04	0.05	0.06	0.07	0.08
P_a	0.949	0.818	0.629	0.436	0.277	0.163	0.090
AOQ (%)	1.90	2.45	<u>2.52</u>	2.18	1.66	1.14	0.72

	Plan 2						
p	0.02	0.03	0.04	0.05	0.06	0.07	0.08
P_a	0.951	0.746	0.450	0.213	0.083	0.027	0.008
AOQ (%)	1.90	<u>2.24</u>	1.98	1.07	0.50	0.19	0.06

some days the buyer will fare better and some days the buyer will fare worse! The interesting fact to note is that if the production line deteriorates and the vendor produces more than 4% defectives, we will detect (and rectify) more of the bad lots, and the overall quality of the product that the buyer actually receives will improve.

We have only discussed one way of defining AOQ. There are other possible scenarios. Our scenario implies that there is a supply of good items that can be substituted for the defectives, which in turn almost implies that the inspection is taking place at the vendor's plant before shipment. A different procedure for testing at the buyer's site would be for the buyer to carry out 100% inspection of the rejected batches and cull the bad items (which would not be paid for). When the percent defectives is small, this procedure gives values of the AOQ and AOQL that are very close to the values for the scenario that we considered.

11.10. OTHER SAMPLING SCHEMES

We have confined our discussions to single sampling plans for sampling by attributes. In the next chapter, we consider multiple sample plans and sequential sampling. In the remaining sections of this chapter, we make a few remarks about four other sampling schemes. For further details, the reader is referred to the books by Duncan (1986) and Schilling (1982) that were cited earlier.

All four of these schemes—rectification sampling plans, chain sampling, skip-lot plans, and continuous sampling—are associated with H. F. Dodge, who pioneered the field of acceptance sampling during his career at Bell Laboratories. In 1977 the *Journal of Quality Technology* devoted an entire issue to his memory, reprinting in that issue nine of his most important papers (Dodge, 1977).

11.11. RECTIFYING INSPECTION PLANS

We assumed in the previous section that every rejected lot would be subject to 100% inspection. We can estimate the cost of inspection using this procedure. Each accepted lot will have n items inspected; in each rejected lot, N items will be inspected. The average total inspection (ATI) is then given by

$$P_a n + (1 - P_a)N = n + (1 - P_a)(N - n) . \qquad (11.11.1)$$

In section 11.8, we chose a sampling plan based on desired values of AQL and LTPD; the lot size N did not appear in the calculations. Dodge and Romig (1959) published tables for choosing values of n and c based on other considerations. They assumed that the objective would be to minimize the ATI for given values of the AOQL, p, and N.

11.12. CHAIN SAMPLING PLANS

We have built up a theory for handling individual lots. That is appropriate if the vendor is erratic. But if he is fairly steady, should, or can, we take that into consideration, and by doing so save money on inspection? This is the rationale for chain sampling, which was introduced by Dodge (1955a).

In chain sampling, we make substantial reductions in n, and set $c = 0$. If there is exactly one defective item in the sample, we will still accept the lot, provided there have been no defective items in the previous i samples. If the number of defectives exceeds one, the lot is rejected.

The constants that define the plan are not n and c, but n and i. A pressure that is, perhaps, not so subtle is put upon the vendor. As long as the vendor's past record is free of blemishes, a single defect will be excused. But if there is a defect, that claim of previous virtue is lost. The vendor must start again from scratch, building up a record of perfect samples until there is again a backlog of at least i without defect.

11.13. SKIP LOT SAMPLING

There is another way of rewarding past good quality. It is skip lot sampling, which also was introduced by Dodge (1955b). The idea here is that after the vendor has established a past record for good quality, we should not insist on inspecting every lot. Rather, we should skip some lots and test only a fraction of the lots that are presented. Which lots are chosen to be tested is decided by a process of randomization. The buyer might agree to test roughly half the lots and choose them by tossing a coin each time a lot arrives. That is very different from telling the vendor that the buyer will, in future, test only the odd-numbered lots that are presented.

11.14. CONTINUOUS SAMPLING

So far, we have envisioned the vendor presenting the items to the buyer in (large) lots. H. F. Dodge (1943) proposed a scheme for sampling by attributes, which could be applied to a situation in which the inspector saw a continuous flow of items passing by on the production line.

These plans were subsequently refined by Dodge and Torrey (1951). A listing of continuous sampling plans is given in MIL-STD-1235B, which appeared in 1981.

11.15. SAMPLING BY VARIABLES

The procedure for sampling by variables is similar to the procedure for sampling by attributes. When it is appropriate, sampling by variables has the advantage of requiring smaller samples. One draws the sample and measures a characteristic, say, for example, the resistance, for each item. The set of n resistances is reduced to a single statistic, such as the mean or standard deviation, which is, in turn, compared to a critical value. There can be difficulties when the conformity of the item calls for its being satisfactory in more than one characteristic, which implies a multivariate testing situation.

EXERCISES

11.1. A single sampling plan has $n = 150$. The engineer accepts the lot if there are four or fewer defectives. What is the probability of accepting a lot with $p = 0.03$? What protection does the buyer have against accepting lots with $p = 0.08$? What is the AOQ for $p = 0.03$? What is the AOQL for this plan? You may use the Poisson approximation.

11.2. Design a single sampling plan that has $\alpha = 5\%$ when $p = 0.03$, and $\beta = 15\%$ when $p = 0.06$.

11.3. A single sampling plan has $n = 200$ and $c = 5$. What is the probability of accepting a lot with $p = 0.03$? What is the probability of failing to reject a lot with $p = 0.06$?

11.4. Find the AOQL for plan 3 of Section 11.6.

CHAPTER TWELVE

Acceptance Sampling II

12.1. THE COST OF INSPECTION

Except for defining ATI and mentioning the Dodge–Romig rectifying inspection plans, thus far we have ignored the cost of inspection in the discussion of sampling plans. In our examples, the buyer could argue in favor of plan 2 because that was the plan that gave him the best protection. On the other hand, it can be argued that it also called for inspecting $n = 200$ items rather than $n = 100$. This is more expensive. If the testing is destructive, the buyer gets 100 fewer items per batch than otherwise. The extra inspection cost will be borne by the buyer, either as a direct charge for the inspection at the buyer's site or as an overhead item figured into the vendor's bid for inspection at the vendor's factory.

It seems a foolish waste to keep on inspecting the whole of a large sample if it is obvious that the vendor is producing at excellent quality with a percentage defective well below the target value. One way of getting around this problem is to adopt a double sampling procedure. We take an initial sample from a lot. If it is a very good sample, we accept the lot. If it is a very bad sample, we reject the lot. If the number of defectives is neither so large nor so small as to make a decision obvious, we take another sample to settle the issue.

In theory, one could continue this procedure to a third sample or even a fourth. The ultimate refinement is the method of sequential analysis developed by Abraham Wald (1947).

12.2. DOUBLE SAMPLING

We will investigate sampling plans with two samples. The first has n_1 items and the second n_2. In many double sampling plans, $n_2 = 2n_1$. This is not a necessary condition. It used to be customary in military procurement, but is not a condition of the modern MIL-STD-105 tables. There are three acceptance numbers: c_1, c_2, and c_3.

We draw the first random sample and note the number, d_1, of defectives. If $d_1 \le c_1$, we accept the batch with no further sampling. If $d_1 > c_2$, we reject the batch outright. If $c_1 < d_1 \le c_2$, we give the lot a second chance. We draw another random sample of n_2 items. Let d_2 denote the total number of defectives in the second sample, and let $d = d_1 + d_2$. We accept the lot if $d \le c_3$. If $d > c_3$, the lot is rejected.

We illustrate this method of double sampling by an example.

Example 12.2.1. Let $n_1 = 100$, $n_2 = 200$, $c_1 = 3$, and $c_2 = c_3 = 9$.

Suppose that $p = 0.02$. For the first sample, we use the binomial tables, with $n = 100$ and $p = 0.02$, and find that the probability of acceptance is 0.859 and the probability of nonacceptance is 0.141. Acceptance after the second sample can take place in several ways. We can have $d_1 = 4$ and $d_2 \le 5$, or $d_1 = 5$, $d_2 \le 4$, and so on. The possibilities are listed in table 12.2.1. We then sum the products of the probabilities of d_1 and d_2 to obtain the desired probability.

Summing, the probability of acceptance on the second round is 0.0989 and the probability of acceptance overall is 0.958, so that $\alpha = 4.2\%$. If $p = 0.05$, we can show in the same way that the probability of acceptance is $0.2579 + 0.0175 = 0.275$. □

A single sample plan with $p_0 = 0.02$, $p_1 = 0.05$, $\alpha = 4.1\%$, and $\beta = 27.5\%$ requires a sample size about 190. Such a plan requires us to take $n = 190$ every time. The double sampling plan may require either 100 or 300 items to be inspected. When $p = 0.02$, there is probability 0.859 that we will only need the first sample. The expected number of items to be tested is

$$E(n) = (0.859)(100) + (0.141)(300) = 128.$$

On the other hand, when $p = 0.05$, the expected number of items is

$$E(n) = (0.258)(100) + (0.742)(300) = 248.$$

The vendor who is producing good quality items is rewarded by a marked decrease in inspection costs, inasmuch as only two-thirds as many items have

Table 12.2.1

First Sample	Probability	Second Sample	Probability	Product
$d_1 = 4$	0.0902	$d_2 \le 5$	0.7867	0.0710
$d_1 = 5$	0.0353	$d_2 \le 4$	0.6288	0.0222
$d_1 = 6$	0.0114	$d_2 \le 3$	0.4315	0.0049
$d_1 = 7$	0.0031	$d_2 \le 2$	0.2351	0.0007
$d_1 = 8$	0.0007	$d_2 \le 1$	0.0894	0.0001
$d_1 = 9$	0.0002	$d_2 \le 0$	0.0176	0.0000

to be sampled. The vendor with poor quality has to pay for inspecting more items.

12.3. SEQUENTIAL SAMPLING

The procedure of sequential sampling was developed by Abraham Wald for the United States during World War II. The items are sampled one by one. After each item, we count the total number of defectives so far and make one of three decisions: accept the lot, reject it, or keep on sampling. The total number of defectives at each stage is plotted against the sample number on a chart that has two lines drawn on it. These lines divide the paper into three zones: acceptance, rejection, and no decision. The plotted points are joined by a line as in a control chart. The line starts in the no-decision region. We keep on sampling until the line crosses into either the acceptance region or the rejection region.

It would seem that with bad luck, we could keep sampling forever; this is theoretically possible, just as it is theoretically possible that we can toss a coin for months before getting the first head, but Wald showed that in practice, his procedure called for testing many fewer items than the traditional fixed-sample-size procedures described earlier.

12.4. THE SEQUENTIAL PROBABILITY RATIO TEST

Wald's sequential probability ratio test was developed during World War II and published in 1947. The basic ideas have been given in the previous section. Suppose that we wish to test a simple null hypothesis, H_0, against a simple alternative, H_1, about a parameter ϕ:

$$H_0: \phi = \phi_0 \quad \text{and} \quad H_1: \phi = \phi_1 .$$

We observe a sequence of independent, identically distributed random variables: $x_1, x_2, \ldots, x_n, \ldots$. Suppose that the density function for X is $f_0(x)$ under the null hypothesis and $f_1(x)$ under the alternative.

After each observation is made, we calculate the likelihood ratio:

$$L = \frac{\Pi\,[f_1(x_i)]}{\Pi\,[f_0(x_i)]} .$$

We calculate two numbers, k_0 and k_1, to correspond to the preassigned values of alpha and beta. If

$$k_0 < L < k_1 , \tag{12.4.1}$$

we take another observation; if $L < k_0$, we accept H_0; if $L > k_1$, we accept

H_1. The calculations of the correct values for k_0 and k_1 are very difficult, so that in practice we use the approximations

$$k_0 = \frac{\beta}{1-\alpha} \quad \text{and} \quad k_1 = \frac{1-\beta}{\alpha},$$

which are very good for small values of alpha and beta. It is also useful in practice to replace the expressions in equation (12.4.1) by their natural logarithms and to write

$$\ln(k_0) < \ln(L) < \ln(k_1). \tag{12.4.2}$$

We now consider two examples. In the first, corresponding to sampling for variables, X will be normally distributed; in the second, X will have a Bernoulli (binomial) distribution.

12.5. THE NORMAL CASE

Suppose that X is normally distributed with known variance and unknown mean. We have

$$H_0: \mu = \mu_0 \quad \text{and} \quad H_1: \mu = \mu_1.$$

For convenience, we assume that $\delta = \mu_1 - \mu_0 > 0$.

We write Y_n for the sum of the first n deviations from μ_0: $Y_n = \Sigma(x_i - \mu_0)$, $1 \le i \le n$. After some algebra, the inequality of equation (12.4.2) is reduced to

$$\frac{\sigma^2 \ln(k_0)}{\delta} < Y_n - \frac{n\delta}{2} < \frac{\sigma^2 \ln(k_1)}{\delta}. \tag{12.5.1}$$

If we plot the data with n along the horizontal axis and Y vertically, the bounds on Y given in equation (12.5.1) become a pair of parallel lines:

$$Y = \frac{\sigma^2 \ln(k_0)}{\delta} + \frac{n\delta}{2} \tag{12.5.2}$$

and

$$Y = \frac{\sigma^2 \ln(k_1)}{\delta} + \frac{n\delta}{2}. \tag{12.5.3}$$

As long as the path of Y remains between the two lines, we continue to take more observations. The testing stops when the path crosses one of the two boundary lines.

AN EXAMPLE

If instead of considering deviations from μ_0, we write

$$z_i = x_i - \frac{\mu_1 + \mu_0}{2} = x_i - \mu_0 - \frac{\delta}{2} \quad \text{and} \quad Z_n = \sum z_i,$$

the two boundaries become the two horizontal lines

$$Z = \frac{\sigma^2 \ln(k_0)}{\delta} \quad \text{and} \quad Z = \frac{\sigma^2 \ln(k_1)}{\delta}, \quad (12.5.4)$$

and the plot looks like a Shewhart control chart. This is particularly true if $\alpha = \beta$. Then the boundary limits are at the same distance above and below the center line $Z = 0$.

12.6. AN EXAMPLE

Suppose that $V(x) = 125$, $\mu_0 = 40$, $\mu_1 = 50$, $\delta = 10$, $\alpha = 5\%$, and $\beta = 10\%$. Then

$$k_0 = \frac{0.10}{0.95} = 0.105, \; k_1 = \frac{0.90}{0.05} = 18.0,$$

$$\ln(k_0) = -2.25, \ln(k_1) = 2.89.$$

The boundary lines of the critical region are

$$Y = 5n - 28 \quad \text{and} \quad Y = 5n + 36.$$

Consider the following sequence of observations:

n	1	2	3	4	5	6	7	8	9
X	47	50	53	46	50	47	54	48	49
Y	7	17	30	36	46	53	67	75	84

$5n - 28$	-23	-18	-13	-8	-3	2	7	12	17
$5n + 36$	41	46	51	56	61	66	71	76	81

The path of Y goes across the upper boundary at the ninth observation, and we, therefore, accept the alternate hypothesis that $\mu = \mu_1 = 50$. Alternatively, we can write

X	47	50	53	46	50	47	54	48	49
z	2	5	8	1	5	2	9	3	4
Z	2	7	15	16	21	23	32	35	39

The upper and lower limits are $Z = 36$ and $Z = -28$. Again, the graph enters the critical region at the ninth observation.

12.7. SEQUENTIAL SAMPLING BY ATTRIBUTES

We now turn to the case of sampling by attributes and let $x_i = 1$ if the ith item is defective, and $x_i = 0$ if the item is nondefective; $Y_n = \Sigma x_i$, $1 \le i \le n$, has a binomial distribution. The two hypotheses are

$$H_0: \quad p = p_0 \quad \text{and} \quad H_1: \quad p = p_1.$$

We continue to take more observations as long as

$$k_0 < \frac{p_1^Y q_1^{n-Y}}{p_0^Y q_0^{n-Y}} < k_1, \tag{12.7.1}$$

which reduces to

$$\frac{\ln(k_0)}{h} < Y + ng < \frac{\ln(k_1)}{h} \tag{12.7.2}$$

where we have written

$$h = \ln\left(\frac{p_1}{p_0}\right) - \ln\left(\frac{q_1}{q_0}\right)$$

and

$$g = \frac{\ln(q_1/q_0)}{h}. \tag{12.7.3}$$

Again, the boundaries are two parallel lines.
If we write $z_i = x_i + g$, the boundaries for Z are horizontal lines.

12.8. AN EXAMPLE OF SEQUENTIAL SAMPLING BY ATTRIBUTES

Suppose that we wish to test

$$H_0: \quad p = 0.02 \quad \text{vs.} \quad H_1: \quad p = 0.06$$

with $\alpha = \beta = 0.10$.

$$\ln(k_0) = -2.197, \quad \ln(k_1) = +2.197$$

$$h = \ln\left(\frac{0.06}{0.02}\right) - \ln\left(\frac{0.94}{0.98}\right) = 1.14,$$

$$g = \frac{\ln(94/98)}{h} = -0.0365,$$

$$\frac{\ln(k_i)}{h} = \pm 1.927.$$

TWO-SIDED TESTS

We continue to take more observations as long as

$$-1.927 + 0.0365n \leq Y \leq 1.927 + 0.0365n \ .$$

The path of Y is a horizontal line when there are no defectives; each defective causes a jump of one unit. The limits, $0.0365n \pm 1.927$, have to be considered in terms of integers.

The lower limit is negative if $n < 52.79$. It is unity for $n = 80.19$. This means that if we have an initial string of nondefectives corresponding to the line $Y = 0$, the path will not cut the lower-limit line until $n = 53$. Thus,

(i) we cannot decide to accept H_0 if $n < 53$;
(ii) if there are no defectives in the first 53 items tested, we accept H_0;
(iii) if there is exactly one defective in the first 53 items, we continue sampling. We accept H_0 if there are no more defectives until $n = 81$, and so on.

For the upper limit, we note that for $n = 2$, the limit is 2.000. If the first two items are defective, we will have $Y = 2$ when $n = 2$, and we reject H_0.

Similarly, $1.93 + 0.0365n = 3$ when $n = 29.39$, and the path of Y will cross the upper line if there are three defectives before $n = 29.39$.

We therefore accept H_1 if

(i) the first two items are defective,
(ii) there are three defectives in the first 29 items,
(iii) there are four defectives in the first 56 items,
(iv) there are five defectives in the first 84 items, and so on.

12.9. TWO-SIDED TESTS

Tests of a simple null hypothesis against a two-sided alternative

$$H_0: \quad \mu = \mu_0 \quad \text{vs.} \quad H_1: \quad \mu \neq \mu_0$$

are broken into two single-sided tests; H_1 becomes

$$\mu > \mu_0 + \delta \quad \text{or} \quad \mu < \mu_0 - \delta$$

with the risk of type I error set at $\alpha/2$ on each side.

The bounds k_0 and k_1 are set at $\beta/(1 - \alpha/2)$ and $(1 - \beta)/(\alpha/2)$, respectively. We again plot $Y = \Sigma (x_i - \mu_0)$. The inequalities for the likelihood ratio define two pairs of parallel lines. Corresponding lines in the pairs intersect on the center line, $Y = 0$. The lines mark off a V-shaped region on the graph paper.

The lines above the center are

$$Y = \left(\frac{\sigma^2}{\delta}\right)\ln(k_0) + \left(\frac{\delta}{2}\right)n \quad \text{and} \quad Y = \left(\frac{\sigma^2}{\delta}\right)\ln(k_1) + \left(\frac{\delta}{2}\right)n.$$

The lines on the bottom half of the paper are their reflections in the center line:

$$Y = -\left(\frac{\sigma^2}{\delta}\right)\ln(k_0) - \left(\frac{\delta}{2}\right)n \quad \text{and} \quad Y = -\left(\frac{\sigma^2}{\delta}\right)\ln(k_1) - \left(\frac{\delta}{2}\right)n.$$

12.10. AN EXAMPLE OF A TWO-SIDED TEST

Suppose that we had elected to carry out a two-sided test in the example of section 12.6, but with $\alpha = 10\%$, corresponding to 5% in each tail. We should have $V(x) = 125$, $\mu_0 = 40$, $\delta = 10$, $\alpha = 10\%$, and $\beta = 10\%$.

Then $\ln(k_0) = -2.25$ and $\ln(k_1) = 2.89$ as before. The pair of parallel lines above the center line are

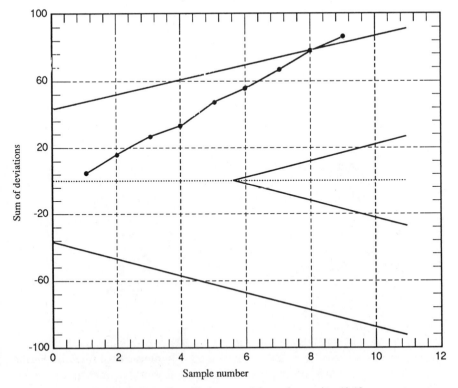

Figure 12.10.1. Two-sided sequential test for section 12.10.

$$Y_n = 5n + 36.1 \quad \text{and} \quad y = 5n - 28.1.$$

The two parallel lines below the center line are

$$Y_n = -5n - 36.1 \quad \text{and} \quad y = -5n + 28.1.$$

The plot of this two-sided test is shown in figure 12.10.1. In it, we see that the cumulative sum of the deviations, Y_n, crosses the line between $n = 8$ and $n = 9$, indicating that the null hypothesis $\mu = 40$ should be rejected in favor of $\mu > 40$.

For the one-sided test of section 12.6, the graph would be the upper half of figure 12.10.1.

12.11. MILITARY AND CIVILIAN SAMPLING SYSTEMS

The standard for attribute sampling used in defense contracts in the United States is the official system called MIL-STD-105D. This sampling system contains a collection of sampling plans. The 105 system first appeared in version A in 1950. It went through two revisions, B and C, before major changes were made in the current standard 105D, which appeared in 1963.

In 1971, the American National Standards Institute and the American Society for Quality Control produced a civilian standard, ANSI/ASQC Z1.4. The 1981 revision of this standard contained the change in terminology for defective to nonconforming and similar rewordings.

The U.S. Department of Defense issued a system for sampling by variables, MIL-STD-414, in 1957. It closely paralleled MIL-STD-105A. It has not been changed to keep pace with the changes in MIL-STD-105 that came with the current version, D. A modern civilian version of MIL-STD-414 was brought out in 1980. This system, called ANSI/ASQC Z1.9, parallels MIL-STD-105D. A complete account of these systems can be found in the latest edition of the book by Duncan (1986).

MIL-STD-1235B is a collection of continuous sampling plans for attributes. It includes the CSP-1 plans of Dodge (1943) and the CSP-2 and CSP-3 plans of Dodge and Torrey (1951).

12.12. MIL-STD-105D

The MIL-STD-105D system contains numerous tables of plans for sampling by attributes. The tables consist of single sampling plans, double sampling plans, and some multiple sampling plans involving more than two stages. Although sequential sampling is mentioned in the accompanying handbook, no tables are included.

The tables are based upon the AQL, which is defined therein as "the maximum percent defective that, for the purpose of sampling inspection, can be considered satisfactory as a process average." The handbook adds that "AQL plans are designed to protect the supplier from having good lots rejected. The consumers' risks of accepting product of inferior quality are only indirectly considered by studying the OC curve for the AQL sampling plans."

The term lot tolerance percent defective (LTPD) is replaced by limiting quality (LQ). A new term, indifference quality, is introduced. This is the quality for which P_a (the probability of acceptance) is 50%. Some people design plans based on the indifference quality, but they are not in common use.

The procedure calls for three levels of inspection: normal, reduced, and tightened. The producer of good quality product is rewarded by reduced inspection costs. The producer of poor quality has higher costs. The implication is that the producer is paying for the inspection. The buyer contracts for a quantity of items that meets a set of standards; reasonable costs for inspection are included in the price agreed upon beforehand. Reduced inspection means increased profits. But if the producer is thus nudged to improve quality, then the buyer profits too by receiving the better quality product.

If the inspection is to be carried out by the producer in his plant, good records must be kept to which the buyer must have access. The buyer's inspectors must also be allowed access for occasional supervision of the actual inspection carried out by the producer. This can produce hard feelings.

12.13. THE STANDARD PROCEDURE

Once the buyer and the producer have agreed on the AQL to be used, the MIL-STD-105D procedure follows several steps. We discuss the situation in which the producer is making batches of the same size, day after day or week after week.

The buyer and seller agree upon the AQL and the general inspection level (which is a function of the importance of the item being inspected). The decision whether to use single or double sampling may be left to the producer. The producer then finds the sample size code from a table and looks in another table to find the critical value, or values, for the given AQL.

Consider, for example, a producer who is producing lots of 1500 items under general inspection level II. That lot size corresponds to sample code K. If single sampling inspection is to be done, the producer begins by looking in the master table for normal inspection. Code letter K calls for a sample of 125 items.

TIGHTENED VERSUS NORMAL INSPECTION   199

Table 12.13.1. **Multiple Sampling for Sample Size Code K (AQL = 2.5%)**

Sample Number	Sample Size	Cumulative Sample Size	Acceptance No. (Ac)	Rejection No. (Re)
1	32	32	0	4
2	32	64	1	6
3	32	96	3	8
4	32	128	5	10
5	32	160	7	11
6	32	192	10	12
7	32	224	13	14

If the AQL is 1%, c (or Ac), the critical number, is three. If there are three or fewer defectives, the lot is accepted; if there are four or more defectives, the lot is rejected. If the AQL is 2.5%, $c = 7$.

If the producer chooses double sampling, $n_1 = n_2 = 80$. Suppose that the AQL is 2.5%. After the first sample is drawn, the lot will be accepted if there are three or fewer defectives and rejected if there are seven or more. After the second sample, the lot will be accepted if there are eight or fewer defectives in the 160 items and reject if there are nine or more. This is a change from the older procedure in which $n_2 = 2n_1$.

The multiple-sampling choice allows up to seven samples, each of 32 items. The acceptance and rejection numbers after each sample are given in table 12.13.1.

This is as close as MIL-STD-105D comes to sequential sampling.

12.14. SEVERITY OF INSPECTION

There are three degrees of severity of inspection—normal, reduced, and tightened. Inspection usually begins at the normal level, which is used when there is no reason to believe that the quality level is either better or worse than has been specified by the AQL. If it becomes apparent that the quality level has deteriorated, tightened inspection is required. If it is clear that the quality level is very good, the supplier may be allowed to go to reduced inspection. Criteria are given for switching from normal to tightened inspection and back, and from normal to reduced inspection and back.

12.15. TIGHTENED VERSUS NORMAL INSPECTION

Tightened inspection usually calls for the same sample sizes, but the requirements for acceptance are more rigorous. Continuing with the example of an AQL of 2.5 and sample size K, the criteria for tightened inspection

Table 12.15.1. Tightened Inspection, Sample Code K, AQL = 2.5%

Type	Cumulative Sample Size	Acceptance No.	Rejection No.
Single	125	5	6
Double	80	2	5
	160	6	7
Multiple	32	*	4
	64	1	5
	96	2	6
	128	3	7
	160	5	8
	192	7	9
	224	9	10

are given in table 12.15.1. In table 12.15.1 the acceptance number for the first sample under multiple sampling is replaced by an asterisk. This denotes that one cannot accept a lot on the basis of the first sample; one can only reject it or take a second sample. This also occurs in table 12.16.1.

12.16. REDUCED INSPECTION

In reduced inspection, the sample size is reduced to approximately 60% of what is required for normal inspection, with a corresponding change in the cost of inspection. We continue with our example and give the criteria for reduced sampling in table 12.16.1. The use of this table and the apparent

Table 12.16.1. Reduced Sampling for Sample Size Code K, AQL = 2.5%

Type	Cumulative Sample Size	Acceptance No.	Rejection No.
Single	50	3	6
Double	32	1	5
	64	4	7
Multiple	13	*	4
	26	0	5
	39	1	6
	52	2	7
	65	3	8
	78	4	9
	91	6	10

ACCEPTANCE SAMPLING BY VARIABLES

discrepancies between the acceptance and rejection numbers are explained in the next section.

12.17. SWITCHING RULES

We begin with normal inspection. If two lots in any sequence of five are rejected, the producer must switch to tightened inspection. We are speaking here of rejection on the original inspection. A lot that is rejected on the original inspection and then accepted after rectification is regarded as a rejected lot.

Tightened inspection then continues until either of the two following events occurs. If five consecutive lots are accepted under tightened inspection, the producer may return to normal inspection. If after ten lots have been inspected, the process has not won its way back to normal inspection, the inspection is stopped, and a search is made for an assignable cause. Inspection and production will resume when it is thought that proper quality has been restored.

The producer is allowed to go from normal to reduced inspection if

(i) the preceding ten lots have been accepted; and
(ii) if the total number of defectives in those ten lots is less than or equal to the acceptance number for the gross sample that is given in a second table.

Notice that in reduced sampling, the acceptance and rejection numbers are not consecutive integers. The producer may continue on reduced sampling as long as the number of defectives in each lot is less than or equal to the acceptance number in the table. If, in any lot, the number of defectives exceeds the acceptance number, the producer must return to normal sampling. If the number of defectives is greater than or equal to the rejection number, the lot is rejected. If the number of defectives falls between the acceptance and rejection numbers, the lot is accepted, but inspection goes back to normal anyway.

12.18. ACCEPTANCE SAMPLING BY VARIABLES

The MIL-STD-414 document has four parts. Part A is the general description. Part B has plans based on the use of the range to estimate the (unknown) standard deviation. Part C has plans for use when σ is estimated from the sample standard deviations. Part D handles the situation when σ is known. There are, therefore, three series of plans covering the situations in B, C, and D. Plans are given for one-sided and two-sided specification limits. The more modern civilian system, ANSI/ASQC Z1.9, has switching rules between normal inspection and reduced or tightened inspection that are similar to those of MIL-STD-105D. For further details, the reader is referred to Duncan (1986).

EXERCISES

12.1. Design a sequential sampling plan for sampling by variables with the following specifications. The target average for the characteristic being measured is 100. We wish to have $\alpha = 0.05$ when $\mu = 100$, and $\beta = 0.15$ when $\mu = 103$. The standard deviation of the characteristic is $\sigma = 6.0$.

12.2. Apply your sequential sampling plan of Exercise 12.1 to the following data sets. When would you make a decision and what would it be?

(i) 102.9 103.2 101.7 109.0 104.3 92.8   105.7
     98.4   99.3   93.6   101.1 107.5 87.3   106.8 107.1
     103.5 108.1 103.3 103.6 108.1 97.6   104.4 111.5
     101.0 95.7   99.4   100.3 103.3 102.3 102.2 108.2
     111.9 87.2   103.8 99.3   103.9 98.9   113.8 103.9
     109.7 102.8 109.1.
(ii) 108.2 101.4 95.4   101.4 97.0   102.3 105.6
     102.0 109.4 102.7 96.9   85.8   98.0   105.2 105.6
     109.0 107.7 98.4   98.3   96.8   108.3 96.9   104.4
     95.8   100.9 96.0   98.6   92.8   99.8   106.2 99.8
     101.5 92.7   109.2 93.0.

12.3. Design a sequential sampling plan for attributes that has $\alpha = 0.05$ when $p = 0.03$, and $\beta = 0.15$ when $p = 0.06$.

12.4. Apply your plan from exercise 12.3 to the following three data sets. When would you make a decision and what would it be?
  (i) The first three items are defective.
  (ii) Items 1, 12, 17, 20, 37, 42, 63, and 67 are defective.
  (iii) Items 75, 91, 135, 174, 175, 190, and 202 are defective.

12.5. A double sampling plan has $n_1 = n_2 = 100$. The engineer accepts the lot on the first sample if $d \le 3$, and rejects it if $d > 7$. The lot is accepted after the second sample if the total number of defects is eight or less. What is the probability of accepting a lot with $p = 0.03$?

CHAPTER THIRTEEN

Further Topics on Control Charts

13.1. INTRODUCTION

In the first example of an x-bar chart in chapter nine, a shift in the mean of the process occurred after the thirtieth run. It became clear that the traditional rule—take action if there is a point outside the outer control lines—did not do a very good job of detecting that shift, since we needed 15 more runs to trigger it. That problem was even more conspicuous in the example of the c-chart in chapter ten. How can we speed up the detection of changes in mean? It is done by looking not just at the individual points on the chart, but at sequences of several points.

The Shewhart control charts are conservative in the sense that as long as the process is rolling along with the process mean equal to the target mean, we will rarely (one time in 400) stop the process when we should not. The alpha risk is kept small. How about the beta risk, which is the risk that we will not stop the process when we should?

In addition to the standard Shewhart charts, we can use several other charts whose purpose is to enable us to detect more quickly changes in the process mean. We discuss three of these charts: the arithmetic moving-average chart, the cumulative sum control chart (Cusum), and the geometric moving-average or exponentially weighted moving-average (EWMA) chart. We are not advocating replacing the Shewhart chart with these new charts—far from it. We are suggesting that the engineer should keep several charts simultaneously.

We begin the chapter with another look at the stop rules that were considered in chapter nine, focusing, in particular, on combining rules 2 and 3 with rule 1. For convenience, we now repeat the three rules that were proposed.

Rule 1. One point outside the three sigma limits. ☐

Rule 2. Two consecutive points either above the upper warning line (two-sigma line) or below the lower warning line. ☐

203

Rule 3. A run of seven points either all above the center line or all below it. □

13.2. DECISION RULES BASED ON RUNS

Rules 2 and 3 are both based on runs of points above or below a line other than the outer control lines. Rule 3 calls for action if there is a run of seven points; rule 2 calls for a run of two points.

Suppose that there has been a shift of $+\sigma/\sqrt{n}$ in the mean, so that the mean changes from μ to $\mu + \sigma/\sqrt{n}$. The probability that a point falls above the center line on the chart is now increased to 0.8413. The probability that the next seven points will constitute a run that triggers rule 3 is

$$P = (0.8413)^7 = 0.30.$$

More generally, if the probability of a point above the line is p, the probability that the next seven in a row will be above the line is p^7.

But we need to ask a different question. Suppose that it takes t points until a run of seven occurs; t is a random variable, whose expectation is the ARL when this rule is used. (Beware of the double use of the word run in this sentence; the word *run* in ARL stands for t.) What can be said about the probability distribution of t? The derivation is beyond the scope of this book, but we can derive a formula for its expected value. Using the results that are derived in the next section, we can show that when there is a shift of $\sigma/\sqrt{n}$ in the mean, the ARL for the rule of seven is given by the formula

$$E(t) = \frac{1 - (0.8413)^7}{(0.1587)(0.8413)^7} = 14.8,$$

which we round to 15.

13.3. COMBINED RULES

Rules 1 and 2 are rarely applied separately. The normal procedure is to combine them into a single rule, rule 2a:

Rule 2a. Take action if either there is one point outside the outer control lines or there are two successive points outside the warning lines and on the same side. □

The table given in chapter nine considered only one rule at a time. In developing the ARL for the combined rule 2a, we follow the argument of Page (1954).

COMBINED RULES

The set of possible values of $\bar{x}$ can be divided into three regions, which we can call good, warning, and action. For rule 2a, the good region consists of values of $\bar{x}$ between the two warning lines, the warning region consists of values between the warning lines and the outer (action) lines, and the action region consists of values outside the action lines. Let the probability that a point falls in each of these three regions be p (good), q (warning), and r (action), where $p + q + r = 1$. The process will stop if there are two successive points in the warning region or one in the action region. Let L denote the ARL when we start the sequence of points from a point in the good region; let L' be the ARL when we start from a point in the warning region.

If we start with a point in the good region, there are three possibilities:

(i) the next point is good, with probability p, and we have the expectation of L more points:
(ii) the next point is in the warning region, and we expect L' more points;
(iii) the next point is in the action region, and we stop.

We combine these possibilities into the equation

$$L = p(1 + L) + q(1 + L') + r. \tag{13.3.1}$$

If we start from a point in the warning region, there are again three possibilities:

(i) the next point will be in the good region, in which case we start again, and $L' = 1 + L$;
(ii) the next point is also in the warning region, and we stop with $L' = 1$;
(iii) the next point is in the action region, and we stop with $L' = 1$. We combine these possibilities into a second equation:

$$L' = p(1 + L) + q + r. \tag{13.3.2}$$

Eliminating L' between the two equations gives

$$L = \frac{1 + q}{1 - p - pq}. \tag{13.3.3}$$

A more general form of the rule would require either a run of m points outside the warning lines or one point outside the action lines. The corresponding formula for the ARL is

$$L = \frac{1 - q^m}{1 - (p + q) + pq^m}$$
$$= \frac{1 - q^m}{r + pq^m}. \tag{13.3.4}$$

We can also combine rules 1 and 3 into rule 3a:

Rule 3a. Take action if there is either a run of seven successvie points on the same side of the center line or a point outside the action lines. □

To calculate the average run length with rule 3a for a positive shift in the mean, we let the "good" region consist of the negative values of $\bar{x}$ and let the warning region run from the center line to the upper three-sigma line. Then we apply equation (13.3.4) with $m = 7$. If we let $r = 0$ and $m = 7$, we derive the formula that was used for rule 3 in section 13.2.

13.4. AN EXAMPLE

Let the shift be $d\sigma/\sqrt{n} = 0.5\sigma/\sqrt{n}$. We calculate the ARL for the two rules in the following way.

For rule 2a,

$$p = F(1.5) = 0.9332, \quad q = F(2.5) - F(1.5) = 0.0606,$$
$$r = 1 - F(2.5) = 0.0062.$$

$$L = \frac{1-q^2}{r+pq^2} = 103.493.$$

For rule 3a,

$$r = 1 - F(3-d); \quad p = 1 - F(d); \quad q = 1 - p - r.$$

In this example, $r = 0.0062$, $q = 0.6853$, and $p = 0.3085$, whence

$$L = \frac{1-q^7}{r+pq^7} = 33.06.$$

13.5. A TABLE OF VALUES OF ARL

Table 13.5.1 gives values of the ARL for various values of the shift in the mean. Rule 3a is triggered by a run of seven or a point outside the outer limits. The values of ARL for rule 3a are shorter than those for rule 2a, but rule 3a has larger values of the alpha risk.

13.6. MOVING AVERAGES

The charts that are presented in this chapter for detecting shifts in the mean come under the general heading of moving averages. We discuss three such charts. The basic Shewhart chart with rule 1 uses only the single latest

ARITHMETIC-AVERAGE CHARTS

Table 13.5.1. Average Run Lengths to Detect a Shift $d(\sigma/\sqrt{n})$ in the Mean. One-Sided Tests

d	Rule 2a	Rule 3a
0.0	571	194
0.2	273	85
0.4	142	44
0.6	77	25
0.8	44	17
1.0	26	13
1.2	17	10
1.4	11	8
1.6	8	7
1.8	6	6
2.0	5	5

observation to make the action decision and pays no heed to any of the previous data. Rule 2 employs only the latest point and its immediate predecessor. The moving-average charts use the earlier data to different extents. The arithmetic moving average uses the average of the k latest observations with equal weights, $1/k$, and neglects observations made before then. The cumulative sum control chart uses all the observations that have been made with the same weight. The geometric moving average (or exponentially weighted moving average, EWMA) chart uses all the previous observations, but with weights that decrease as the points grow older.

Arithmetic and EWMA charts have applications beyond traditional quality assurance, notably in econometrics and in engineering process control. For this reason, through the rest of the chapter, we change our notation to make it more general in its application. We plot the sample number as discrete time, t, along the horizontal axis. The ith sample is the sample drawn at t_i; the observed response is denoted by y_i; the variance of y is denoted by σ^2. In the special case of Shewhart charts, $y_i = \bar{x}_i$, and $V(y_i) = \sigma_{\bar{x}}$.

13.7. ARITHMETIC-AVERAGE CHARTS

The first of our charts for detecting shifts in the mean is the (arithmetic) moving-average chart. Sample number, or time, is measured along the horizontal axis. Against this, we plot not the value of y_i for the ith sample, but z_i, the mean of that and the previous $k - 1$ values of y_i. The value of k used in the chart is the choice of the engineer; $k = 4$ and $k = 5$ are commonly used. If $k = 4$, the ordinate for $t = 5$ is

$$z_5 = \frac{y_2 + y_3 + y_4 + y_5}{4}.$$

Figure 13.7.1 shows a moving-average chart with $k = 4$ for the data of figure 9.8.1. There are 47 averages on the chart, beginning with

$$\frac{\bar{x}_1 + \bar{x}_2 + \bar{x}_3 + \bar{x}_4}{4} = \frac{10.28 + 10.70 + 9.50 + 10.13}{4}$$

$$= 10.1525 .$$

Because we are now plotting averages of four samples, the outer control lines are pulled in to

$$\mu \pm \frac{3\sigma_{\bar{x}}}{\sqrt{k}} = 10.13 \pm 3\left(\frac{0.442}{2}\right) .$$

The shift in the mean that took place after the thirtieth sample is now more apparent than it was with the traditional x-bar chart. The twenty-seventh average is $(\bar{x}_{27} + \bar{x}_{28} + \bar{x}_{29} + \bar{x}_{30})/4$; this is the first average that contains a point with the "new" mean.

For low values of k, these charts detect large changes in the mean quickly, but they are slower to detect smaller changes. On the other hand, the smaller changes are easier to detect with higher values of k, but at a

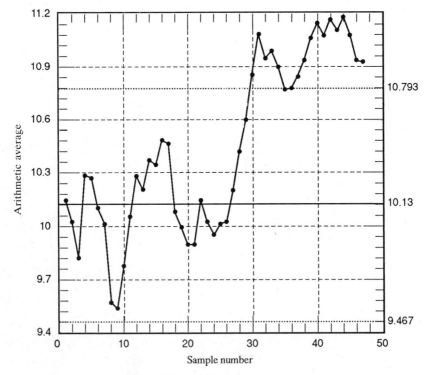

Figure 13.7.1. Arithmetic moving average.

price. The damping effect of using high k values slows the detection of the larger shifts. It has also been shown by Nelson (1983b) that in their effort to smooth out actual time trends in the data, these charts may accidentally introduce some specious trends.

13.8. CUMULATIVE SUM CONTROL CHARTS

Cumulative sum control charts were introduced by E. S. Page (1961); they are commonly called by the shorter name, Cusum charts. In Cusum charts, we plot along the vertical axis the sum of all the deviations of the sample average from some predetermined value since an arbitrary starting time. Thus,

$$z_t = \sum (\bar{x}_i - \mu_0), \quad i = 1, \ldots, t, \quad (13.8.1)$$

where μ_0 is sometimes called the target value; it is usually the value of the process mean under steady state, corresponding to the center line on the Shewhart chart.

The procedure for using these charts amounts essentially to performing a sequential probability ratio test backwards. The engineer uses a V-mask, which is a piece of cardboard or plastic with a large V-shaped notch cut in it. The mask is placed on the chart with the V opening backwards (>). The apex of the V is at the same level (z value) as the latest point on the chart with a lead distance, d. The latest point to be plotted on the chart is (t_i, z_i); the apex of the V-mask is placed at ($d + t_i, z_i$). The horizontal line joining those two points bisects the angle of the V. The process is declared to be out of control if the cumulative sum disappears beneath the mask (crosses the edge of the V), which will happen if $\bar{x}$ has been rising or falling sharply. The similarity to the two-sided sequential probability ratio test with its V-shaped critical region will be apparent.

Two parameters are involved in the test procedure: the lead time, d, and the angle, 2θ, between the edges of the V. The choice of the angle depends on how large a deviation from the target value the engineer is prepared to tolerate. The lead time is a function of the values of the alpha and beta risks that are acceptable. The values of these parameters can be calculated as one would with a sequential test procedure. Often, they are established by cut-and-try methods, using previous data on the performance of the plant. Lucas has written several papers on the use of Cusum charts including parabolic masks (1973, 1976, 1982).

The use of the V-mask, which is symmetric about the horizontal line, implies that the engineer is carrying out a two-sided test. Hence, the two critical values, i.e., the two values that are used to decide that there has been a shift in the mean, are symmetric about the target value, μ_0, and are at, say, $\mu_0 \pm \delta$. The engineer should then choose the angle θ so that the edges of the V have slopes corresponding to $\pm \delta/2$. If the scales on the two coordinate axes are the same, $\tan \theta = \delta/2$ will be taken; otherwise, $\tan \theta$ will be modified accordingly.

In calculating the lead distance, we recall from section 12.9 that the outer lines of the acceptance region in the two-sided sequential test are

$$Z = \left(\frac{\sigma_{\bar{x}}^2}{\delta}\right) \ln(k_1) + \left(\frac{\delta}{2}\right) n ; \qquad Z = -(\sigma_{\bar{x}} \delta) \ln(k_1) - \left(\frac{\delta}{2}\right) n ,$$

where $k_1 = (1 - \beta)/(\alpha/2)$. The lines intersect at

$$n = -\left(\frac{2}{\delta}\right)\left(\frac{\sigma_{\bar{x}}^2}{\delta}\right) \ln(k_1) .$$

The V in the Cusum chart is facing backwards, and so we take

$$d = \left(\frac{2}{\delta}\right)\left(\frac{\sigma_{\bar{x}}^2}{\delta}\right) \ln(k_1) .$$

For a cumulative sum chart that is equivalent to a one-sided test, we can follow the argument of section 12.5.

Example 13.8.1. Suppose an engineer makes a cumulative sum chart of the data in table 9.8.1 and wishes to check for a shift in the mean after 15 points. The chart, which is shown as figure 13.8.1, is made with the following parameters: $\mu_0 = 10.13$ and $\sigma_{\bar{x}} = 0.442$, which were the process values calculated earlier in section 9.6, and $\delta = 0.5$, $\alpha = 0.05$, and $\beta = 0.10$. The data points and the Cusums are as follows:

Sample number	1	2	3	4	5	6
Sample average	9.25	10.13	10.58	9.60	10.30	9.65
Cusum $\Sigma (\bar{x} - 10.13)$	−0.88	−0.88	−0.43	−0.96	−0.79	−1.27

7	8	9	10	11	12	13	14	15
10.25	9.88	10.35	10.30	11.08	10.70	11.28	11.23	10.60
−1.15	−1.40	−1.18	−1.01	−0.06	0.51	1.66	2.76	3.23

The lead distance, d, is calculated as

$$d = \frac{2(0.442)^2}{(0.50)^2} \ln(36) = 5.60 .$$

The engineer takes a mask with half angle $\theta = \arctan(0.25)$ and places it on the chart with the apex at a distance d to the right of the fifteenth point. The

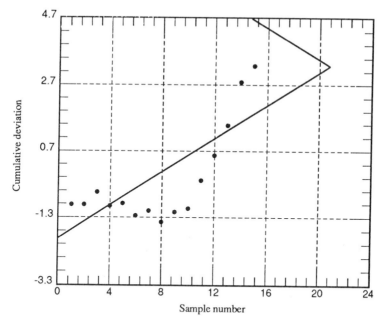

Figure 13.8.1. Cumulative sum chart.

lower arm of the V intersects the plot of the Cusums between samples 12 and 13. The engineer concludes that there has indeed been a shift in the mean.

The fifteenth point has $z = 3.23$, and so the equation of the lower arm of the mask is the line that passes through the point $(15.00 + 5.60, 3.23)$ with slope 0.25, i.e.,

$$y_t = 0.25t - 1.92 .$$

When the V-mask is positioned with its apex a distance d ahead of the fifteenth point, its arm will intersect the Cusum graph if there is a value of z_t below $0.25t - 1.92$. When $t = 12$, we have $0.25t - 1.92 = 1.08$; when $t = 13$, $0.25t - 1.92 = 1.33$. Since $z_{12} = 0.51 < 1.08$ and $z_{13} = 1.66 > 1.33$, the arm of the mask intersects the graph between $t = 12$ and $t = 13$.

When the mask is used at the nth point, the arm of the mask is the line

$$y = y_n + \left(\frac{\delta}{2}\right)(t - n - d) . \tag{13.8.2}$$

For this example, it is

$$y = 0.25t + y_n - 1.40 - \frac{n}{4} . \qquad \square$$

13.9. A HORIZONTAL V-MASK

If in the example of section 13.8, we let

$$z_t = \sum \left(x - \mu_0 - \frac{\delta}{2}\right) = \sum (\bar{x} - 10.38) = y_t - 0.25t,$$

the arm of the V-mask becomes the horizontal line

$$z = z_n - 1.40. \tag{13.9.1}$$

The mask cuts the graph if at some previous time t,

$$z_t < z_n - 1.40.$$

In this example, $z_{15} = -0.52$, and

$$z_{12} = -2.49 < -1.92.$$

The general form of equation (13.9.1) is

$$z_t = z_n - \frac{\delta d}{2}. \tag{13.9.2}$$

Example 13.9.1. The data in table 13.9.1 are daily production rates (conversion percentages) of product in a chemical plant. The plant has been running at an average rate $\mu = 43.0$ and standard deviation $\sigma = 1.1$. The plant manager wishes to be able to detect an increase in production rate of 0.5 with $\alpha = 0.05$ and $\beta = 0.10$.

Table 13.9.1. Data for Example 13.9.1

	1	2	3	4	5	6
Observation	42.7	43.9	42.4	43.7	44.4	42.3
Deviation	−0.55	0.65	−0.85	0.45	1.15	−0.95
Cusum	−0.55	0.10	−0.75	−0.30	0.85	−0.10
	7	8	9	10	11	12
Observation	42.0	44.2	43.4	44.7	44.3	44.7
Deviation	−1.25	0.95	0.15	1.45	1.05	1.45
Cusum	−1.35	−0.40	−0.25	1.20	2.25	3.70
	13	14	15	16	17	18
Observation	42.7	43.7	43.0	45.9	45.4	43.7
Deviation	−0.55	0.45	−0.25	2.65	2.15	0.45
Cusum	3.15	3.60	3.35	6.00	8.15	8.60

GEOMETRIC MOVING AVERAGES

The plant manager calculates deviations $z_t = \Sigma (x - 43.25)$. Equation (13.9.2) becomes

$$z_t = z_n - \left(\frac{1.21}{0.5}\right) \ln(18) = z_n - 6.99.$$

The plant manager will report the shift in the rate after the sixteenth run.

□

13.10. GEOMETRIC MOVING AVERAGES

Geometric, or exponentially weighted, moving-average charts were introduced by S. W. Roberts (1959). In these charts, we plot the ordinate

$$z_t = ry_t + (1 - r)z_{t-1} \tag{13.10.1}$$

against time.

The choice of r lies with the engineer. Values in the range

$$0.10 \leq r \leq 0.30 \tag{13.10.2}$$

are usual.

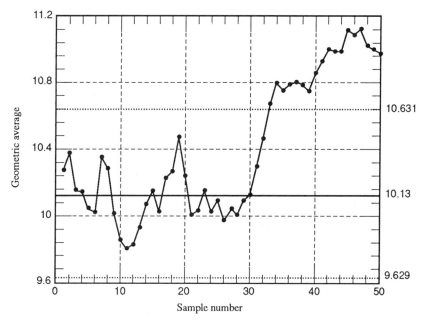

Figure 13.10.1. Geometric average ($r = 0.25$).

The weights given to earlier observations drop as t increases, like the terms in a geometric series; hence the name. These charts are usually preferred to the arithmetic moving-average charts. Figure 13.10.1 shows a geometric chart with $r = 0.25$ for the data of figure 9.8.1. Again, one notices that this chart detects the shift in the mean faster than the Shewhart chart.

Equation (13.10.1) may be transformed to give z_t in terms of the responses:

$$z_t = ry_t + r(1-r)y_{t-1} + r(1-r)^2 y_{t-2} + \cdots \qquad (13.10.3)$$

To compute the control lines for the EWMA chart, we note that the variance of z_t is given by

$$V(z_t) = (\sigma^2)[r^2 + r^2(1-r)^2 + r^2(1-r)^4 + \cdots].$$

The sum of this series for large t is

$$V(z_t) = \sigma^2 \left(\frac{r}{2-r}\right). \qquad (13.10.4)$$

13.11. EWMA AND PREDICTION

J. S. Hunter (1986) approaches the EWMA from the point of view of prediction in a stochastic process, or time series. He lets y_t denote the response that is observed from the tth sample, and he estimates the response at time $t + 1$ by the prediction formula

$$\hat{y}_{t+1} = y_t + r\hat{e}_t \qquad (13.11.1)$$

where $\hat{e}_t$ is the deviation, $y_t - \hat{y}_t$, of the predicted response at time t from the observed value. Substituting for $\hat{e}_t$ in equation (13.11.1), we get

$$\hat{y}_{t+1} = ry_t + (1-r)\hat{y}_t. \qquad (13.11.2)$$

This is the same as equation (13.10.1) with $\hat{y}_{t+1}$ instead of z_t, and $\hat{y}_t$ instead of z_{t-1}.

In this formulation, r measures the depth of memory that the process itself has, and may be estimated from the data by choosing the value of r that minimizes $\Sigma \hat{e}_i^2$. Table 13.11.1 shows these calculations for a set of 20 points for $r = 0.1, 0.2,$ and 0.3. The lowest value of $\Sigma \hat{e}_i^2$ occurs at $r = 0.2$. The sum of squares increases as r becomes larger. The value 0.2 is good enough for most purposes. The engineer who wishes for a value with more decimal places can use quadratic interpolation with the values 0.1, 0.2, and 0.3. The first observation is 23.6; the predicted value of the first observation

Table 13.11.1. EWMA Forecast Computations for Estimating r

Data	$r = 0.1$	$y_i - \hat{y}_i$	$r = 0.2$	$y_i - \hat{y}_i$	$r = 0.3$	$y_i - \hat{y}_i$
23.6	23.80	−0.20	23.80	−0.20	23.80	−0.20
23.5	23.78	−0.28	23.76	−0.26	23.74	−0.24
24.1	23.75	0.35	23.71	0.39	23.67	0.43
23.4	23.79	−0.39	23.79	−0.39	23.80	−0.40
23.3	23.75	−0.45	23.71	−0.41	23.68	−0.38
24.7	23.70	1.00	23.63	1.07	23.56	1.14
23.1	23.80	−0.70	23.84	−0.74	23.91	−0.81
24.0	23.73	0.27	23.69	0.31	23.66	0.34
23.8	23.76	0.04	23.75	0.05	23.76	0.04
23.9	23.76	0.14	23.76	0.14	23.78	0.12
24.1	23.78	0.32	23.79	0.31	23.81	0.29
24.3	23.81	0.49	23.85	0.45	23.90	0.40
24.5	23.86	0.64	23.94	0.56	24.02	0.48
24.7	23.92	0.78	24.05	0.65	24.16	0.54
24.4	24.00	0.40	24.18	0.22	24.32	0.08
23.9	24.04	−0.14	24.23	−0.33	24.35	−0.45
24.1	24.03	0.07	24.16	−0.06	24.21	−0.11
24.3	24.03	0.27	24.15	0.15	24.18	0.12
24.1	24.06	0.04	24.18	−0.08	24.22	−0.12
24.3	24.06	0.24	24.16	0.14	24.18	0.12
$\Sigma \hat{e}_i^2$		3.84		3.62		3.68

is the target value, 23.8. We are seeing a drift upwards in both the observed and predicted values.

The change of emphasis is important. It turns our attention to the use of the quality control chart as a tool for process control. Hunter (1986) suggests that we plot both y_t and $\hat{y}_t$ on the same piece of paper. The reader is referred to his article for his insights into the use of EWMA for control purposes.

EXERCISES

13.1. For the data set in exercise 9.1 make the following charts:

(a) an arithmetic average using three samples in each average;
(b) an arithmetic moving average using four samples;
(c) a geometric (EWMA) chart with $r = 0.25$.

Note that in the arithmetic mean chart with span 3 and the EWMA chart the last point on the chart is close to the upper control line and the graph is heading upwards. This is not true for the arithmetic averages with span 4. Note also that the one sample that falls above the UCL in the x-bar chart does not cause an outside point on the average charts.

13.2. The following data set consists of 40 sample averages. The process mean was 0.825; the standard deviation of the sample averages was 0.03.

0.868	0.808	0.786	0.797	0.850	0.792	0.882	0.754
0.837	0.831	0.801	0.797	0.855	0.819	0.803	0.841
0.905	0.865	0.828	0.838	0.827	0.834	0.811	0.871
0.806	0.863	0.841	0.892	0.800	0.862	0.850	0.910
0.846	0.827	0.837	0.842	0.909	0.871	0.855	0.878

Make the same three charts as in exercise 13.1.

13.3. Make the same three charts for the data in exercise 9.3.

13.4. Make the same three charts for the data in exercise 9.4.

13.5. Make the same three charts for the data in exercise 9.5, assuming that $\mu = 39.85$ and $\sigma = 0.78$.

13.6. We wish to detect a shift in the mean of as much as 1.0 from the target value of 64.0 in the data of exercise 9.3. Make a Cusum chart to test for a shift with $\alpha = 5\%$ and $\beta = 10\%$, using a V-mask, and also without a mask. Start testing at the 15th sample, and continue with the 16th, etc., until you spot a significant shift. How many samples will you need before you detect a shift?

13.7. Repeat exercise 13.6 for the data in exercise 9.4. In this example assume that you wish to detect a shift of 0.8 from an original mean of $\mu = 17.3$.

13.8. Repeat exercise 13.6 for the data in exercise 9.5, assuming that you wish to detect a shift of 0.8 from a target value of 39.85 with $\sigma = 0.78$.

13.9. Treat the data of exercise 13.2 as if they were single observations and make an I chart.

CHAPTER FOURTEEN

Bivariate Data: Fitting Straight Lines

14.1. INTRODUCTION

In the previous chapters, we have dealt primarily with univariate data. Each observation was a single measurement—the length of a rod, the breaking strength of a strand of fiber. In the next two chapters, we turn to multivariate data: bivariate data in this chapter, and more than two variables in the next. The general topic of fitting lines and planes to data by the method of least squares is called regression. We can only give an introduction to the topic here. The reader who wishes to go further will find numerous books on the topic. Prominent among them are *Applied Regression Analysis* by N. R. Draper and H. Smith (1981) and *Fitting Equations to Data*: *Computer Analysis of Multifactor Data* by C. Daniel and F. Wood (1980).

In this chapter, our data will be a set of n points, each one of which consists of observations on a pair of variables. We may have taken n male undergraduate students and for the ith student, measured his height, x_i, and his weight, y_i, or else, perhaps, his scores in the Scholastic Aptitude Test (SAT) or the Graduate Record Examination (GRE) for English (verbal) and mathematics (quantitative).

In both these examples, the variables are obviously connected, but not by some algebraic function. By and large, tall men are heavier than short men and heavy men are taller than light men. Of course, there are exceptions—the short, fat man and the long, thin beanpole. Similarly, when we look at the whole range of students, the better students tend to have high marks on both parts of the test, but this is not true over a narrow range. When the range of students is truncated and consists only of those who are accepted to study for master's or doctoral degrees, the connection is not so clear. We do have a considerable number of arts students who are weak quantitatively when measured against the average for graduate students, and science students whose verbal skills are low compared to the graduate student average. That does not mean that they necessarily rank below the average for the man in the street, just that they rank below the average for the subgroup that we have chosen. If we take our sample of undergraduates and make a scatter plot of height against weight, we will get an elliptical, or cigar-shaped, cloud of points running from the southwest to the northeast. Often, it looks as if there is a straight line around which the points fall.

Our basic problem in this chapter is to fit a straight line to bivariate data, with the aim of being able to use it to predict future values of Y given X. We would like to be able, if we are told that a student is six feet tall, to predict his weight. A more practical use would be to predict the yield of a chemical process given the percentage of impurity in the feedstock.

In this chapter and the next, we use in the illustrations edited printouts from the Minitab software package. Every standard software package has subroutines for fitting lines and planes. They vary only slightly in detail and in the availability of optional features. The reader should check with the local computer center to find out what is available.

14.2. THE CORRELATION COEFFICIENT

In section 4.15, we introduced the concept of the covariance of two random variables:

$$\text{Cov}(X, Y) = E\{[X - E(X)][Y - E(Y)]\} .$$

When the covariance is estimated from a data set, we substitute $\bar{x}$ for $E(X)$ and $\bar{y}$ for $E(Y)$. The covariance is then estimated by

$$\frac{\sum (x_i - \bar{x})(y_i - \bar{y})}{n - 1} . \qquad (14.2.1)$$

Suppose that we draw coordinate axes through the centroid $(\bar{x}, \bar{y})$ of the data. If there is a strong tendency for Y to increase as X increases, as with height and weight, the data will tend to fall in the southwest and northeast quadrants. In each of these quadrants, the cross product is positive, and so the covariance is positive. On the other hand, if Y is the yield of a product and X is the percentage of impurity, Y will tend to decrease as X increases. The data will tend to fall in the northwest and southeast quadrants where the cross products are negative, and the covariance will be negative.

The value of the covariance depends upon the scale of the variables. If the scale of X is changed from meters to feet, and from kilograms to pounds, the covariance is multiplied by $(3.28)(2.205) = 7.23$.

The correlation, or correlation coefficient, of X and Y is a scale-free version of the covariance. It is defined by

$$\frac{\text{Cov}(X, Y)}{\sigma_X \sigma_Y} . \qquad (14.2.2)$$

CORRELATION AND CAUSALITY

It is estimated from a sample by r, where

$$r^2 = \frac{\left[\sum (x_i - \bar{x})(y_i - \bar{y})\right]^2}{\sum (x_i - \bar{x})^2 \sum (y_i - \bar{y})^2} \quad (14.2.3)$$

and r has the same sign as the covariance.

It can be shown algebraically that r must lie in the interval $-1.0 \leq r \leq +1.0$. If the points lie on a straight line, the correlation is ± 1.0, the sign of r being the same as the sign of the slope of the line. If the line is horizontal, the correlation is zero.

On the other hand, if X and Y are independent, then when $Y > 0$, we are as likely to have $X > 0$ as to have $X < 0$. The covariance of X and Y is zero, and their correlation is zero also.

One must be careful of converse arguments here. It does not follow that if r is close to unity, the relation between X and Y is close to linear. Suppose that X takes the values 1, 2, 3, 4, and 5, and that $Y = X^2$. Then we have

x	1	2	3	4	5
y	1	4	9	16	25

Applying equation (14.2.3),

$$r^2 = 0.9626 \quad \text{and} \quad r = 0.98.$$

Nor does it follow that if $r = 0$, the variables are independent. We have already seen this in example 4.16.1.

The correlation coefficient is invariant under changes of scale and changes of origin. In the case where we switched from meters and kilograms to feet and pounds, the covariance was multiplied by (3.28)(2.205), but the standard deviation of X was multiplied by 3.28 and the standard deviation of Y by 2.205. The multipliers cancel out in the correlation coefficient. More generally, it is a simple algebraic exercise to show that if $W = a + bX$ and $U = c + dY$, then

$$\text{corr}(W, U) = \text{corr}(X, Y).$$

14.3. CORRELATION AND CAUSALITY

One should avoid the trap of confusing correlation with causality. It does not follow that if X and Y are highly correlated, then X must cause Y, or vice versa. The population of Russia and the output of cups and saucers in the United States have both increased, by and large, over the past half century. They are both positively correlated with time, and, hence, with one another, but we cannot rationally argue that the one caused the other. There are numerous examples of these nonsense, or spurious, correlations.

University enrollment in the United States has increased steadily since the end of World War II. The numbers of both men and women have increased. It would not be appropriate here to speculate on whether the increase of women caused an increase in male enrollment, or vice versa, but we could use the enrollment figures for men to estimate the female enrollment. On a more practical note, if X and Y are two chemical measurements that are highly correlated, and if X is much cheaper to obtain than Y, we may, perhaps, feel comfortable about using the cheaper X as a proxy, or surrogate, for Y.

14.4. FITTING A STRAIGHT LINE

Suppose that we have a set of n bivariate data points. We propose to fit a straight line to the data by the method of least squares, with the primary idea of predicting the value of Y given the value of X.

If we take several observations at the same value of X, the observed values of Y will vary. The weights of students who are 70 inches tall vary about a value $E(Y|X=70)$. Our model implies that the expectations $E(Y|X)$ lie on a straight line. This line is called the regression line of Y on X. The noted English biometrician Sir Francis Galton fitted a straight line to the plot of the heights of sons versus the heights of their fathers in his studies of inheritance, in particular of the "law of universal regression." The name *regression* has stuck to the technique and is generally used by statisticians to describe fitting lines and planes to data.

Let the ith data point be (x_i, y_i). The model is

$$y_i = \alpha + \beta x_i + e_i . \qquad (14.4.1)$$

The coefficients α and β are unknown constants to be estimated from the data; the values of x_i are taken as given; e_i represents the random noise in the system. We can write

$$\mu_i = \alpha + \beta x_i , \qquad (14.4.2)$$

and then

$$y_i = \mu_i + e_i . \qquad (14.4.3)$$

We now make our first assumption about the behavior of e_i, namely, that e_i is a random variable with zero expectation. Then

$$\mu_i = E(Y|X=x_i) .$$

AN EXAMPLE   221

It is an unfortunate accident of history that we usually refer to e_i as the error. It is not error in the sense that someone has made a mistake. If our straight-line model is correct, e_i represents random variation; if our model is wrong, as it would be if we were fitting a straight line to points that lie on or about a parabola, e_i will be a combination of both noise and "badness of fit." We are not going to discuss that situation yet. For the moment, we assume that our model is correct.

14.5. AN EXAMPLE

The data of table 14.5.1 appeared in the *Austin American Statesman* newspaper in the fall of 1978. The data are the number of calories, Y, per 12-ounce can, and the percentage of alcohol, X, for the regular and light types of beer brewed by eight different producers. The first eight beers were classified by the brewers as light beers and the last eight as regular.

To fit the line on Minitab, we entered the data in columns 1 and 2 of their work sheet and gave the command REGRESS C1 1 C2, telling the program to regress the Y variable that was in C1 against one predictor variable, X, in C2. An edited version of the printout is given in table 14.5.2. The details of the printout are discussed as this chapter proceeds. A plot of the data is shown in figure 14.5.1.

The printout begins with a statement of the fitted equation. Then the coefficients in the equation are repeated and their t values are given. The t value for the slope of the line, 5.14, is comfortably larger than 2.0, and we reject the hypothesis that the slope is zero. On the other hand, the t value for the constant term is small. This suggests that it might be appropriate to fit a line through the origin, and we do that later. The calculation of those t values is discussed in section 14.12.

The standard deviation of Y refers to the standard deviation of the error terms. It is discussed in section 14.11. The derivation of the square of the correlation coefficient and the table of sums of squares that follows it are in section 14.13.

Table 14.5.1. Beer Data

	Y	X		Y	X		Y	X
1.	96	3.52	2.	68	2.43	3.	110	3.61
4.	96	3.22	5.	70	2.32	6.	70	2.27
7.	134	3.40	8.	97	3.17	9.	124	3.89
10.	140	3.79	11.	134	3.76	12.	130	3.81
13.	140	3.08	14.	140	3.36	15.	160	3.90
16.	129	3.89						

Table 14.5.2. Printout of Minitab Regression for the Beer Data (Edited)

THE REGRESSION EQUATION IS Y = −23.3 + 41.4 X

	COLUMN	COEFFICIENT	ST. DEV. OF COEF.	T-RATIO = COEF/S.D.
	—	−23.30	27.25	−0.86
X	C2	41.384	8.054	5.14

THE ST. DEV. OF Y ABOUT REGRESSION LINE IS S = 17.56, 14 d.f.

R-SQUARED = 0.654   R = 0.8087

ANALYSIS OF VARIANCE

DUE TO	DF	SS	MS = SS/DF
REGRESSION	1	8138.5	8138.5
RESIDUAL	14	4315.2	308.2
TOTAL	15	12453.8	

(BACK SOLUTION FOR SPECIAL POINTS)

ROW	X C2	Y C1	PRED. Y VALUE	ST. DEV. PRED. Y	RESIDUAL	ST.RES.
5X	2.32	70.00	72.71	9.30	−2.71	−0.18
6X	2.27	70.00	70.65	9.66	−0.65	−0.04
13R	3.08	140.00	104.17	4.86	35.83	2.12

The printout ends with a listing, with some details, of three "special" points (see sections 14.14 and 14.15). The point with the R rating, beer 13, has a bad fit to the data. It seems to have too many calories for that amount of alcohol. The fitted line predicts only 104 calories for beer with 3.08% alcohol; number 13 had 140 calories. The X ratings denote beers that have especially large influence on the fit. There are two of them, numbers 5 and 6 (number 2 comes very close to earning an X rating). They are very weak beers; they stand a long way from the main group; they pull the line down.

For the whole group of 16 beers, the correlation coefficient between calories and alcohol content is $r = 0.81$. One might argue that the three weak beers should be excluded. One argument would be that we wish to confine our investigation to beers of a certain minimum strength, say, 3%; those three beers are below specification. When the three lightest beers are excluded, the correlation coefficient falls to 0.41, and the t value for the slope falls to a nonsignificant 1.47.

THE METHOD OF LEAST SQUARES

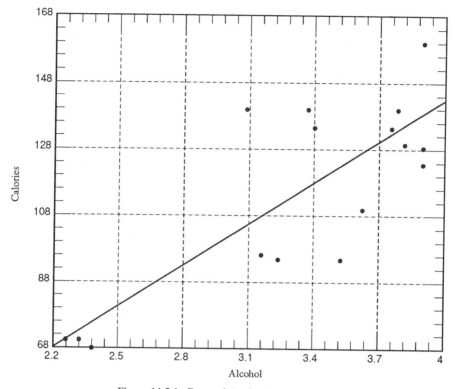

Figure 14.5.1. Regression of calories on alcohol.

14.6. THE METHOD OF LEAST SQUARES

Suppose that we estimate α and β by $\hat{\alpha}$ and $\hat{\beta}$. Substituting those estimates in equation (14.4.2), we have an estimate of μ_i,

$$\hat{y}_i = \hat{\alpha} + \hat{\beta} x_i. \tag{14.6.1}$$

The difference between this estimated value and the observed value is the deviation

$$\hat{e}_i = y_i - \hat{y}_i. \tag{14.6.2}$$

This difference is called the ith residual. The method of least squares is to take for $\hat{\alpha}$ and $\hat{\beta}$ the values that minimize the sum of the squares of the residuals:

$$S_e = \sum \hat{e}_i^2. \tag{14.6.3}$$

The quantity S_e is called the sum of squares for error, or the residual sum of squares.

Substituting the estimates $\hat{\alpha}$ and $\hat{\beta}$ in equation (14.6.2), we have

$$S_e = \sum (y_i - \hat{\alpha} - \hat{\beta} x_i)^2 . \qquad (14.6.4)$$

Differentiating S_e, in turn, with respect to $\hat{\alpha}$ and to $\hat{\beta}$, and equating the derivatives to zero, gives two equations

$$2 \sum (y_i - \hat{\alpha} - \hat{\beta} x_i) = 0$$
$$2 \sum x_i (y_i - \hat{\alpha} - \hat{\beta} x_i) = 0 ,$$

which simplify to

$$\sum y_i = n\hat{\alpha} + \hat{\beta} \sum x_i , \qquad (14.6.5)$$
$$\sum x_i y_i = \sum x_i \hat{\alpha} + \hat{\beta} \sum x_i^2 , \qquad (14.6.6)$$

and dividing equation (14.6.5) by n,

$$\bar{y} = \hat{\alpha} + \hat{\beta} \bar{x} . \qquad (14.6.7)$$

This says that $(\bar{x}, \bar{y})$ lies on the fitted line, i.e., the fitted line passes through the centroid of the data.

Equations (14.6.5) and (14.6.6) are called the normal equations.

If we multiply equation (14.6.7) by $\sum x_i$ and subtract it from equation (14.6.6), we have

$$\sum x_i (y_i - \bar{y}) = \hat{\beta} \sum \left(x_i^2 - \bar{x} \sum x_i \right) . \qquad (14.6.8)$$

We rewrite this equation as

$$S_{xy} = \hat{\beta} S_x , \qquad (14.6.9)$$

so that

$$\hat{\beta} = \frac{S_{xy}}{S_x} . \qquad (14.6.10)$$

We then substitute this value in equation (14.6.7) and obtain

$$\hat{\alpha} = \bar{y} - \hat{\beta} \bar{x} . \qquad (14.6.11)$$

The prediction line can now be written as

$$\hat{y} = \bar{y} + \hat{\beta}(x - \bar{x}) . \qquad (14.6.12)$$

A SIMPLE NUMERICAL EXAMPLE   225

For the beer data, $\bar{x} = 3.389$, $\bar{y} = 114.875$, $S_{xy} = 196.658$, and $S_x = 4.752$. Then $\hat{\beta} = 196.658/4.752 = 41.3844$, $\hat{\alpha} = -23.297$.

14.7. AN ALGEBRAIC IDENTITY

$$\sum (x_i - \bar{x})(y_i - \bar{y}) = \sum x_i y_i - \bar{x}\sum y_i - \bar{y}\sum x_i + n\bar{x}\bar{y}$$
$$= \sum x_i y_i - n\bar{x}\bar{y} - n\bar{x}\bar{y} + n\bar{x}\bar{y}$$
$$= \sum x_i y_i - n\bar{x}\bar{y}\;.$$

We can show in the same way that

$$\sum (x_i - \bar{x})y_i = \sum (y_i - \bar{y})x_i = \sum x_i y_i - n\bar{x}\bar{y}\;,$$

and so the term on the left of equation (14.6.9) is

$$S_{xy} = \sum (x_i - \bar{x})(y_i - \bar{y})\;. \tag{14.7.1}$$

If we set $x_i = y_i$ in the algebraic identity, we also see that the multiplier on the right of the equation is

$$S_x = \sum (x_i - \bar{x})^2\;. \tag{14.7.2}$$

14.8. A SIMPLE NUMERICAL EXAMPLE

This example has been constructed to illustrate the ideas of the previous section.

Example 14.8.1

x	y	$x - \bar{x}$	$y - \bar{y}$
1.0	13.3	−3.0	−8.8
2.0	16.1	−2.0	−6.0
3.0	19.0	−1.0	−3.1
4.0	21.9	0.0	−0.2
5.0	25.2	+1.0	+3.1
6.0	27.7	+2.0	+5.6
7.0	31.5	+3.0	+9.4

$\bar{x} = 4.0$, $\bar{y} = 22.1$, $S_{xy} = 84.0$, and $S_x = 28.0$, so that $\hat{\beta} = 3.0$, and $\hat{\mu} = 22.1 - 12.0 = 10.1$. The prediction line is

$$y = 10.1 + 3.0x\;.$$

Table 14.8.1

x	y	$\hat{y}$	$\hat{e}$
1.0	13.3	13.1	+0.2
2.0	16.1	16.1	0.0
3.0	19.0	19.1	−0.1
4.0	21.9	22.1	−0.2
5.0	25.2	25.1	+0.1
6.0	27.7	28.1	−0.4
7.0	31.5	31.1	+0.4

The predicted values and the residuals are shown in table 14.8.1. The residual sum of squares is $S_e = 0.42$.

Note that the sum of the residuals is zero. That follows from equation (14.6.12) since

$$\hat{e}_i = y_i - \hat{y}_i = (y_i - \bar{y}) - \hat{\beta}(x_i - \bar{x}),$$

and so

$$\sum \hat{e}_i = \sum (y_i - \bar{y}) - \hat{\beta} \sum (x_i - \bar{x}) = 0. \qquad (14.8.1)$$

Also, from equation (14.6.8),

$$\sum x_i \hat{e}_i = 0. \qquad (14.8.2)$$

So there are two restrictions on the residuals, and the sum of squares of the residuals has $n - 2$ degrees of freedom. One can as well say that 2 d.f. have been lost because two parameters, α and β, have been estimated from the data. □

14.9. THE ERROR TERMS

So far we have only made one assumption about the nature of the error terms:

$$E(e_i) = 0. \qquad (i)$$

This assumption is enough to make $\hat{\beta}$ an unbiased estimate of the slope of the line. Three more assumptions are usually added. The observations are uncorrelated:

$$E(e_h e_i) = 0 \quad \text{for } h \neq i. \qquad (ii)$$

The observations have the same variance:

$$V(y_i) = E(e_i^2) = \sigma^2 \quad \text{for all } i. \tag{iii}$$

Assumption (iii) is called the assumption of homoscedasticity. The last assumption is that the errors are normally distributed. This implies, in conjunction with the other three assumptions, that

$$e_i \sim N(0, \sigma^2) \quad \text{and} \quad y_i \sim N(\mu_i, \sigma^2), \tag{iv}$$

and that the observations are independent.

14.10. THE ESTIMATE OF THE SLOPE

In this section, we develop the properties of the estimate, $\hat{\beta}$, of the slope of the line. The estimate, $\hat{\beta}$, is a linear estimate of the slope, i.e., it is a linear combination of the observations, y_i. For convenience, we write

$$\hat{\beta} = \sum w_i y_i, \quad \text{where} \quad w_i = \frac{x_i - \bar{x}}{S_x}.$$

We now show that $\hat{\beta}$ is an unbiased estimate of the slope.

$$\begin{aligned} E(\hat{\beta}) &= E\left(\sum w_i y_i\right) \\ &= \sum w_i E(y_i) \\ &= \sum w_i \alpha + \beta \sum w_i x_i \\ &= \beta. \end{aligned}$$

Note that in showing that $\hat{\beta}$ is an unbiased estimate of the slope, we have only used the first assumption about the error terms.

The variance of $\hat{\beta}$ is computed with assumptions (i), (ii), and (iii).

$$\begin{aligned} V(\hat{\beta}) &= \sum w_i^2 V(y_i) = \frac{S_x \sigma^2}{S_x^2} \\ &= \frac{\sigma^2}{S_x}. \end{aligned} \tag{14.10.1}$$

This is intuitively reasonable. The wider the spread in the values of x, the more precise is the estimate of the slope.

Under the normality assumption, $\hat{\beta}$ is a linear combination of normal random variables and is itself normally distributed. Even if the observations are not normal, the central limit theorem justifies us in acting as if $\hat{\beta}$ were approximately normally distributed. The reader should note that this is the first time that we have made use of the normality assumption. It leads to the tests that are derived in the next few sections.

14.11. ESTIMATING THE VARIANCE

If our original model is correct, and the locus of $E(Y|X)$ is a straight line, $E(Y) = E(\hat{Y})$ at each point. The residuals reflect only the noise in the system. With homoscedasticity (homogeneous variance), the residuals provide an estimate of the error variance. We have seen in equations (14.8.1) and (14.8.2) that there are two linear constraints on the residuals. It can be shown, after some algebra, that

$$E(S_e) = (n-2)\sigma^2,$$

so that

$$s^2 = \frac{S_e}{n-2}$$

is an unbiased estimator of σ^2.

In the printout for the beer data, the sum of squares of the residuals appears on the second line of the section headed by "Analysis of Variance." It is 4315 with 14 d.f., which leads to the estimate $4315/14 = 308$ for the variance. The square root of that quantity, 17.6, appears earlier in the table as the estimate of the standard deviation.

14.12. t-TESTS

If the errors are normally distributed, the quotient S_e/σ^2 has a chi-square distribution with $n-2$ d.f. We have already seen that the estimate, $\hat{\beta}$, of the slope is a normal random variable. It follows that

$$t = \frac{\hat{\beta} - \beta}{\sqrt{s^2/S_x}} \qquad (14.12.1)$$

has a t distribution with $n-2$ d.f.

This t statistic can be used to compute confidence intervals for the "true" slope, β. The statistic

$$t = \frac{\hat{\beta}}{\sqrt{s^2/S_x}} \qquad (14.12.2)$$

can be used to test the hypothesis that the "true" slope is zero. It is called the t-value for $\hat{\beta}$. Similar calculations lead to the t-value for the other coefficient, $\hat{\alpha}$.

A common rule of thumb is to regard values of t that exceed 2.0 in absolute value as significant. Smaller values of t would lead to accepting the hypothesis that the true coefficient is zero, or, at least, to arguing that the evidence is not strong enough to reject that null hypothesis. In the beer data, the t-value for the slope is

$$t = \frac{41.3844}{17.6/\sqrt{4.752}} = 5.14,$$

which confirms that the line of best fit is not horizontal. The t-value for $\hat{\alpha}$ is only -0.86, which suggests refitting a line through the origin.

14.13. SUMS OF SQUARES

Another approach to testing the significance of the slope of an estimated line is through the analysis of variance. We argue that if $\beta = 0$, which means that X is of no use as a predictor of Y, the model becomes

$$E(Y) = \text{constant}, \quad (14.13.1)$$

and the constant is estimated by $\bar{y}$.

The scatter, or sum of squares of residuals, about the horizontal line is $S_y = \Sigma (y_i - \bar{y})^2$. Fitting the predictor, X, reduces the scatter to S_e. The amount of the reduction is called the sum of squares for regression and is denoted by

$$S_R = S_y - S_e. \quad (14.13.2)$$

The total sum of squares, S_y, has $n - 1$ d.f. The reduction is obtained at the cost of 1 d.f., corresponding to the slope, β. That leaves $n - 2$ d.f. for S_e. If the "true" line is indeed horizontal, the apparent reduction should just be equivalent to 1 d.f. worth of noise, i.e., to the error variance. The effectiveness of the reduction, therefore, can be measured by comparing S_R to the estimate of the variance. This is the F-test. The analysis of variance describes this division of S_y into the two components:

$$S_y = S_R + S_e. \quad (14.13.3)$$

We now look at the sum of squares for regression more closely. We recall from equation (14.6.12) that

$$\hat{y}_i = \bar{y} + \hat{\beta}(x_i - \bar{x}),$$

so that

$$\hat{e}_i = y_i - \hat{y}_i = (y_i - \bar{y}) - (\hat{y}_i - \bar{y})$$

and

$$S_e = S_y - 2\hat{\beta}S_{xy} + \hat{\beta}^2 S_x$$
$$= S_y - \hat{\beta}S_{xy},$$

whence

$$S_R = \hat{\beta}S_{xy}. \qquad (14.13.4)$$

Computationally, the easiest way to compute S_e is to compute S_y and S_R and subtract.

The F statistic, described earlier, is

$$F = \frac{S_R}{s^2}. \qquad (14.13.5)$$

Under the null hypothesis that $\beta = 0$, F has the F distribution with $(1, n-2)$ d.f. But we can also write

$$F = \frac{\hat{\beta}S_{xy}}{s^2} = t^2, \qquad (14.13.6)$$

and so the F-test is equivalent to the t-test that was described earlier.

The analysis of variance also leads to the correlation coefficient. We have derived the sum of squares for regression as the reduction in scatter, which is explained by adding X as a predictor. This quantity depends on the scale of Y. However, the fractional reduction is scale-free:

$$\frac{S_R}{S_y} = \frac{S_{xy}^2}{S_x S_y} = r^2. \qquad (14.13.7)$$

This gives a physical interpretation of the correlation coefficient in terms of the fit of the linear model.

For the beer data,

$$S_R = (41.3844)(196.658) = 8138.5,$$

whence $r^2 = 8138.55/12{,}453.8 = 0.654$,

$$S_e = 12{,}453.8 - 8138.5 = 4315.2 \quad \text{and} \quad s^2 = 4315.2/14 = 308.2.$$

INFLUENTIAL POINTS 231

14.14. STANDARDIZED RESIDUALS

It follows from equation (14.10.1) that

$$V(\hat{y}_i) = V(y) + (x_i - \bar{x})^2 V(\hat{\beta})$$
$$= \sigma^2 \left[\frac{1}{n} + \frac{(x_i - \bar{x})^2}{S_x} \right]. \qquad (14.14.1)$$

A similar formula can be derived for $V(\hat{e}_i)$.

In the last part of the printout—the back solution for the "special" points—there is a column for the standard deviation of $\hat{y}_i$ obtained from equation (14.14.1) with s substituted for the unknown σ. The last column contains the standardized residuals. In this column, each residual is divided by the estimate of its standard deviation. The quotient is essentially a t-value for that residual. A value greater than 2.0 for the standardized residual indicates a point whose fit is questionable, and it is flagged by the letter R. In the beer data, point 13 earns an R rating.

14.15. INFLUENTIAL POINTS

In the beer data, the two weakest beers received X ratings to indicate that they had a large influence on the slope of the prediction line. They earned these ratings because they were a long way from the centroid of the data base. The actual criterion of how far is a long way is a matter of choice. This computer run was made with an earlier version of Minitab. Its choice of cutoff point gave X ratings to beers 5 and 6. Beer 2 missed an X rating by a whisker. In the current version of the Minitab package, the authors have changed the cutoff value and those two beers no longer earn that rating!

One should be careful not to fall into the trap of automatically dropping from the data any points that have X ratings, on the argument that they must surely be extreme. Such action can sometimes be justified. In the particular example of the beer data, a case could be made for setting those points aside if, in retrospect, they were considered to be outside the limits within which the scientists wished to work in future. Then they might be justified in using a prediction line based only on the stronger beers. This kind of action has to be taken on a case-to-case basis, and only with due consideration.

A similar rating can be used in multiple regression, when Y is predicted by several X variables. In that situation, it is not always easy to spot the influential points with only a visual examination of the data.

14.16. CONFIDENCE INTERVALS FOR A PREDICTED VALUE

Suppose that we predict the value of Y when $X = x_0$. The variance of the predicted value is

$$V(\hat{y}_0) = V(\bar{y}) + \hat{\beta}(x_0 - \bar{x})$$
$$= V(y) + (x_0 - \bar{x})^2 V(\hat{\beta})$$
$$= \sigma^2 \left[\frac{1}{n} + \frac{(x_0 - \bar{x})^2}{S_x} \right]. \qquad (14.16.1)$$

This is the formula that is used to calculate the entries in the column headed "ST. DEV. PRED. Y" in the printout. Clearly, $V(\hat{y}_0)$ increases as x_0 moves away from the center, $\bar{x}$, of the data base. It takes its minimum value when $x_0 = \bar{x}$. A confidence interval for the "true" value, μ_0, when $x = x_0$, can be obtained in the usual way by

$$\hat{y}_0 - t^* s^* \le \mu_0 \le \hat{y}_0 + t^* s^*, \qquad (14.16.2)$$

where t^* is the tabulated value of t with $n - 2$ d.f., and S^* is the standard deviation of $\hat{y}_0$, obtained from equation (14.16.1).

If we now take an observation, y_0, at $X = x_0$, we may wish to compare the observed value with our estimate. It is important to remember that there will be noise in our observation too. The variance of the difference is

$$V(y_0 - \hat{y}_0) = \sigma^2 \left[1 + \frac{1}{n} + \frac{(x_0 - \bar{x})^2}{S_x} \right]. \qquad (14.16.3)$$

This can be appreciably larger than $V(\hat{y}_0)$, and the confidence interval is correspondingly wider.

14.17. LINES THROUGH THE ORIGIN

In the printout for the beer data, the t-value for the constant term is -0.86. We cannot reject the hypothesis that the constant term is zero, and so we fit a new line, constrained to pass through the origin. This is done in Minitab by two commands, first REGRESS c1 1 c2;, followed by NOCONSTANT. The printout appears in table 14.17.1.

Notice the change in the analysis of variance table. The sum of squares for Y is now the raw sum of squares, Σy_i^2; it has not been adjusted by taking deviations about the mean value, y. The sum of squares of residuals has increased from 4315.2 to 4540.6; but there has been an increase in the number of d.f. from 14 to 15, so that the estimate of the variance has actually fallen from 308 to 302.7. No reference is made to the correlation

RESIDUAL PLOTS

Table 14.17.1. Beer Data (Line through the Origin)

REGRESS c1 1 c2;
NOCONSTANT.

THE REGRESSION EQUATION IS Y = +34.6 X1

	COLUMN	COEFFICIENT	ST. DEV. OF COEF.	T-RATIO = COEF./S.D.
NOCONSTANT				
X1	C2	34.588	1.286	26.90

ANALYSIS OF VARIANCE

DUE TO	DF	SS	MS = SS/DF
REGRESSION	1	219053.4	219053.4
RESIDUAL	15	4540.6	302.7
TOTAL	16	223594.0	

coefficient. We are no longer asking whether including X as a predictor is an improvement over the line $y = \bar{y}$. Now we ask if it is an improvement over the line $y = 0$.

The least squares estimate of the slope is obtained from the single normal equation. Minimizing the residual sum of squares

$$\sum (y_i - \hat{\beta} x_i)^2$$

leads to

$$\sum x_i y_i = \hat{\beta} \sum x_i^2 . \qquad (14.17.1)$$

It is easily shown that $\hat{\beta}$ is an unbiased estimate of the slope, and that

$$V(\hat{\beta}) = \frac{\sigma^2}{\sum x_i^2} . \qquad (14.17.2)$$

The residuals are no longer constrained to add up to zero, because there is no normal equation that corresponds to a constant term. It is the loss of that constraint that adds one more degree of freedom to the error term.

14.18. RESIDUAL PLOTS

We have already mentioned that the residuals (when there is a constant term) are constrained to sum to zero. That is the only constraint, and so we should expect them to be more or less randomly and normally distributed. Some like to make a normal plot of the residuals and see whether they lie, more or less, on a straight line.

Roughly half the residuals should be positive and half negative. If you have one large positive residual and the others are small and negative, perhaps there really is a straight line passing through, or close to, all the points but one; that one point is an outlier somewhere above the line.

Suppose that the data are entered in ascending order of X, the observation with the smallest value of X being number 1 and so on. A glance at the list of residuals should show a random ordering of plus and minus signs. If the residuals are alternately plus and minus, we may have fitted a straight line to a sine curve! If there is a string of positive residuals followed by a string of negatives and again a string of positives, we may have fitted a straight line to a parabola or a hyperbola (see the example in the next section). Before plotting on the computer was as convenient as it is now, these precautions of checking the randomness of the plus and minus signs were standard practice.

Nowadays, it is easy to plot the residuals against X. (A plot of the residuals for the beer data is shown as figure 14.18.1.) One can notice the phenomena that have just been mentioned, and perhaps something else. Do the residuals seem to get bigger as X increases? Do they get bigger as Y increases? Does this mean that, perhaps, we were wrong to think that the variance was constant? Maybe the variance is a function of Y. In that case,

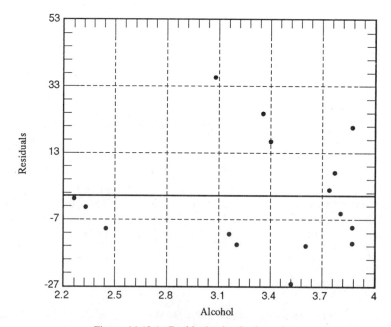

Figure 14.18.1. Residuals plot for beer data.

OTHER MODELS 235

one should be using weighted regression or a transformation. The consideration of that complication is beyond the scope of this book. The reader may wish to refer to a book that specializes in regression, such as Draper and Smith (1981).

14.19. OTHER MODELS

So far, we have assumed that our model is correct, and that there is some underlying linear connection between the two variables—not necessarily a functional relationship. A large value of the correlation coefficient tends to give credence to that idea. In the following example, we fit a straight line where a hyperbola is more appropriate.

Example 14.19.1. A company sets up a new process for producing parts. After two weeks, they start to take a sample of 1000 parts each week and to record the number of defectives. The management is delighted to see that the number of defectives decreases. The data set is given in table 14.19.1.

Table 14.19.2 shows a summary version of the printout when a straight-line model is fitted to the data.

With so large a value of r as 0.875, one might be tempted to think that the straight line is the "correct" model. A plot of the data, figure 14.19.1, reveals this to be incorrect. The points appear to lie on a fairly smooth curve that looks like a rectangular hyperbola. A plot of the residuals against X, figure 14.19.2, shows that the end points lie above the fitted line and the points in the middle lie below the line—a distinct pattern.

Does this mean that fitting a straight line to this data is useless? Not necessarily. There is no question that Y decreases as X increases. The straight-line model gives a reasonable approximation to reality as long as one stays within limited ranges of the X variable.

There is enormous potential for trouble if one tries to extrapolate—to use the equation to predict values of Y outside the range of the data. If one extends the line far enough, it will cross the X axis and will predict a negative number of defectives, which is better than even the most enthusiastic quality engineer would predict. In the 1970s, some economists plotted

Table 14.19.1

Week	Def.	Week	Def.	Week	Def.	Week	Def.
3	65	4	50	5	41	6	33
7	27	8	30	9	21	10	18
11	16	12	14	13	15	14	11
15	10	16	9	17	8	18	10
19	7	20	6	21	6	22	4
23	5	24	4	25	3	26	3

Table 14.19.2

The regression equation is Y = 46.36 − 2.00 X.

Predictor	Coef	Stdev	t-ratio
Constant	46.3586	3.792	12.22
X	−2.00174	0.236	−8.48

s = 8.004 $R^2 = 0.766$ R = −0.875

the price per barrel of crude oil against time. That straight line extrapolation forecasted that oil would cost $80 a barrel by 1990; some of us are finding that hard to believe now.

Extrapolation, and, in particular, linear extrapolation, is very easy—you just keep drawing the line. It is also very, very dangerous.

Engineers are used to converting curves to straight lines by using special paper in which X and/or Y are replaced by their logarithms, square roots, reciprocals, etc. We can fit a hyperbola to the data by introducing a new variable, $1/X$, and fitting a straight line with $1/X$ as the predictor instead of X. The new "line" is

$$Y = -3.94 + 216.81\left(\frac{1}{X}\right)$$

with $s = 1.948$ and $r = 0.993$.

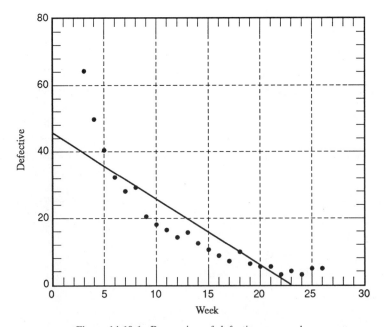

Figure 14.19.1. Regression of defectives on week.

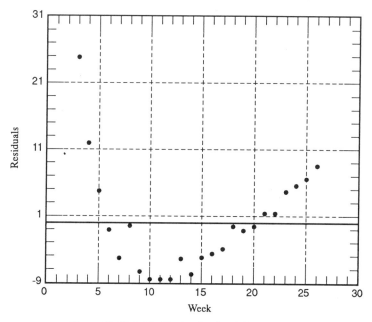

Figure 14.19.2. Residuals plot for defectives data.

This is obviously a good fit to the data. None of the usual assumptions has been violated by using $1/X$ instead of X as the predictor of Y. Multiplying both sides of the equation by x gives the rectangular hyperbola:

$$xy + 3.94x = 216.81 \, .$$

□

14.20. TRANSFORMATIONS

Another model that might be appropriate for a data set that looks like this is the exponential model:

$$Y = \alpha e^{\beta X} \, . \qquad (14.20.1)$$

As it stands, this model is not a linear model, by which we mean that the right-hand side is not a linear function of the two unknown parameters, α and β. We can, however, make it linear by a transformation. We take natural logarithms of both sides of the equation and get

$$Z = c + dX \qquad (14.20.2)$$

where $Z = \ln(Y)$, $c = \ln(\alpha)$, and $d = \beta$.

This procedure does some violence to our assumptions about the error term, but it does enable us to fit the model. We fit the straight line and then take exponents to return to equation (14.20.1). The new regression equation is

$$\ln(Y) = 4.28 - 0.124x,$$

which corresponds to

$$y = 72.24e^{-0.124x}.$$

The fit that we have made is equivalent to plotting the data on semilog paper and drawing the best line through the points. We can also get the equivalent of log–log paper and fit the model

$$y = ax^b$$

by taking logarithms of both sides and fitting the line

$$w = c + bz$$

where $w = \ln(Y)$, $c = \ln(a)$, and $z = \ln(X)$.

The results that we get with these transformations are only approximations to the best fit. It does not follow that the pair of coefficients, $a = \exp(c)$, b, that minimizes the residual sum of squares in the log–log world is the same as the pair that minimizes the sum in the untransformed world. The problem of finding the best fit in these nonlinear problems without using linearizing transformations is full of mathematical complexities, and the reader is again referred to Draper and Smith (1981).

EXERCISES

14.1. For a set of 20 observations, we have $\Sigma x = 200$, $\Sigma y = 810$, $\Sigma (x - \bar{x})^2 = 92$, $\Sigma (y - \bar{y})^2 = 104$, and $\Sigma (x - \bar{x})(y - \bar{y}) = 87$. Fit the line $y = \alpha + \beta x$ to the data and test the hypothesis $\beta = 1.0$ against the alternative $\beta \neq 1.0$.

14.2. Fit a straight line, $y = \alpha + \beta x$, to the following data:

X =	1	2	3	4	5	6	7	8	9	10
Y =	13	17	19	12	14	29	34	27	36	33

Plot Y against X and plot the residuals against the predicted values, $\hat{y}_i$. Compute the correlations between y_i, $\hat{y}_i$, and $\hat{e}$. Note that the correlation between the predicted values and the residuals is zero. Test the hypothesis $\beta = 2$ against the alternative $\beta \neq 2$.

14.3. Fit a straight line to the following data. Note that $R^2 = 0.915$, which is very impressive. Make plots of Y against X and of the residuals against the predicted values. You will see that for $x > 4$, the line is not an implausible model at first glance. The residual plot shows a clear parabolic pattern. Verify that for each point $y = x^2 - 3x + 5$. You have fitted a straight line to a parabola! Estimate from your fitted line the values of Y when $X = 13$, 14, and 15, and compare them with the actual values from the quadratic. This illustrates how extrapolation from a model that seems to be plausible can be a trap.

$X =$	1	2	3	4	5	6	7	8	9	10	11	12
$Y =$	3	3	5	9	15	23	33	45	59	75	93	113

14.4. The table shows the miles per gallon for the ten most economical cars in the 1990 EPA fuel economy survey, as reported in the *Austin American Statesman*.

m.p.g. city	m.p.g. highway
53	58
49	52
46	50
46	50
46	50
43	49
40	44
38	42
38	39
37	43

Obtain a least squares prediction line to predict the highway mileage from the city mileage:

$$\text{highway} = \beta_0 + \beta_1(\text{city})$$

It could be argued that the line should pass through the origin. Test that hypothesis.

14.5. Since the data in exercise 14.4 has repeated observations at city = 46 and 38, there is an estimate of pure error between those duplicates. This can be used to check your model, i.e., to test whether the straight-line model is adequate. Test that hypothesis. Readers who are using Minitab should use the commands

REGRESS C1 1 C2 ;
PURE .

14.6. In a study of the growth of InP, engineers recorded the van der Pauw measurements of mobility at both 77K and 300K. Fit a straight line to predict the mobility at 300K from the mobility at 77K. A plot of the data will explain why the twelfth sample has large influence and also a large residual. Interpret this.

	X (77K)	Y (300K)		X (77K)	Y (300K)
1.	3990	9020	7.	3270	3760
2.	3940	8620	8.	3960	10460
3.	3980	9670	9.	3920	9970
4.	3930	9830	10.	3860	8110
5.	3980	11200	11.	3330	4890
6.	3770	7880	12.	2330	2200

14.7. Suppose that you felt comfortable in arguing that for the data in exercise 14.6, the twelfth sample was far below the range of interest and that you might, therefore, reasonably drop it. Do that and fit a new line. Is the fit markedly better?

14.8. In exercise 14.2, you showed that $\text{corr}(\hat{e}, \hat{y}_i) = 0$. Prove that this is always true when there is a constant term, α, in the model, i.e., when the line is not forced to pass through the origin. It can also be shown that $\text{corr}(\hat{e}, y_i) = \sqrt{1 - r^2}$. In exercise 14.1, $r^2 = 0.718$ and so $\text{corr}(\hat{e}, y_i) = 0.531$, which is not trivial. This a reason for some statisticians to prefer to plot $\hat{e}$ against $\hat{y}_i$ rather than y_i.

14.9. Two scientists wish to develop a formula for predicting the height of an aspen tree (Y) standing in a forest from its diameter at breast height. The measurements in meters for 32 trees follow. Fit a line $Y = \alpha + \beta x$ to the data and comment about your results. It is suggested that they should use $\sqrt{x}$ rather than x as the predictor. Fit that model and compare its fit to that of the first model.

	1	2	3	4	5	6	7	8	9	10
X =	0.90	1.20	1.45	1.80	2.00	2.20	3.40	3.40	3.50	7.30
Y =	2.20	2.80	3.20	3.78	4.60	3.10	5.35	5.70	5.35	9.20

	11	12	13	14	15	16	17	18	19
	9.10	10.50	13.00	13.70	15.10	15.40	15.80	17.30	19.40
	9.40	11.50	16.10	15.90	16.70	17.40	15.60	15.50	23.00

	20	21	22	23	24	25	26	27	28
	19.50	21.50	22.50	22.60	22.80	23.00	25.10	25.20	27.80
	19.35	23.10	22.50	18.10	22.40	22.50	23.80	22.50	23.50

	29	30	31	32
	30.20	32.10	32.40	35.40
	23.50	23.80	23.50	22.50

14.10. Suppose that for a set of 12 observations, $\Sigma x = 21$, $\Sigma y = 60$, $\Sigma x^2 = 91$, $\Sigma xy = 200$, and $\Sigma y^2 = 801$. Fit the model $y = \beta x$. Test the hypothesis $\beta = 2.0$ against the alternative $\beta \neq 2.0$.

14.11. Another estimate of the slope of a line through the origin is $\tilde{\beta} = \bar{x}/\bar{y}$, when $\bar{x} \neq 0$. Show that $\tilde{\beta}$ is an unbiased estimate of the slope. Calculate its variance and show that it is greater than the variance of the least squares estimate. Do the calculations for the data of exercise 14.10.

14.12. A purchasing agent compares the prices of a certain kind of wire and fits a regression line

$$y = 0.52 + 0.83x,$$

where y is the cost in U.S. dollars, and x is the length in feet. Suppose that the price had been computed in Swiss francs (1 S. Fr. = $0.62) per meter. Express the regression line in terms of the new units and prove that the correlation between price and length is unchanged.

14.13. In an octane blending study that will be mentioned again in the next chapter, R. D. Snee (1981) determined the octane ratings of 32 blends of gasoline by two standard methods—motor ($F1$) and research ($F2$). The data follow with $F1$ as Y and $F2$ as X. Fit a regression line to enable an engineer to predict the $F1$ number from the $F2$ number.

	1	2	3	4	5	6	7	8
X =	105.0	81.4	91.4	84.0	88.1	91.4	98.0	90.2
Y =	106.6	83.3	99.4	94.7	99.7	94.1	101.9	98.6

	9	10	11	12	13	14	15	16
	94.7	105.5	86.5	83.1	86.2	87.7	84.7	83.8
	103.1	106.2	92.3	89.2	93.6	97.4	88.8	85.9

	17	18	19	20	21	22	23	24
	86.8	90.2	92.4	85.9	84.8	89.3	91.7	87.7
	96.5	99.5	99.8	97.0	95.3	100.2	96.3	93.9

	25	26	27	28	29	30	31	32
	91.3	90.7	93.7	90.0	85.0	87.9	85.2	87.4
	97.4	98.4	101.3	99.1	92.8	95.7	93.5	97.5

Blends 1 and 10 stand out from the others. They have by far the highest octanes because of a high aromatic content, and it is reasonable to set them aside. Drop them from the data set and rerun the regression. Do you get a better fit to the data?

14.14. In a survey of roofing asphalts, 22 different asphalts were compared. Y is the percentage of asphaltene in each asphalt; X is the percentage of resin. Fit a straight line to predict the percent asphaltene from the percent resin. It is suggested that number 22 is somewhat different from the others. Is this suggestion plausible? Is there is a marked change in the prediction line if number 22 is deleted? When that is done, number 11 has a large residual. Is this plausible? Is there a marked change in fit when number 11, too, is deleted?

	1	2	3	4	5	6	7	8	9	10	11
$X =$	22	21	25	23	29	26	25	27	25	21	24
$Y =$	35	35	29	32	27	29	28	31	30	36	39

	12	13	14	15	16	17	18	19	20	21	22
	26	23	24	22	27	29	24	24	27	24	34
	33	31	31	36	26	32	31	29	27	27	23

14.15. These are measurements on 11 samples of asphalt. Y is the penetration at 115°F; X is the softening point in degrees F. Fit a straight line to predict Y from X. Asphalt 4 is seen to be very influential. Make a plot of the data and explain this. Does the fit improve when point 4 is dropped?

When you fit the data without number 4, the old asphalt, number 5, stands out as influential. Suppose that you drop it, too; does that improve the fit? What justification would you have for dropping one, or both, of those asphalts in practice?

$$Y = \quad 77 \quad 149 \quad 111 \quad 49 \quad 57 \quad 76$$
$$X = \quad 158 \quad 136 \quad 153 \quad 194 \quad 172 \quad 154$$

$$106 \quad 119 \quad 93 \quad 153 \quad 112 \quad 100$$
$$150 \quad 142 \quad 156 \quad 145 \quad 147 \quad 145$$

14.16. Prove that if $W = a + bX$ and $U = c + dY$, then $\operatorname{corr}(W, U) = \operatorname{corr}(X, Y)$.

CHAPTER FIFTEEN

Multiple Regression

15.1. INTRODUCTION

In this chapter, we progress from the straight line, or fitting Y to a single predictor variable, to the more complicated problem of handling several predictors—multiple regression. Before the advent of electronic computers, the sheer weight of the calculations involved prevented this technique from being used very much. In order to handle a problem with p predictor variables, one has to invert a $(p+1) \times (p+1)$ matrix. Thirty years ago, inverting a 5×5 matrix on a desk calculator took several hours, even if no errors crept into the calculation, but nowadays, multiple regression is one of the most commonly used statistical tools.

We start with a manufactured example to illustrate the underlying theory. We fit the model

$$y = \beta_0 + \beta_1 x_1 + \beta_2 x_2 + e$$

to the set of seven points shown in table 15.2.1. As in chapter fourteen, the term e denotes the random error; it is subject to the same conditions. We use two methods. At first, we make a two-stage fit using straight lines. In that procedure, we begin by adjusting X_1 for X_2 and Y for X_2, and then fit a line to the adjusted values; finally, we express the equation of the last line in terms of the original variables.

The second approach is the method of least squares. Later, we use matrices to develop the least squares method for the general case with more than two predictor variables.

Sometimes Y is called the dependent variable and the X variables are called the independent variables. This goes back to the terminology of elementary algebra. When one took the equation $y = 2x + 3$ and substituted a value for x to obtain y, x was called independent and y dependent; when the equation was transformed to $x = (y-3)/2$, the labels changed; x became the dependent variable and y the independent variable. There is no requirement that the X variables be independent of each other in any broad sense. Our procedure accommodates the polynomial equation

$$Y = \beta_0 + \beta_1 X + \beta_2 X^2$$

by setting $X = X_1$, and $X^2 = X_2$.

It also handles the trigonometric model

$$Y = \beta_0 + \beta_1 \sin(wt) + \beta_2 \cos(wt)$$

by setting $\sin(wt) = X_1$ and $\cos(wt) = X_2$. We do require that X_1 and X_2 be linearly independent, which means that we cannot have x_1 being a linear function of x_2; we cannot have, for example, $X_1 = 2X_2 + 6$. More is said about this later.

15.2. THE METHOD OF ADJUSTING VARIABLES

This is an old-fashioned method that has only pedagogical value in this age of computers. The procedure for fitting Y to two predictor variables, X_1 and X_2, involves fitting three lines:

1. Y in terms of X_2 with residual $u = y - \tilde{y}$, where $\tilde{y}$ is the fitted value;
2. X_1 in terms of X_2 with residual $v = x_1 - \tilde{x}_1$;
3. $u = \beta v$.

The data for an example are shown in table 15.2.1. The three lines, calculated in this example, are

$$\tilde{y} = 5.0 + 0.35 x_2 ; \qquad \tilde{x}_1 = 4.0 + 0.30 x_2 ; \qquad u = 0.819 v .$$

Substituting $y - \tilde{y}$ for u and $x_1 - \tilde{x}_1$ for v in the last line gives the final fitted model:

$$\begin{aligned} y &= \tilde{y} + 0.819(x_1 - \tilde{x}_1) \\ &= (5.0 + 0.35 x_2) + 0.819(x_1 - 4.0 - 0.30 x_2) \\ &= 1.724 + 0.819 x_1 + 0.104 x_2 . \end{aligned}$$

In this procedure, we chose to "adjust" Y and X_1 for X_2, and then to fit a line to the adjusted values. It is left to the reader to show that we would have arrived at the same equation if we had started by "adjusting" Y and X_2 for X_1 instead. If there were three predictor variables, we could start by adjusting for X_3, and then adjust for X_2. This is a process that soon becomes tedious.

THE METHOD OF LEAST SQUARES 245

Table 15.2.1. Adjusting Variables

Y	X_1	X_2	$\tilde{y}$	u	$\tilde{x}_1$	v
15	14	23	13.05	1.95	10.9	3.1
12	12	18	11.30	0.70	9.4	2.6
14	11	19	11.65	2.35	9.7	1.3
13	10	23	13.05	−0.05	10.9	−0.9
13	9	18	11.30	1.70	9.4	−0.4
9	7	22	12.70	−3.70	10.6	−3.6
8	7	17	10.95	−2.95	9.1	−2.1

15.3. THE METHOD OF LEAST SQUARES

We denote the estimates of β_0, β_1, and β_2 by $\hat{\beta}_0$, $\hat{\beta}_1$, and $\hat{\beta}_2$, respectively, as before. The estimated value of Y when $X_1 = x_{i1}$ and $X_2 = x_{i2}$ is

$$\hat{y}_i = \hat{\beta}_0 + \hat{\beta}_1 x_{i1} + \hat{\beta}_2 x_{i2}.$$

The corresponding residual is $\hat{e} = y_i - \hat{y}_i$, and we choose the estimates so as to minimize the sum of squares of the residuals:

$$S_e = \sum (y_i - \hat{\beta}_0 - \hat{\beta}_1 x_{i1} - \hat{\beta}_2 x_{i2})^2.$$

The summation is made over all n observations.

We differentiate S_e with respect to each $\hat{\beta}_i$ in turn, and set the derivatives equal to zero. This gives us a set of three equations that are called the normal equations:

$$\sum y_i = n\hat{\beta}_0 + \hat{\beta}_1 \sum x_{i1} + \hat{\beta}_2 \sum x_{i2},$$

$$\sum x_{i1} y = \hat{\beta}_0 + \hat{\beta}_1 \sum x_{i1}^2 + \hat{\beta}_2 \sum x_{i1} x_{i2},$$

$$\sum x_{i2} y = \hat{\beta}_0 \sum x_{i2} + \hat{\beta}_1 \sum x_{i1} x_{i2} + \hat{\beta}_2 \sum x_{i2}^2.$$

For the data in our example, the equations are

(1) $\qquad 84 = 7\hat{\beta}_0 + 70\hat{\beta}_1 + 140\hat{\beta}_2,$

(2) $\qquad 874 = 70\hat{\beta}_0 + 740\hat{\beta}_1 + 1412\hat{\beta}_2,$

(3) $\qquad 1694 = 140\hat{\beta}_0 + 1412\hat{\beta}_1 + 2840\beta_2.$

We subtract $10 \times (1)$ from (2) and $20 \times (1)$ from (3) to eliminate $\hat{\beta}_0$, and obtain

(4) $\qquad 34 = 40\hat{\beta}_1 + 12\hat{\beta}_2,$

(5) $\qquad 14 = 12\hat{\beta}_1 + 40\hat{\beta}_2,$

whence

$$\hat{\beta}_1 = 0.819, \quad \hat{\beta}_2 = 0.104, \quad \text{and} \quad \hat{\beta}_0 = 1.72,$$

so that the fitted equation is indeed the same as the equation that was obtained earlier by the first method. The estimated values, $\hat{y}_i$, and the residuals are shown in table 15.3.1.

We note that $\Sigma \hat{e}_i = 0.01$ (it should be identically zero but for roundoff error). Also, $\Sigma \hat{e}_i^2 = 10.72$. The variance, σ^2, is estimated by

$$s^2 = \frac{S_e}{n-3} = \frac{10.72}{4} = 2.68$$

with $s = 1.64$.

We will see later that in this example,

$$V(\hat{\beta}_1) = V(\hat{\beta}_2) = \frac{40\sigma^2}{1456} = 0.0275\sigma^2,$$

which is estimated by $(0.2715)^2$. Thus, for $\hat{\beta}_1$, $t_1 = 0.819/0.2715 = 3.02$, and for $\hat{\beta}_2$, $t_2 = 0.39$.

In this case, it would be reasonable to conclude that β_2 could be zero, and to drop x_2 from the prediction equation for y. When this is done, the new least squares equation is the straight line:

$$y = 3.50 + 0.870 x_1.$$

The coefficients β_0 and β_1 changed when x_2 was dropped. The sum of squares for the residuals about this line is 11.10, which is only a 3.5% increase over the sum of squares for the model with two predictor variables. The addition of x_2 to the model with x_1 alone does not improve the fit significantly.

Table 15.3.1

y	$\hat{y}$	d
15	15.59	−0.59
12	13.43	−1.43
14	12.71	1.29
13	12.31	0.69
13	10.97	2.03
9	9.75	−0.75
8	9.23	−1.23

15.4. MORE THAN TWO PREDICTORS

We now turn to the general case, in which there are p predictors. Suppose that there are N data points, each of which contains $p+1$ coordinates (variables), $x_1, x_2, \ldots, x_p, y$. We wish to fit the linear model

$$y_i = \beta_0 + \beta_1 x_{i1} + \cdots + \beta_p x_{ip} + e_i, \qquad (15.4.1)$$

with the usual assumptions about the error terms, e_i, namely, that they are independent normal variables with zero expectations and the same variance, σ^2. This model is called a linear model because it is a linear expression in the unknown coefficients. As we said before, it matters little to us whether the x variables are squares or cubes, sines or cosines; the important point is that the model is a linear expression in the unknown betas.

In matrix notation, we write this model as:

$$\mathbf{Y} = \mathbf{X}\boldsymbol{\beta} + \mathbf{e}, \qquad (15.4.2)$$

where $\mathbf{Y}$ is the N-dimensional vector of the observations, and $\mathbf{e}$ is the corresponding vector of errors; $\boldsymbol{\beta}$ is a vector that consists of the $p+1$ coefficients. The matrix $\mathbf{X}$ is called the design matrix. It has N rows, one for each data point, and $p+1$ columns, numbered 0 through p, corresponding to the coefficients. The zero (leftmost) column of $\mathbf{X}$ is a column of ones; this corresponds to a dummy variable, X_0, which is always unity, and is associated with the constant term; otherwise, the element x_{ij} in the ith row and the jth column is the value of X_j for the ith data point.

We wish to obtain a vector, $\hat{\boldsymbol{\beta}}$, of estimates of the coefficients, β_j, by the method of least squares. Let $\hat{\mathbf{Y}} = \mathbf{X}\hat{\boldsymbol{\beta}}$ be the vector of estimated values of the dependent variable. The vector of residuals is

$$\hat{\mathbf{e}} = \mathbf{Y} - \hat{\mathbf{Y}} = \mathbf{Y} - \mathbf{X}\hat{\boldsymbol{\beta}}. \qquad (15.4.3)$$

The estimation procedure calls for minimizing the residual sum of squares,

$$S_e = \sum \hat{e}_i^2 = \hat{\mathbf{e}}'\hat{\mathbf{e}}$$
$$= \mathbf{Y}'\mathbf{Y} - 2\mathbf{Y}'\mathbf{X}\hat{\boldsymbol{\beta}} + \hat{\boldsymbol{\beta}}'\mathbf{X}'\mathbf{X}\hat{\boldsymbol{\beta}}.$$

It can be shown that differentiating S_e with respect to each of the $\hat{\beta}_i$, and equating the derivatives to zero, leads to the normal equations

$$\mathbf{X}'\mathbf{Y} = (\mathbf{X}'\mathbf{X})\hat{\boldsymbol{\beta}}, \qquad (15.4.4)$$

from which we obtain the estimates

$$\hat{\boldsymbol{\beta}} = (\mathbf{X}'\mathbf{X})^{-1}\mathbf{X}'\mathbf{Y}. \qquad (15.4.5)$$

The reader should confirm equation (15.4.4) for the example in section 15.3.

It is obviously necessary that the matrix $\mathbf{X}'\mathbf{X}$ should not be singular, in order for its inverse to exist. This implies that the X variables must be linearly independent, which means that they must *not* have a linear relationship among themselves. An example of this problem is given in section 15.13.

15.5. PROPERTIES OF THE ESTIMATES

The least squares estimates of the coefficients are unbiased, $E(\hat{\boldsymbol{\beta}}) = \boldsymbol{\beta}$. Their covariance matrix is

$$\mathrm{Cov}(\hat{\boldsymbol{\beta}}) = (\mathbf{X}'\mathbf{X})^{-1}\sigma^2 = \mathbf{V}\sigma^2. \qquad (15.5.1)$$

In particular, if v_{jj} is the jth diagonal element of $\mathbf{V}$, the variance of $\hat{\beta}_j$ is $v_{jj}\sigma^2$.

Each of the estimates, $\hat{\beta}_j$, is a linear combination of the observations, $y_1, y_2, \ldots, y_N$. Such estimates are called linear estimates. The mathematical justification for the use of the method of least squares is found in the Gauss–Markov theorem. This states that of all possible linear estimates of the unknown coefficients, β_j, that are unbiased, the least squares estimates are the (unique) estimates with the smallest variances. They are sometimes called the BLUE estimates (best linear unbiased estimates).

The error variance, σ^2, is estimated by the unbiased estimator

$$s^2 = \frac{S_e}{N - p - 1}. \qquad (15.5.2)$$

This estimate has $N - p - 1$ degrees of freedom. It is used to compute t-values for the coefficients. For β_j, we have

$$t_j = \frac{\hat{\beta}_j}{\sqrt{s^2 v_{jj}}}. \qquad (15.5.3)$$

15.6. MORE ABOUT THE EXAMPLE

The $\mathbf{X}'\mathbf{X}$ matrix and its inverse for the example in section 15.2 are as follows:

$$\mathbf{X}'\mathbf{X} = \begin{bmatrix} 7 & 70 & 140 \\ 70 & 740 & 1412 \\ 140 & 1412 & 2840 \end{bmatrix},$$

and

$$10192 \mathbf{X}'\mathbf{X}^{-1} = \begin{bmatrix} 107856 & -1120 & -4760 \\ -1120 & 280 & -84 \\ -4760 & -84 & 280 \end{bmatrix}.$$

Multiplying $\mathbf{X}'\mathbf{X}$ by the vector $\mathbf{X}'\mathbf{Y}$, which is $(84, 874, 1694)'$, we get the vector of coefficients $(1.72, 0.82, 0.10)$. We also see from the matrix $\mathbf{X}'\mathbf{X}^{-1}$ that $v_{11} = v_{22} = 280/10{,}192 = 0.0275$, so that, as mentioned earlier,

$$V(\hat{\beta}_1) = V(\hat{\beta}_2) = 0.0275\sigma^2.$$

15.7. A GEOLOGICAL EXAMPLE

A geologist sinks several deep holes into the ground. A sample of the composition of the earth is taken at various depths and analyzed. Much of the analysis consists of noting the percentages of various geologic components that are present in the sample. Table 15.7.1 contains such a set of data. There are 37 samples (rows) and six measurements on each (columns); X1 is the depth in meters at which the sample was taken. The remaining columns contain the percentages of the components. We wish to obtain a linear prediction equation to predict the amount of spherulites from the other variables. The next section consists of an edited version of the printout. The data was read into the Minitab worksheet in columns 1 through 7, as in table 15.7.1. The first column is the observation number and plays no further part in the discussion.

15.8. THE PRINTOUT

REGRESS c7 5 c2–c6

This command means regress the Y variable, which is in c7, against five predictor variables that are in columns 2 through 6. The regression equation is

Y = 102 − 0.0121 X1 − 1.58 X2 − 0.816 X3 − 0.993 X4 − 1.03 X5

Predictor	Coef.	St.dev.	t-ratio
Constant	101.720	2.462	41.31
X1	−0.012137	0.005373	−2.26

Table 15.7.1. Analysis of Geological Samples

	X1	X2	X3	X4	X5	Y
1	155.6	0.7	0.3	2.4	4.8	89.6
2	185.8	0.8	0.0	5.1	0.0	93.6
3	226.8	1.0	0.5	3.5	0.3	94.1
4	254.8	0.6	3.9	9.3	0.0	83.1
5	258.5	1.1	1.5	11.2	0.0	84.8
6	272.5	0.6	0.2	7.1	1.0	90.0
7	285.7	1.5	0.7	10.2	0.3	83.4
8	308.4	1.1	0.2	9.8	0.7	85.4
9	323.3	0.7	0.4	13.2	1.2	81.7
10	153.6	1.7	0.0	6.6	5.8	81.0
11	188.7	0.5	0.1	2.3	4.6	90.6
12	220.1	0.6	0.6	1.0	1.1	94.0
13	230.7	0.3	0.2	5.6	5.4	86.2
14	254.6	1.1	0.9	10.2	1.1	84.9
15	266.6	1.3	0.2	13.6	0.6	84.2
16	283.8	1.1	0.4	6.6	2.0	86.1
17	303.4	1.0	0.9	4.5	2.9	89.8
18	319.8	1.0	6.3	6.9	2.7	82.8
19	339.3	0.9	0.8	6.9	2.4	86.6
20	350.6	0.8	1.3	10.4	2.3	83.9
21	289.9	1.1	1.4	6.4	1.2	89.1
22	314.6	1.2	0.0	4.0	1.7	92.9
23	326.7	1.0	0.3	7.5	0.0	83.3
24	359.1	0.5	0.4	10.4	1.7	83.6
25	376.1	0.8	0.0	7.3	1.5	89.3
26	405.7	1.3	6.8	9.6	0.7	77.4
27	432.8	1.1	2.8	4.7	2.1	87.0
28	445.3	1.1	1.1	6.8	1.0	88.2
29	474.3	0.8	1.3	4.8	0.9	90.0
30	483.1	1.0	1.4	3.6	0.8	80.7
31	193.0	1.0	0.1	7.2	4.9	86.7
32	193.1	0.8	0.0	7.2	3.5	88.0
33	212.9	1.5	0.6	3.8	0.2	93.6
34	277.8	1.0	4.2	1.2	2.9	90.1
35	291.0	0.9	0.5	5.0	1.0	91.6
36	291.0	1.0	1.0	3.5	2.6	91.2
37	310.8	1.0	2.7	0.8	3.1	91.6

X1 = depth in meters, X2 = % phenocrysts, X3 = % lithic fragments, X4 = % granophyre, X5 = % amygdules, Y = % spherulites.

X2	−1.578	1.373	−1.15
X3	−0.8161	0.2544	−3.21
X4	−0.9935	0.1270	−7.82
X5	−1.0314	0.2781	−3.71

$s = 2.347$ $R^2 = 0.735$ $R = 0.86$

We note that the coefficients of all the predictors are negative. This is reasonable, at least for X2 − X6; when the percentages of the other components increase, the amount of spherulite decreases. All except the coefficient of X2 have t-values that exceed 2.0 in absolute value. Later, we make a second run, dropping X2 from the model.

The analysis of variance table shows how the total scatter of the observations has been divided.

Analysis of Variance

SOURCE	DF	SS	MS
Regression	5	472.491	94.498
Error	31	170.749	5.508
TOTAL	36	643.240	

The total sum of squares, $\Sigma (y_i - \bar{y})^2$, has $37 - 1 = 36$ d.f. The error sum of squares has $36 - 5 = 31$ d.f. The sum of squares for regression is obtained by subtraction. It has 5 d.f., equal to the number of predictors. The ratio $472.491/643.240 = 0.735 = R^2$; $R = 0.86$ is called the coefficient of multiple correlation.

The column headed MS contains the mean squares. They are the sums of squares divided by their degrees of freedom. The mean square for error is s^2. The F test for the hypothesis that

$$\beta_1 = \beta_2 = \beta_3 = \beta_4 = \beta_5 = 0$$

uses for the test statistic the ratio of the mean square for regression to the mean square for error,

$$F = \frac{94.498}{5.508} = 17.16.$$

This is compared to the tabulated value of F with p d.f. in the numerator and $n - p - 1$ d.f. in the denominator.

In this example, the value of F is clearly significant. However, that tells us nothing new because we have already established by the t-values that several of the coefficients are not zero.

15.9. THE SECOND RUN

Since the t-value for the coefficient of X_2 is not significant, we can conclude that X_2 is not needed as a predictor. More formally, we test and accept the hypothesis that $\beta_2 = 0$.

REGRESS c7 4 c2 c4–c6

The regression equation is

$Y = 100 - 0.0117\ X1 - 0.856\ X3 - 1.01\ X4 - 0.987\ X5$

Predictor	Coef.	St.dev.	t-ratio
Constant	100.120	2.041	49.04
X1	−0.011673	0.005385	−2.17
X3	−0.8556	0.2533	−3.38
X4	−1.0069	0.1271	−7.92
X5	−0.9866	0.2768	−3.56

$s = 2.359$ $R^2 = 0.723$ $R = 0.85$

Analysis of Variance

SOURCE	DF	SS	MS
Regression	4	465.22	116.30
Error	32	178.02	5.56
TOTAL	36	643.24	

All the t-values are now significant. There is a trivial increase in s and a trivial decrease in R^2. There are small changes in the remaining coefficients (but not all in the same direction) and in their t-values. Three points have large standardized residuals:

Unusual Observations

Obs.	X1	Y	Fit	St.dev.Fit	Residual	St.Resid.
10	154	81.000	85.959	1.137	−4.959	−2.40R
23	327	83.300	88.498	0.657	−5.198	−2.29R
30	483	80.700	88.869	1.120	−8.169	−3.94R

The residual for observation 30 is about 10% of the observed value. The geologist should check this observation if possible; there may have been an error in the analyis. The presence of a large residual should prompt an engineer to question an observation and to check it, or even to repeat it if possible. One should be wary of discarding an observation just because it has a large residual. If we regard the standardized residual as a t-value with 2.0 corresponding roughly to one chance in 20, we should not be surprised if about 5% of the residuals earn R ratings.

15.10. FITTING POLYNOMIALS

The data shown in table 15.10.1 come from a chemical engineering process for polymerization. The X variable is the reactor temperature; Y is the yield of tetramer. At lower temperatures, increasing temperature increases the yield; at higher temperatures, there is a tendency to overdo the polymerization and to produce pentamer instead of the desired product, tetramer. A plot of the data is shown in figure 15.10.1.

When the model

$$Y = \beta_0 + \beta_1 X$$

is fitted to the data, the regression line is

$$Y = -47.93 + 0.5094 X$$

with $s = 4.270$ and $R^2 = 0.862$. It is clear that fitting a straight line explains a large fraction of the scatter in the data and one might be tempted to stop there, but that could be a mistake.

To fit a quadratic model, we let $X_1 = X$, $X_2 = X^2$, and carry out the usual multiple-regression procedure with two predictor variables. The regression equation is

$$Y = -853.3 + 7.116X - 0.013482 X^2.$$

Predictor	Coef.	St.dev.	t-ratio
Constant	−853.3	127.4	−6.70
X	7.116	1.043	6.82
X^2	−0.013482	0.002128	−6.34

$s = 1.300$ $\quad R^2 = 0.99 \quad$ $R = 0.995$

The significant t-value for the quadratic term shows that it was a good idea to fit the quadratic and not to stop with the straight line.

What we have just done is theoretically correct, but there are problems. The first is the size of the numbers involved. Suppose that we wish to predict the yield at a temperature of 255 degrees. We substitute 255 for X in the fitted equation. However, $X^2 = 65{,}025$, and we are interested in changes in the yield of a few percent, which means that we have to carry several places of decimals in our estimate of the coefficient β_2. This becomes even more of

Table 15.10.1. Polymer Yields

Temperature (X)	220	230	240	250	260	270
Ave. yield (Y)	58.9	71.6	78.0	81.6	86.0	85.2

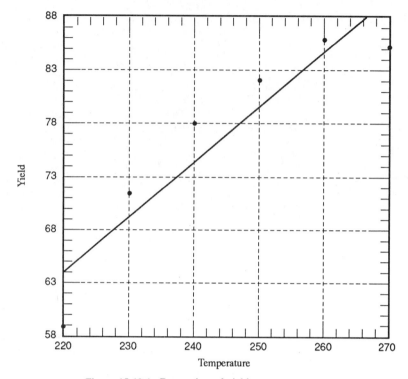

Figure 15.10.1. Regression of yield on temperature.

a problem when the X variable is time and is recorded as $X = 1982, 1983, 1984, \ldots$.

The more serious problem lies in the fact that when the values of X are positive integers, X and X^2, while "independent" in the sense that neither is a *linear* function of the other, are highly correlated. This causes the matrix $\mathbf{X'X}$ to be badly conditioned. Thirty years ago this would have been a problem because of the inability of the small "mainframe" computers to actually invert the matrix and solve the normal equations. Even though nowadays we can do the inversion on PCs, the estimates are highly correlated and, therefore, of dubious value.

The problem can be mitigated by changing the coordinates. Instead of using as the predictor variable $X = $ temp., we use $Z_1 = $ (temp. $-245)/5 = -5, -3, -1, +1, +3,$ and $+5$. In this transformation, 245 is the average of the X values; the denominator 5 was chosen to make the values of Z_1 small integers, which is merely a matter of convenience. Then Z_1 and Z_1^2 will have zero correlation, i.e., they will be orthogonal.

It is more convenient computationally to modify Z_1^2. The average value of Z_1^2 is 35/3. Subtracting this from Z_1^2 gives $40/3, -8/3, \ldots$, and so we elect to work with the more convenient numbers.

$$Z_2 = \frac{3Z_1^2 - 35}{8} = 5, -1, -4, -4, -1, 5;$$

Z_1 and Z_2 are a pair of first- and second-degree *orthogonal* polynomials. Using the orthogonal polynomials has two consequences; the calculations are easier, and we can assess the contributions of the linear and quadratic terms independently.

The fit to this quadratic model is

$$Y = 76.8833 + 2.5471 Z_1 - 0.8988 Z_2$$

with $s = 1.300$ and $R^2 = 0.99$ as before.

We can substitute back for X in terms of Z_1 and Z_2 to obtain the earlier equation. However, if we only want to predict the yields for particular temperatures, we can more easily work with Z_1 and Z_2. The predicted yield for 235 degrees is obtained by substituting $Z_1 = -2$ and $Z_2 = 5$ to obtain $\hat{y} = 74.4$.

15.11. A WARNING ABOUT EXTRAPOLATION

Table 15.11.1 shows the predicted values of the yields at three temperatures by the two models.

The two estimates differ by 2.6 at 235 degrees, but we note that the observed yield for 230 degrees was 71.6 and for 240 degrees 78.0; both estimates fall between those two observations. A similar remark can be made about the estimates at 245 degrees. However, at 280 degrees, the estimates are far apart; they differ by 12.6. As the temperature increases, the linear model keeps climbing at a steady rate of 0.51 per degree, and will continue to do so. The quadratic model, on the other hand, is beginning to lose steam; the estimates are starting to drop.

Here lies the trap of extrapolation. For some sets of data with relatively small variance, the linear and quadratic models give predictions of Y, within the ranges of the independent variables for which data were taken, that are rather close to one another; the engineer might not get into serious difficulty by using either model. However, the engineer who extrapolates is asking for trouble. It is then that the linear and curved models diverge faster and faster. Unfortunately, the simple first-degree model is a very appealing model. Sooner or later, somebody, at a safe distance from the plant, is going

Table 15.11.1. Predicted Yields

Temperature	235	245	280
Linear	71.8	76.9	94.7
Quadratic	74.4	80.8	82.1

to notice that for the reported data, yield increases at a rate of five points per ten degrees, and will surely draw a straight line and expect you to match that line outside the range!

15.12. TESTING FOR GOODNESS OF FIT

In the previous sections, we fitted our polynomials to the means of observations at six temperatures. There were actually four observations at each temperature, and we can use this data to illustrate a method of testing the adequacy of the fitted model.

The residual sum of squares is the sum of the squares of the deviations, $\Sigma (y_i - \hat{y}_i)^2$. If the model is correct, this deviation is due entirely to noise, both the noise in the observation y_i itself and the combination of noises in the observations that went into computing $\hat{y}_i$. If the model is incorrect, the deviation also contains a component because $E(y_i)$ is not the same as $E(\hat{y}_i)$. As we square and sum the deviations, these discrepancies from the inadequate model accumulate. The residual sum of squares thus consists of two components: noise and a lack of fit. In the usual situation, we cannot separate the two. However, there are two cases in which we can examine the residual sum of squares and obtain a test for "goodness of fit," or, more appropriately, lack of fit.

The first case appears in the next section in an example on octane blending. With octane measurements, experience with the testing procedure has given the engineer a good estimate of σ^2, both from previous work and from the work of others. The engineer can compare the mean square for residuals, s^2, to the prior value, σ^2. If s^2 is considerably larger than σ^2, that is evidence of a lack of fit in the model.

In the case at hand, we can use the variation between the duplicate observations to obtain what is called an estimate of "pure error," the true value of σ^2. The actual data are shown in table 15.12.1; some minor adjustments have been made to eliminate roundoff problems.

The differences between the four observations at 220 degrees have nothing to do with whether we chose the right regression model. They are due entirely to noise. Their sum of squares, $\Sigma (y_i - \bar{y}_i)^2$, is an estimate of

Table 15.12.1. Polymer Yields

Temperature	220	230	240	250	260	270
Yields	60.2	72.6	77.1	80.7	86.4	84.4
	59.2	70.3	77.5	80.7	85.5	86.6
	58.9	72.3	77.5	81.9	85.6	84.1
	57.3	71.2	79.9	83.1	86.5	85.7

$3\sigma^2$ with 3 d.f. Pooling this with the similar estimates from the other temperatures provides a sum of squares for "pure error" with 18 d.f.

When a straight line is fitted for yield against temperature using all 24 observations, we obtain the same fitted line as before when we used the means. The sum of squares for error is 313.2 with 22 d.f.; the sum of squares for "pure error" is 21.44 with 18 d.f. The difference, $313.2 - 21.4 = 291.8$, measures error and a lack of fit with $22 - 18 = 4$ d.f. We divide each sum of squares by its d.f. If there is no lack of fit, we should have two estimates of σ^2. We compare the ratio of these estimates to the F statistic. In the present example, we have

$$F = \frac{291.8/4}{21.44/18} = 61.2$$

to be compared to $F^*(4, 18) = 2.93$.

In the general case, if there are q d.f. for pure error, we denote the sums of squares for error and for pure error by S_e and S_{pe}, respectively; the test statistic is

$$F = \frac{(S_e - S_{pe})/(n - 1 - p - q)}{S_{pe}/q}$$

with $n - 1 - p - q$ and q d.f.

15.13. SINGULAR MATRICES

We have already mentioned that the matrix $X'X$ must be nonsingular (a singular matrix is one that does not have an inverse). This means that the columns of X must be linearly independent. The following example illustrates how such a singularity can occur in practice. Suppose we are mixing three component gasolines to make blends, and that the volume fractions of the components in any blend are denoted by x_1, x_2, and x_3; the sum $x_1 + x_2 + x_3$ is identically equal to unity. If we try to predict some response, such as the octane number, by the equation

$$y = \beta_0 + \beta_1 x_1 + \beta_2 x_2 + \beta_3 x_3 ,$$

the X matrix will have four columns, but its rank will be three. The sum of columns 1 through 3 is a column of unit elements, which is identical with column 0. In this case, we will not be able to find a unique solution to the normal equations. If $\beta = (\beta_0, \beta_1, \beta_2, \beta_3)'$ is a solution, so is the vector $(\beta_0 + 3t, \beta_1 - t, \beta_2 - t, \beta_3 - t)'$ for any value of t. There is clearly one variable too many. The solution to this difficulty is to drop one of the columns from the model.

15.14. OCTANE BLENDING

The following data are taken from an octane blending study reported by R. D. Snee (1981). The complete study involved six components. We confine ourselves to the blends involving only three components: alkylate (ALK), light straight run (LSR), and light cat-cracked (LCC). The response recorded is the research octane number of the blend in the presence of an additive. The data are given in table 15.14.1.

The usual approach, recommended by Scheffé (1958) is to drop the constant term from the model and fit the new model:

$$Y = \beta_1 X_1 + \beta_2 X_2 + \beta_3 X_3 .$$

Minitab achieves this by adding the subcommand NOCONSTANT. The data Y, X_1, X_2, and X_3 are read into columns 1, 2, 3, and 4, respectively. The variables are denoted by $Y = $ OCT, $X_1 = $ ALK, $X_2 = $ LSR, and $X_3 = $ LCC.

REGRESS c1 3 c2–c4;

NOCONSTANT.

The regression equation is

OCT = 106 ALK + 84.1 LSR + 101 LCC

Predictor	Coef.	St.dev.	t-ratio
Noconstant			
ALK	106.054	0.779	136.15
LSR	84.0741	0.6873	122.33
LCC	100.876	0.686	146.95

s = 0.9608

Table 15.14.1. Octane Blends

Y	X_1	X_2	X_3
106.6	1.00	0.00	0.00
83.3	0.00	1.00	0.00
99.7	0.00	0.00	1.00
94.1	0.50	0.50	0.00
103.1	0.50	0.00	0.50
93.6	0.00	0.50	0.50
97.4	0.00	0.25	0.75
88.8	0.00	0.75	0.25
97.4	0.33	0.33	0.34

THE QUADRATIC BLENDING MODEL

This formulation is the usual linear blending model. The individual coefficients are estimates of the octanes of the pure components. For the prediction of the octane of future blends, we can act as if the octane number of the light straight run is 84.1.

We could, instead, have dropped one of the gasoline columns and fitted, for example,

$$Y = \beta_0 + \beta_1 + \beta_3.$$

With this model, the fitted equation is

$$\text{OCT} = 100.9 + 5.2 \text{ ALK} - 16.8 \text{ LSR}.$$

This model amounts to using the light cat-cracked as a base fuel; its octane is estimated by the constant term. The other coefficients represent the differences between the component and LCC. Thus, the estimated octane of pure alkylate is $100.9 + 5.2 = 106.1$, and of light straight run $100.9 - 16.8 = 84.1$.

15.15. THE QUADRATIC BLENDING MODEL

The fit of the linear blending model to the data in table 15.14.1 had $s = 0.96$. The standard deviation of an octane determination is about three-tenths of a number. Clearly, there is strong evidence that the linear model is not adequate, and that there is some curvature, or synergism. The engineer should try a quadratic model. The complete quadratic model, under the restriction that $\Sigma x_i = 1.00$, is

$$Y = \beta_1 X_1 + \beta_2 X_2 + \beta_3 X_3 + \beta_{12} X_1 X_2 + \beta_{13} X_1 X_3 + \beta_{23} X_2 X_3. \tag{15.15.1}$$

A term such as βX_1^2 is redundant, because X_1^2 can be written as

$$X_1^2 = X_1(1 - X_2 - X_3) = X_1 - X_1 X_2 - X_1 X_3;$$

its inclusion would only lead to another singular matrix.

Let us continue with the octane blending example. The fitted quadratic equation is

$$Y = 107.0 \text{ ALK} + 83.2 \text{ LSR} + 99.8 \text{ LCC}$$
$$- 2.84 X_1 X_2 + 0.050 X_1 X_3 + 8.73 X_2 X_3$$

with $s = 0.1955$. This is clearly a more satisfactory fit.

EXERCISES

15.1. The data set used in exercise 14.4 contained a third column. This was a composite mileage, obtained by combining the city and highway mileages:

$$\text{comp.} = \beta_0 + \beta_1(\text{city}) + \beta_2(\text{highway})$$

M.P.G. CITY	M.P.G. HIGHWAY	M.P.G. COMP.
53	58	55
49	52	45
46	50	47
46	50	47
46	50	47
43	49	45
40	44	42
38	42	39
38	39	38
37	43	40

Use least squares to calculate the coefficients β_0, β_1, and β_2. In retrospect, the composite value for the second car is clearly wrong, probably a misprint; why? Drop that car and recalculate the values.

15.2. The plot of the data in exercise 14.6 shows that the appropriate model is a parabola rather than a straight line. Fit a parabolic model

$$y = \beta_0 + \beta_1 x + \beta_2 x^2$$

to the data.

15.3. The values of x and x^2 in exercise 15.2 are highly correlated. What is their correlation? 3700 is a good approximation to $\bar{x}$. Fit the parabolic model:

$$y = b_0 + b_1(x - 3700) + b_2(x - 3700)^2 .$$

Notice that the t-value for the quadratic term is the same as it was in exercise 15.2, and that the fitted curve and s^2 are, of course, the same. Why do the t-values for the first-degree term differ in exercises 15.2 and 15.3?

15.4. An engineer wishes to predict the mean annual runoff of rainwater. He has observations from thirteen regions for the following variables:

EXERCISES

Y: mean annual runoff in inches
X_1: mean annual preciptiation in inches
X_2: area of watershed (square miles)
X_3: average land slope (%)
X_4: stream frequency (square miles)$^{-1}$

No.	Y	X_1	X_2	X_3	X_4
1	17.38	44.37	2.21	50	1.36
2	14.62	44.09	2.53	7	2.37
3	15.48	41.25	5.63	19	2.31
4	14.72	45.50	1.55	6	3.87
5	18.37	46.09	5.15	16	3.30
6	17.01	49.12	2.14	26	1.87
7	18.20	44.03	5.34	7	0.94
8	18.95	48.71	7.47	11	1.20
9	13.94	44.43	2.10	5	4.76
10	18.64	47.72	3.89	18	3.08
11	17.25	48.38	0.67	21	2.99
12	17.48	49.00	0.85	23	3.53
13	13.16	47.03	1.72	5	2.33

Fit the linear model

$$y = \beta_0 = +\beta_1 X_1 + \beta_2 X_2 + \beta_3 X_3 + \beta_4 X_4.$$

Is it reasonable to regard point 7 as a special case? Assuming that it is, refit the model, omitting that point.

15.5. Coking coals were investigated by Perch and Bridgewater (*Iron and Steel Engineering*, vol. 57 [1980], pp. 47–50). They observed $X_1 =$ percent fixed carbon, X_2 percent ash, X_3 percent sulfur, and $Y =$ coking heat in BTU per pound. Fit the prediction model

$$y = \beta_0 + \beta_1 x_1 + \beta_2 x_2 + \beta_3 x_3.$$

The first coal seems to differ from the others, and to earn an X rating from Minitab. How does it differ? Omit the first coal from the data and rerun the regression. Do you get a better fit? In what way?

	X_1	X_2	X_3	Y		X_1	X_2	X_3	Y
1.	83.8	11.2	0.61	625	14.	55.6	6.6	0.90	715
2.	78.9	5.1	0.60	680	15.	54.7	6.5	1.54	705
3.	76.1	5.3	1.65	680	16.	53.4	4.5	1.10	730

4.	72.2	8.1	1.06	710	17.	60.4	9.9	1.08	725
5.	73.2	7.0	1.02	710	18.	60.8	8.1	1.41	710
6.	73.8	6.0	0.75	685	19.	61.9	6.8	1.03	710
7.	70.6	8.6	0.74	705	20.	61.8	6.8	0.99	700
8.	68.4	9.8	1.09	685	21.	61.9	6.6	0.90	715
9.	70.5	6.7	0.76	680	22.	61.1	6.4	0.91	710
10.	63.2	11.8	1.85	700	23.	59.0	7.6	1.36	740
11.	55.8	10.4	0.71	720	24.	59.3	7.0	1.31	730
12.	56.3	8.8	1.70	705	25.	56.6	7.6	1.07	730
13.	57.1	5.3	0.93	730					

15.6. These are measurements of properties of twelve samples of asphalt. The measurements are

Y: penetration at 115°F
X_1: softening point (°F)
X_2: penetration at 32°F
X_3: penetration at 77°F
X_4: percentage of insoluble n-pentane
X_5: flash point (°F)

Y	X_1	X_2	X_3	X_4	X_5
77	158	21	35	28.0	635
149	136	18	38	25.8	590
111	153	20	34	26.4	550
49	194	16	26	31.1	625
57	172	14	27	33.6	595
76	154	12	26	27.0	625
106	150	17	38	27.9	625
119	142	10	26	26.3	590
93	156	18	32	33.9	550
153	145	18	41	30.2	550
112	147	14	29	30.3	545
100	145	12	32	27.8	595

We want to predict the penetration at 115°F from the other variables. Fit Y to all five predictors. Then eliminate the predictor that has the smallest *t*-value and fit again. Continue this process, called backwards elimination, until all the predictors left in the model have significant *t*-values. What is your final prediction equation?

15.7. (A continuation of exercise 15.6.) Find the correlations of each of the predictors with Y. You will notice that X_1 is the predictor with the highest correlation. Yet the best subset of four predictors does not

EXERCISES

contain X_1. Find R^2 and s^2 when Y is fitted to X_1 alone and compare them to the corresponding values for the best subset of four. In the previous exercise, you might have decided that X_1 was of no use as a predictor. This peculiar behavior does sometimes happen.

15.8. (A continuation of exercise 15.7.) When you fitted Y to X_2, X_3, X_4, and X_5, point 3 had a large residual: $y = 111.0$ and $\hat{y} = 124.76$. Drop that point from the data set and repeat the backwards elimination procedure used in exercise 15.6.

15.9. The method used in section 15.2 and the method of least squares give the same fitted equation for the data in table 15.2.1. Prove that they give the same equation for any linear model with two predictor variables.

15.10. Prove that the least squares estimates are unbiased (section 15.5).

CHAPTER SIXTEEN

The Analysis of Variance

16.1. INTRODUCTION

In chapter eight, we considered experiments to compare two populations. The basic idea of the chapter was that the engineer took two samples, one from each population, and compared the sample means by the two-sample t-test. That procedure included obtaining an estimate of the variance from the experimental data. In this chapter, we extend those ideas to the situation in which we want to compare more than two populations in the same experiment. This is done by the analysis of variance. A fuller treatment of this subject can be found in John (1971).

16.2. THE ANALYSIS OF VARIANCE

The analysis of variance was introduced by R. A. Fisher about 60 years ago. The essence of the procedure is that we take the sum of squares of the deviations of all the observations in an experiment about the overall, or grand, mean; this is called the total sum of squares. We then subdivide the total sum of squares into components, which correspond to different sources of variation, and test certain hypotheses using F-tests and t-tests. The reader will notice the very strong connections between the analysis of variance and multiple regression through the underlying principle of least squares.

The analysis of variance was originally used in agronomic experiments, and so the jargon of agronomy continues in the field. An agronomist plants crops in a field divided into small plots. In those *plots*, different *varieties* of some crop are sown, usually one variety to a plot. At harvest time, the yields of each plot are recorded; these are the data. Alternatively, the same variety might be planted in each plot, and then different *treatments* applied, such as different fertilizers or different methods of cultivation. It is convenient to continue to use the word treatments in this chapter as a general expression. The two different catalysts compared in chapter eight can be called two treatments; the six temperatures in the example in chapter fifteen can be regarded as six treatments. The agricultural word "plot" is often replaced by the word *run*, usually denoting a "run" made in a chemical or other plant under a certain set of operating conditions. It is now common to replace the name analysis of variance by the acronym ANOVA.

16.3. THE ONE-WAY LAYOUT

One-way layout is jargon for the simplest experiment in which we compare t different treatments. Later in the chapter, more complicated experiments—two-way layouts and three-way layouts—are introduced. Suppose that the engineer has samples of observations from each of t treatments and that the sample from the ith treatment contains n_i observations with a total of $N = \Sigma n_i$ observations. The jth observation on the ith treatment is denoted by y_{ij}. We can call y_{ij} the response observed on the jth run with the ith treatment. Table 16.3.1 shows a set of data for an experiment with six observations on each of five treatments.

The mathematical model for the data is

$$y_{ij} = \mu_i + e_{ij}. \qquad (16.3.1)$$

The parameters μ_i are the (unknown) treatment means that we wish to compare; the e_{ij} terms are the random noise with the usual assumption that they are statistically independent normal variables with the same (also unknown) variance σ^2.

Let $\bar{y}_i$ denote the average of the observations on the ith treatment and $\bar{y}$ denote the average of all N observations. The n_i observations on the ith treatment provide a sum of squares of deviations, $\Sigma (y_{ij} - \bar{y}_i)^2$, with expected value $(n_i - 1)\sigma^2$ and $n_i - 1$ d.f. Summing these terms, we have

$$S_e = \sum_i \sum_j (y_{ij} - \bar{y}_i)^2. \qquad (16.3.2)$$

Table 16.3.1

	A	B	C	D	E
	45.1	42.3	44.9	42.8	43.5
	44.8	41.8	42.1	44.0	42.6
	46.2	42.3	43.9	43.3	42.5
	45.9	41.9	43.8	43.0	43.3
	46.1	41.1	45.7	42.1	43.3
	45.5	42.4	45.1	41.1	42.6
Averages	45.60	41.97	44.25	42.72	42.97

We now write $S_y = \Sigma (y_i - \bar{y})^2$, and

$$S_y = S_t + S_e. \qquad (16.3.3)$$

The term S_t is called the sum of squares for treatments. It can be shown, with a little algebra, that

$$S_t = \Sigma n_i (\bar{y}_i - \bar{y})^2, \qquad (16.3.4)$$

and that its expectation is

$$E(S_t) = \Sigma n_i (\mu_i - \bar{\mu})^2 + (t-1)\sigma^2, \qquad (16.3.5)$$

where $\bar{\mu}$ is the average of the μ_i. The sum of squares, S_t, has $t-1$ d.f.

These sums of squares are combined in the analysis of variance table 16.3.2.

The mean square for treatments is $M_t = S_t/(t-1) = 49.33/4 = 12.333$; $s^2 = S_e/(N-t) = 17.05/25 = 0.682$. Their ratio is the F statistic: $F = 12.333/0.682 = 18.08$. The F statistic tests the hypothesis that all the treatment means, μ_i, are equal (in which case, the expected value of the numerator of F is σ^2). We compare the value 18.08 to the tabulated value of F for $(4, 25)$ d.f., which is 2.76 at $\alpha = 0.05$, and reject the null hypothesis.

The formulas that were given before are not the formulas that are used in practice. It is easier to proceed as follows. Suppose that there are n_i observations on the ith treatment and that their total is T_i. Let G be the grand total of the N observations; $G = \Sigma T_i = N\bar{y}$. We compute $C = G^2/N$, which is called the correction for the mean. Then we compute the sums of squares:

The total sum of squares $= S_y = \Sigma_i \Sigma_j y_{ij}^2 - C$;
SS between treatments $= S_t = \Sigma (T_i^2/n_i) - C$;
SS for error $= S_e = S_y - S_t$ with $n - t$ d.f.

The F statistic,

$$\frac{M_t}{s^2} = \frac{S_t/(t-1)}{s^2},$$

has $(t-1, N-t)$ d.f.

Table 16.3.2. ANOVA Table

SOURCE	DF	SS	MS	F
treat.	4	49.330	12.333	18.08
error	25	17.050	0.682	
TOTAL	29	66.380		

Table 16.4.1. Sample Means

B	D	E	C	A
41.97	42.72	42.97	44.25	45.60

16.4. WHICH TREATMENTS DIFFER?

We have just reached the conclusion that it is not true that all the population means, μ_i, are equal. This does not imply that they all differ from each other. Table 16.4.1 lists the treatment averages, $\bar{y}_i$, in increasing order.

We would reject the hypothesis $\mu_h = \mu_i$ if

$$t = \frac{|(\bar{y}_h - \bar{y}_i)|}{s\sqrt{1/n_h + 1/n_i}} > t^*,$$

where t^* is the tabulated value of t with probability $\alpha/2$ in each tail and $N - t$ d.f. Equivalently, we declare that $\mu_h \neq \mu_i$ if, and only if,

$$|\bar{y}_h - \bar{y}_i| > t^*(s\sqrt{1/n_h + 1/n_i}). \qquad (16.4.1)$$

In the particular case, such as the example of the previous section, where the sample sizes are all equal to n, equation (16.4.1) reduces to

$$|\bar{y}_h - \bar{y}_i| > t^*\sqrt{2s^2/n}. \qquad (16.4.2)$$

The term on the right side of equation (16.4.2) applies to all pairs of treatments. It is called Fisher's least significant difference (LSD).

In the example, we have for $\alpha = 5\%$ with 25 d.f., $t^* = 2.06$. With $s^2 = 0.682$ and $n = 6$, the LSD is 0.98. Applying this LSD to the averages in table 16.4.1, we conclude that

$$A > B, C, D, E; \quad C > B, D, E; \quad E > B;$$

we cannot conclude that $\mu_4 \neq \mu_5$ or that $\mu_4 \neq \mu_2$.

Some statisticians argue that Fisher's LSD is too liberal and that it declares too many differences to be significant. Other formulas for significant differences have been derived by Tukey, Newman and Keuls, and Duncan. Some software packages compute these diferences automatically or on request. Details are given in the manuals.

16.5. THE ANALYSIS OF COVARIANCE

We can sometimes sharpen our ability to distinguish between treatments by taking into account and adjusting for covariates, or, as they are often called,

Table 16.5.1. Yields and Impurities

A		B		C	
Y	X	Y	X	Y	X
54.4	2.29	54.6	2.19	51.6	3.43
51.1	2.73	54.4	2.45	54.1	2.41
53.7	2.17	50.2	3.42	52.9	2.72
51.1	3.23	49.7	3.31	52.3	3.48
47.3	3.79	52.5	2.78	53.0	3.17
50.1	3.40	53.7	2.57	55.1	2.48
51.6	2.99	51.9	3.38	51.6	3.16
51.9	2.76	50.3	3.22	56.9	2.16

concomitant variables. A common situation occurs when each run is made on a separate batch of raw material and we may want to "adjust" Y for the amount of some impurity in the batch. In animal nutrition experiments where some young animals are fed ration A for a few months and others are fed ration B, and the weight gains are compared, one might take the initial weight of each animal as a covariate.

The adjustment for the covariate is linear. We add a term to our earlier model in equation (16.3.1) and write

$$y_{ij} = \mu_i + \beta x_{ij} + e_{ij}, \qquad (16.5.1)$$

where x_{ij} is the value of the covariate for the ijth observation.

This could be called a first-order adjustment. The model implies two things. The plot of Y versus X is a straight line (except for random noise) for each treatment and, furthermore, the slope of the line is the same (β) for every treatment. We illustrate the analysis of covariance with the following example.

Example 16.5.1. Three reagents for a chemical process were compared in a pilot plant. Eight runs were made with each reagent (treatment). The raw material, unfortunately, contained an impurity. The data are shown in table 16.5.1; Y denotes the yield and X the percentage of the impurity in the raw material for each run.

Figure 16.5.1 is a plot of Y versus X. It is clear that the yield decreases as the amount of impurity increases, and that, by and large, the yields with treatment A are lower than the yields with C. Nevertheless, the one–way ANOVA in table 16.5.2, which ignores the levels of the impurity, does not

Table 16.5.2. One-Way ANOVA Ignoring Covariate

SOURCE	DF	SS	MS	F
Treatments	2	16.96	8.48	2.12
Error	21	83.86	3.99	
TOTAL	23	100.81		

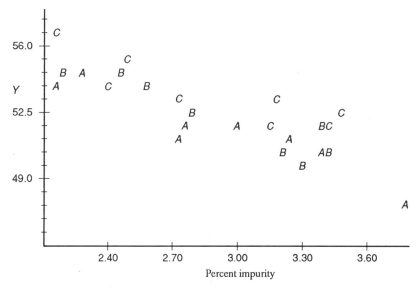

Figure 16.5.1. Plot of yield vs. impurity.

give a significant value of F, and we cannot conclude on the basis of that analysis that the treatments differ. The two smallest yields for C are 51.6, a value that is exceeded by three runs with A and by five runs with B, but those values of Y occur with relatively high values of X, namely, 3.43 and 3.16.

If, however, we adjust for the amount of impurity on each run by the analysis of covariance, we can establish that there are differences between treatments. Table 16.5.3 shows the printout from an analysis of covariance program. In the next section, we show how the figures in table 16.5.3 are obtained.

The treatment averages were

$$A = 51.400, \quad B = 52.162, \quad C = 53.437.$$

After adjusting for the covariate, $S_t = 14.32$, which is somewhat lower than before, and $s^2 = 0.6512$, which is much lower. The F ratio increased to $(14.32/2)/0.6512 = 10.99$, which is significant, and we are able to conclude that the three catalysts are not all equal.

Table 16.5.3. Analysis of Covariance

SOURCE	DF	ADJ SS	MS	F
Covariate	1	70.83	70.8331	108.77
Treatments	2	14.32	7.1589	10.99
Error	20	13.02	0.6512	
TOTAL	23	100.81	4.3832	

16.6. THE ANALYSIS OF COVARIANCE AND REGRESSION

The rationale for the analysis of covariance table 16.5.3 can be seen if we approach the problem from the point of view of regression. Let X_1 denote the percentage of the impurity present in a run. We first fit a single straight line through all 24 observations with the model

$$y = b_0 + b_1 x_1 . \qquad (16.6.1)$$

The sum of squares for error about this line is 27.342.

We next introduce two indicator variables: $X_2 = 1$ if the observation is made with catalyst B and zero otherwise; $X_3 = 1$ if the observation is made with catalyst C and zero otherwise. The following model fits three parallel lines to the data, one for each treatment.

$$y = b_0 + b_1 x_1 + b_2 x_2 + b_3 x_3 + e . \qquad (16.6.2)$$

The common slope of the three lines is $b_1 = \beta$; $b_0 = \mu_1$, the intercept for the line for the first treatment; $b_2 = \mu_2 - \mu_1$; and $b_3 = \mu_3 - \mu_1$.

With this model, equation (16.6.2), S_e drops to 13.02; see the accompanying table 16.6.1. The reduction in S_e achieved by considering three separate parallel lines rather than just one line is $27.34 - 13.02 = 14.32$. This is the (adjusted) sum of squares for treatments in table 16.5.3.

If, on the other hand, we refer back to the one-way analysis of variance, table 16.5.2, the sum of squares for error when we ignored the covariate was 83.86. The reduction in S_e that occurs when we include the covariate is

Table 16.6.1. Regression

The equation is $Y = 61.92 - 3.60 \, X1 + 0.744 \, X2 + 1.88 \, X3$.
$X1 = $ Impurity, $X2 = 1$ for B, $X3 = 1$ for C.

Predictor	Coef.	St.dev.	t-ratio
Constant	61.919	1.048	59.07
X1	−3.6023	0.3454	−10.43
X2	0.7445	0.4035	1.85
X3	1.8799	0.4038	4.66

$s = 0.8070 \qquad R^2 = 0.871$

Analysis of Variance,

SOURCE	DF	SS	MS	F	p
Regression	3	87.789	29.263	44.94	0.000
Error	20	13.024	0.651		
TOTAL	23	100.813			

83.86 − 13.02 = 70.84. This quantity (allowing for some roundoff error) appears in table 16.5.3 as the (adjusted) sum of squares for the covariate.

We notice that the t-values for the two indicator variables are 1.85 for B and 4.66 for C. Since $1.85 < 2.00$, we cannot conclude that $b_1 \neq 0$, or that $\mu_2 - \mu_1 \neq 0$. We conclude that catalysts A and B do not differ significantly, but that C gives significantly higher yields than A. We can also establish by similar calculations that C gives significantly higher yields than B.

16.7. QUANTITATIVE FACTORS

In section 15.10, we introduced an example in which a quadratic polynomial was fitted to the yields of a polymer plant. We look at it again as an example of one-way analysis of variance with a factor, temperature, that has six levels. The term quantitative is used for a factor whose levels are expressed as quantities, as opposed to a qualitative factor whose levels are not numerical, e.g., make of car with levels Ford, Plymouth, and Chevrolet.

The data were shown in table 15.12.1, which is repeated as table 16.7.1 for convenience. The analysis of variance table for all 24 observations is shown in table 16.7.2.

Out of the sum of squares between temperatures, we can take components with single degrees of freedom for the linear and quadratic trends. We recall, from section 15.12, that the values taken by Z_1, the linear polynomial, were $-5, -3, -1, 1, 3$, and 5, with $\Sigma z_1^2 = 70$.

The coefficient in the regression equation was $\Sigma (z_{1i} \bar{y}_i)/70 = 2.5471$. The corresponding sum of squares is

$$S_L = \frac{4(\text{linear contrast})^2}{\Sigma z_1^2} = 1816.62 \ ;$$

the factor 4 in the numerator is the number of observations that were made at each level of temperature.

A similar argument defines the quadratic contrast as $\Sigma (z_{1i} \bar{y}) = -75.5$ and the corresponding sum of squares as

$$S_Q = \frac{4(\text{quadratic contrast})^2}{\Sigma z_2^2} = \frac{22{,}801.00}{84} = 271.44 \ .$$

Table 16.7.1. Polymer Yields

Temperature	220	230	240	250	260	270
Yields	60.2	72.6	77.1	80.7	86.4	84.4
	59.2	70.3	77.5	80.7	85.5	86.6
	58.9	72.3	77.5	81.9	85.6	84.1
	57.3	71.2	79.9	83.1	86.5	85.7

Table 16.7.2. Analysis of Variance

SOURCE	DF	SS	MS	F
Temperature	5	2108.35	421.67	354.01
Residual	18	21.44	1.19	
TOTAL	23	2129.79		

We can now amend the analysis of variance table in table 16.7.3.

The F test statistic, $F(3, 18) = 5.68/1.19 = 5.68$, indicates that the model with only the quadratic trend is not entirely adequate and that there is still some unexplained curvature. The reader will notice that this is the same as the pure error test for goodness of fit that was used in the previous chapter to test the adequacy of the simpler first-degree model.

Single degrees of freedom for orthogonal contrasts can also be isolated with qualitative factors. Two contrasts, $A = \Sigma a_i \bar{y}_i$ and $B = \Sigma b_i \bar{y}_i$, are orthogonal if $\Sigma a_i b_i = 0$. Suppose that we have as the levels of a qualitative factor six makes of car: Cadillac, Chevrolet, Plymouth, Mercedes, Mazda, and Toyota, with n observations on each make.

The contrast

$$A = \bar{y}_1 + \bar{y}_2 + \bar{y}_3 - \bar{y}_4 - \bar{y}_5 - \bar{y}_6$$

compares the American cars as a group to the foreign cars. Its sum of squares is

$$S_A = \frac{nA^2}{6} = \frac{nA^2}{\Sigma a_i^2}.$$

The orthogonal contrast

$$B = 2\bar{y}_1 - \bar{y}_2 - \bar{y}_3 + 2\bar{y}_4 - \bar{y}_5 - \bar{y}_6$$

compares the luxury cars to the ordinary cars. Its sum of squares is

$$S_B = \frac{nB^2}{12} = \frac{nB^2}{\Sigma b_i^2}.$$

Table 16.7.3. Analysis of Variance

SOURCE	DF	SS	MS	F
Lin. Temp.	1	1816.62	1816.62	
Quad. Temp.	1	271.44	271.44	
Cubic etc.	3	20.29	6.76	5.68
Residual	18	21.44	1.19	
TOTAL	23	2129.79		

THE TWO-WAY LAYOUT

Indeed, B can be divided into two contrasts:

$$2\bar{y}_1 - \bar{y}_2 - \bar{y}_3 \quad \text{(luxury vs. ordinary for American cars)}$$

and

$$2\bar{y}_4 - \bar{y}_5 - \bar{y}_6 \quad \text{(luxury vs. ordinary for imported cars)}.$$

The two new contrasts are orthogonal, both to one another and to the first contrast, A.

16.8. THE TWO-WAY LAYOUT

Suppose that a chemical engineer wants to investigate two factors, temperature and pressure, at four levels each, and makes one run with each combination of temperature and pressure—16 runs in all. The engineer can arrange the data in a rectangular array, with the columns corresponding to the four levels of temperature and the rows to the levels of pressure. This is called a two-way layout. The fact that in this example the two factors each have the same number of levels is irrelevant. We could just as well have five temperatures and six pressures. We can analyze the data as if there were two one-way layouts—one for rows and one for columns. The analysis is illustrated in the following example.

Example 16.8.1. The data set for an experiment like that which has just been described is shown in table 16.8.1.

If we were to make a one-way analysis for the temperatures, ignoring the fact that the observations were made at different pressures, we would get $S_t = 8.52$ with 3 d.f. and $S_e = 20.06$ with 12 d.f. The corresponding F statistic is

$$F(3, 12) = \frac{2.84}{1.67} = 1.70,$$

which is not significant; $F^*(3, 12) = 3.49$. Similarly, a one-way analysis for

Table 16.8.1. A Two-Way Layout

Pressure	Temperature				Sum
	1	2	3	4	
1	8.95	9.27	12.08	9.76	40.06
2	10.54	11.44	11.17	9.96	43.11
3	12.59	12.89	13.03	11.48	49.99
4	9.47	11.28	11.32	8.90	40.97
Sum	41.55	44.88	47.60	40.10	174.13

Table 16.8.2. Two-Way ANOVA

SOURCE	DF	SS	MS	F
temp.	3	8.518	2.839	5.180
press.	3	15.126	5.042	9.200
error	9	4.936	0.548	
TOTAL	15	28.580		

the pressures, ignoring temperatures, also does not give a significant F value.

However, we can take out of the total sum of squares *both* the sum of squares for temperatures *and* the sum of squares for pressures and obtain the following ANOVA table, table 16.8.2.

Notice that the value of s^2 has dropped to 0.548 from its earlier value of 1.67 when we looked at temperature alone. The F statistics are compared to the tabulated value $F^*(3, 9) = 3.86$, and we now conclude that there are differences both between temperatures and between pressures. We can go ahead and apply the methods of section 16.4 to see which temperatures differ, or we can, if appropriate, apply the methods of section 16.7 and take out single d.f. contrasts for linear temperatures, quadratic temperatures, and cubic temperature, and similar contrasts for the pressures.

In the general case, where there are two factors, A and B, with a and b levels, respectively, there will be $a - 1$ d.f. for the sum of squares for A, $b - 1$ d.f. for B, and $(a - 1)(b - 1)$ d.f. for the sum of squares for error.

To compute the sums of squares for A and for B, we let $T_{i\cdot}$ denote the sum of the observations at the ith level of A and $T_{\cdot j}$ denote the sum of the observations at the jth level of B. Then the sums of squares for A and B are

$$S_a = \sum \left(\frac{T_{i\cdot}^2}{b} \right) - C \quad \text{and} \quad S_b = \sum \left(\frac{T_{\cdot j}^2}{a} \right) - C.$$

The sum of squares for error is obtained by subtraction as

$$S_e = S_y - S_a - S_b.$$

□

16.9. ORTHOGONALITY

The reason that we were able to take out the sums of squares for each of the two factors in the prior example is that the experiment was balanced. Each temperature appeared exactly once with each pressure, and so when two temperatures are compared, the pressures "cancel out," and vice versa. This balance has the technical name of orthogonality; the two factors are said to be orthogonal to one another. Orthogonality is a very important property in

Table 16.9.1. Residuals

−0.570	−1.082	1.048	0.603
0.257	0.325	−0.625	0.040
0.587	0.055	−0.485	−0.160
−0.277	0.701	0.061	−0.484

an experimental design, and we will see more of it in the next three chapters.

We can use the following model for the observation in the ith row and the jth column:

$$y_{ij} = m + r_i + c_j + e_{ij}, \qquad (16.9.1)$$

where m represents an overall mean, r_i is the extra contribution for being in the ith row, and c_j is the extra contribution for being in the jth column. The sum of the r_i and the sum of the c_j are both identically zero.

In our example, m is estimated by the overall, or grand, average, $174.13/16 = 10.883$. For $\hat{r}_1$, we subtract the grand average from the average of the observations in the first row:

$$\hat{r}_1 = -0.868, \qquad \hat{r}_2 = -0.105, \qquad \hat{r}_3 = +1.615, \qquad \hat{r}_4 = -0.641;$$

similarly,

$$\hat{c}_1 = -0.495, \qquad \hat{c}_2 = +0.337, \qquad \hat{c}_3 = +1.017, \qquad \hat{c}_4 = -0.858.$$

We can go further and calculate an estimated value for the observation in row i and column j as

$$\hat{y}_{ij} = \hat{m} + \hat{r}_i + \hat{c}_j,$$

so that $\hat{y}_{11} = 10.883 - 0.868 - 0.495 = 9.520$, and the residual in that cell is $8.95 - 9.520 = -0.570$. The complete set of residuals is shown in table 16.9.1. The sums of the residuals in each row and in each column are identically zero.

The sum of the squared residuals is indeed 4.936, as was shown in table 16.8.2. We can now see why the sum of squares for error has nine d.f. In each of the first three columns, the first three entries are random, but the fourth is fixed because the column total must be zero. Similarly, given the first three entries in each row, the fourth is fixed. Thus, there is a total of nine cells that can be filled freely; the entries in the other seven cells are forced by the conditions that the rows and the columns sum to zero.

16.10. RANDOMIZED COMPLETE BLOCKS

In our discussion in chapter eight of the two-sample t-test, we noted that the precision of the experiment can be improved by pairing. Instead of taking $2n$ units of raw material and assigning n to each treatment, we should, if possible, take n units, split each unit in half, and run one-half with one treatment and the second half with the other. A semiconductor manufacturer could take a boat that holds 32 wafers in an oven and load it with 16 wafers that have received one treatment and 16 that have received another; he would call it a split lot.

An agronomist who is carrying out an experiment to compare v varieties of barley would divide a field into plots. Then the agronomist would take n plots and sow variety A in them, n for variety B, and so on. One experimental design would be to choose the n plots for A *at random* from the total number, vn, of plots available; similarly for each of the other varieties. But often there is considerable variation in fertility over the whole field, and so the field might be divided into strips of v plots each that are fairly homogeneous and each variety planted in one plot chosen at random from each strip. The technical term used for the strips is *blocks*, and the experiment is called a randomized complete block experiment.

The name carries over to engineering applications. Our engineer, who wants to compare t treatments, might take n batches of raw material, divide each batch into t equal portions, and assign one of the portions to each treatment. The portions of raw material play the role of the agronomist's plots, and the batches are the blocks. In the analysis, we treat the blocks as if they were the levels of a second factor. We can return to our example and replace the word pressure by the word block and go through exactly the same calculations. Again, it is important that each treatment appear the same number of times in each block in order to achieve orthogonality—to make treatments orthogonal to blocks.

16.11. INTERACTION

The model, equation (16.9.1), that we have used so far is a simple additive model; we add to the mean first a row term, then a column term, and,

Table 16.11.1. Filling Weights of Cans

68.0	65.0	75.0	57.0	32.0	70.0
56.0	52.0	55.0	48.0	65.0	47.0
40.0	51.0	52.0	36.0	49.0	45.0
84.0	87.0	88.0	73.0	34.0	70.0
50.0	52.0	52.0	50.0	45.0	61.0

INTERACTION

Table 16.11.2. ANOVA Table

SOURCE	DF	SS	MS	F
times	4	2601.13	650.28	5.00
heads	5	1225.37	245.07	1.88
error	20	2602.47	130.12	
TOTAL	29	6428.97		

finally, an error term. It is also rather restrictive because it implies that the amount added to each observation because it is in the first column is the same in every row. This implies that the improvement in yield by increasing the temperature from level 1 to level 2 is the same, no matter what the pressure. The failure of the data to follow this simple, but rigid, pattern is called interaction, or nonadditivity.

When there are several observations in each cell of our array, we have a test for the existence of interaction, which is discussed in the next section. When there is only one observation in each cell, we have no such test. The sum of squares for error measures both interaction and error, and is correspondingly inflated. We can sometimes, however, spot interaction by examining the table of residuals. The residuals in the table, like the residuals in a regression, should not show any pattern.

Example 16.11.1. E. R. Ott and R. D. Snee (1973) reported an investigation of the heads of a multihead machine for filling cans of a soft drink. They took samples of cans filled by six heads of the machine every hour for five hours. The data shown in table 16.11.1 are the coded weights of the amount of drink that the head put into the can.

The analysis of variance table is table 16.11.2. There is no significant difference between heads. But when we look at the residuals, table 16.11.3, we see that head 5 exhibits some very erratic behavior, because its column contains the four largest residuals. It can be argued, with the benefit of hindsight, that we could have spotted this peculiar behavior by looking carefully at table 16.11.1. Perhaps so, but it is easier to spot this kind of anomaly from the table of residuals.

Table 16.11.3. Residuals

4.20	−0.60	6.40	0.00	−17.20	7.20
−0.47	−6.27	−6.27	−1.67	23.13	−8.47
−8.13	1.07	−0.93	−5.33	15.47	−2.13
8.70	9.90	7.90	4.50	−26.70	−4.30
−4.30	−4.10	−7.10	2.50	5.30	7.70

□

Table 16.12.1. Data

	I	II	III	IV	Row sum
1.	83	86	85	90	679
	81	84	82	88	
2.	77	80	80	87	656
	78	82	83	89	
3.	80	81	88	86	657
	78	78	85	81	
Col. totals	477	491	503	521	1992

16.12. INTERACTION WITH SEVERAL OBSERVATIONS IN EACH CELL

When there are several observations per cell, we can split the residual sum of squares into two components, one due to interaction and the other to error. This is done in the same way as in regression, where we divided the residual sum of squares into two portions: pure error and goodness of fit.

Example 16.12.1. Consider an experiment with two factors, A (rows) with three levels and B (columns) with four levels, and two observations in each cell. The data are shown in table 16.12.1. The analysis of variance table is table 16.12.2.

To compute the sums of squares for A and for B, we let T_{ij} denote the sum of the observations at the ith level of A and the jth level of B, i.e., in the ijth cell. Then the sums of squares are the following:

$$S_a = \sum \left(\frac{T_{i.}^2}{br}\right) - C; \qquad S_b = \sum \left(\frac{T_{.j}^2}{ar}\right) - C;$$

$$S_{ab} = \sum_i \sum_j \left(\frac{T_{ij}^2}{r}\right) - C - S_a - S_b.$$

The sum of squares for error is then obtained by subtraction as

$$S_e = S_y - S_a - S_b - S_{ab}.$$

Table 16.12.2. Analysis of Variance

SOURCE	df	SS	MS	F
A	2	42.25	21.13	
B	3	174.00	58.00	
A × B	6	70.75	11.79	3.29
Error	12	43.00	3.58	
TOTAL	23	330.00		

Table 16.12.3. Cell Means

82	85	83.5	89
77.5	81	81.5	88
79	79.5	86.5	83.5

The value of the F ratio, 3.29, is significant, $F^*(6, 12) = 3.00$, and there is clear evidence of interaction. To see this more clearly, we look at the cell means, table 16.12.3. We note that in every column save the third, the mean in the first row is at least three points higher than the mean in the last row. In the first two rows, the last cell is appreciably higher than the others. One suspects that the observations 86.5 (row 3, column 3) and, perhaps, 83.5 (row 3, column 4) should be looked at carefully.

In the general case of two factors, A and B, with a and b levels, and r observations in each cell, there will be $abr - 1$ d.f. for the total sum of squares. This will be divided by the ANOVA calculations into $a - 1$ d.f. for A, $b - 1$ for B, as before, together with $(a - 1)(b - 1)$ d.f. for the AB interaction and the remaining $ab(r - 1)$ d.f. for the sum of squares for error. □

16.13. THREE FACTORS

With three factors, A, B, and C at a, b, and c levels each, the analysis proceeds in the same way as for two factors except for some extra terms in the table. The observation at the ith level of A, the jth level of B, and the kth level of C is denoted by y_{ijk}. The sums of squares for A, B, and C are computed as before, and have $a - 1$, $b - 1$, and $c - 1$ d.f., respectively; we have a third subscript on the totals, which are now denoted by $T_{i..}$, $T_{.j.}$ and $T_{..k}$. Then

$$S_a = \frac{\sum T_{i..}^2}{bc} - C, \text{ and so on}.$$

Similarly, the interaction sums of squares are given by

$$S_{ab} = \frac{\sum T_{ij.}^2}{c} - C - S_a - S_b,$$

$$S_{ac} = \frac{\sum T_{i.k}^2}{b} - C - S_a - S_c,$$

$$S_{bc} = \frac{\sum T_{.jk}^2}{a} - C - S_b - S_c.$$

If there are m observations per cell, with $m > 1$, we can also take out a component for the three-factor interaction, S_{abc}, which will have $(a-1)(b-1)(c-1)$ d.f. The residual sum of squares is then

$$S_e = \sum \sum \sum (y_{ijkm} - \bar{y}_{ijk.})^2$$

with $abc(m-1)$ d.f. In the formulas for the sums of squares, terms such as $T_{i..}^2/br$ become $T_{i...}^2/brm$.

After all these six terms are subtracted from the total sum of squares, the remainder is the sum of squares for error. Years ago, these calculations were tedious. Nowadays, they are easily done for us on a PC.

16.14. SEVERAL FACTORS

The methods presented in this chapter for the two- and three-way layout can be extended to several factors. The large factorial experiment with several factors at four, five, or six levels is more appropriate to agriculture than to engineering. There are two reasons for this. The cost per plot in the agricultural experiment is usually much cheaper than the cost per run for an engineer. Furthermore, the agronomist has to plant all the plots together in the spring and the plants grow side by side; then the agronomist has to wait until the end of the season to harvest the data. Anything omitted has to wait until next year. The engineer makes runs, one after another, in sequence, and is therefore more able to conduct a series of small experiments, pausing after each to decide where to go next. These remarks are, of course, generalities, but they point out a fundamental difference in experimental strategies in the two fields.

In the next chapter, we turn to a group of factorial experiments that are commonly used in industry—experiments with several factors at only two or three levels—and develop some of the strategy of sequential experimentation.

EXERCISES

16.1. A research engineer for an oil company wants to compare four gasolines. The engineer makes five determinations of the octane number of each gasoline. The data are given in the following. Do the gasolines differ significantly in their octane numbers? Can you recognize any of the gasolines as being different from the others?

EXERCISES

	A	B	C	D
	81.5	80.6	79.0	80.8
	81.2	80.4	78.6	80.4
	80.3	80.3	80.0	80.1
	81.7	80.4	79.9	80.7
	80.6	80.3	79.2	80.4

16.2. In another octane study, the engineer wants to compare five gasolines. The engineer takes five cars and makes one determination of the octane number (on the road) for each gasoline in each car. Is there a significant difference between cars? Which gasolines differ?

Gasolines	Cars				
	I	II	III	IV	V
A	88.2	86.6	88.1	86.7	87.5
B	86.6	87.6	86.8	89.0	88.6
C	86.9	86.1	87.3	86.9	87.0
D	86.1	86.2	86.8	86.6	87.5
E	85.7	85.9	86.1	86.0	85.6

16.3. An engineer compared acid catalysts from three suppliers in a chemical process at the pilot plant stage. Unfortunately, the engineer was not able to make the same number of plant runs with each of the acids. Carry out the analysis of variance calculations and find which acid differs from the others (if any). The figures shown are the percentages of raw material converted to product.

A	B	C
73	75	78
75	79	74
78	78	79
73	80	71
71	81	76
	77	78
		74

16.4. The percentages of product from three sources of raw material follow. There were seven samples from the first source and four from each of the other two. Do the sources differ in the amount of product?

$A =$	72.5	72.2	73.0	72.0	71.8	72.7	71.2
$B =$	71.1	69.5	70.9	72.0			
$C =$	72.5	71.4	72.5	71.7			

16.5. In a two-way layout, the rows correspond to four machines and the columns to five levels of operating temperature. Analyze the data.

	\multicolumn{5}{c}{Temperature}
Machine	A	B	C	D	E
I	21.6	21.7	21.9	23.3	23.1
II	22.7	21.6	23.6	24.2	24.7
III	21.2	24.8	23.0	25.5	26.2
IV	21.9	22.5	22.3	22.5	24.7

16.6. Five wafers, fabricated under different conditions, were compared for thickness. Five measurements were made on each wafer. Do the wafers differ significantly in thickness? (Wafers are rows.)

1736	1725	1726	1730	1659
1665	1657	1674	1693	1637
1623	1647	1685	1721	1738
1696	1725	1735	1735	1735
1711	1712	1742	1744	1701

16.7. In exercise 16.6, the columns represent different positions on the wafer. Positions (columns) 1, 2, 3, and 4 are 90 degrees apart on the edge of the wafer. The measurements in column 5 are made at the center of the wafer. Is there cause from this data to think that there are significant differences between positions on the wafer? Does the thickness at the center differ from the average thickness on the edge?

16.8. A company wishes to consider changing the way in which one of the materials that it uses in its fabrication process is made. Two subcontractors are invited to make five samples of the material prepared by the old method and five samples prepared by the new method. The response is a measure of the tensile strength of the material and a higher value is desirable. Is there a difference between the methods? Is there a significant difference between suppliers? Is there significant evidence of interaction?

Method	Supplier A	Supplier B
Old	130, 124, 128, 125, 122	125, 129, 127, 127, 126
New	128, 127, 133, 126, 129	135, 137, 134, 136, 128

16.9. The following data are the lifetimes of components. The components come from two different manufacturers, A. They are tested at four operating temperatures, B, and two power ratings, C. Three components are tested under each set of conditions. We may assume that the lifetimes are normally distributed. Which are the important factors? Make a two-way table of means to explain the $B*C$ interaction.

Manufacturer A_1

	C_1			C_2		
B_1	1252	2544	1848	2664	3024	2908
B_2	2952	2288	2092	1960	3360	2212
B_3	2892	3076	2328	2896	3108	3368
B_4	2184	2828	2416	3200	2668	3148

Manufacturer A_2

	C_1			C_2		
B_1	2088	2316	1796	4068	3448	3472
B_2	2452	2124	2088	2188	2548	2392
B_3	2824	1836	2780	2520	3200	2764
B_4	3172	2412	2528	4176	3236	3064

CHAPTER SEVENTEEN

Design of Experiments: Factorial Experiments at Two Levels

17.1. INTRODUCTION

Chapter sixteen was a general survey of the analysis of variance. It ended with a discussion of factorial experiments. The factorial experiments are designed experiments. The engineer carefully decides beforehand to make, for example, 20 observations, one at each of five levels of A and at each of four levels of B. That design gives a good experiment with the important property of orthogonality.

We turn now more specifically to applications of experimental design to engineering problems, especially in manufacturing engineering, using factorial experiments or fractional factorials. In this chapter, we consider factors at two levels; in chapter eighteen, factors with more than two levels; and in chapter nineteen, applications that have come to be associated with the Japanese engineer and statistician G. Taguchi. We only have space for an introduction to factorial experiments and their fractions. The reader who wishes to go further will find fuller discussions in the books *Statistics for Experimenters* by G. E. P. Box, J. S. Hunter, and W. G. Hunter, and *Statistical Design and Analysis of Experiments* by P. W. M. John and in articles in such journals as *Technometrics* and the *Journal of Quality Technology*.

An experiment with n factors, each of which has exactly two levels, is called a 2^n factorial experiment. These experiments play a very important part in modern engineering experimental design. They were first discussed in a systematic way by F. Yates (1937), who was using them for experiments at the Rothamsted Agricultural Station near London. His first example dealt with using fertilizers to help in growing potatoes. His three factors were nitrogen, N, phosphorus, P, and potassium, K, which were applied at two levels: zero and four pounds per plot. During and after World War II, applications to chemical engineering were developed by George Box and others at Imperial Chemical Industries (ICI). Now they are widely used in chemical engineering and in the semiconductor industry throughout the world. In Japan, they have been adapted successfully for many years by G. Taguchi and his associates. His ideas are now being applied extensively in the West.

A chemical engineering example of a 2^3 factorial would be an experiment in a pilot plant with three factors: temperature at 200 or 300 degrees, pressure at 500 or 550 pounds per square inch, and flow rate at 8 or 12 gallons per minute.

With four factors, the complete factorial calls for $2 \times 2 \times 2 \times 2 = 16$ runs; with five factors, 32 runs; and with six factors, 64 runs. That many runs may tax the budget of time or money too highly, so we turn to fractional factorials, which were introduced by David Finney (1945). Instead of running all 64 points in the factorial, we can run a well-planned half fraction, or half replicate, sometimes denoted by 2^{6-1}, with 32 points, or a quarter replicate with only 16 points. We discuss a one-sixteenth replicate of a 2^7 factorial, which we denote by 2^{7-4} or $2^7//8$, the latter standing for seven factors at two levels each in eight points. This reduction in the number of points does not come without cost. We will see in section 17.12 that when we use fractions, it is necessary to assume that some of the interactions are zero.

As we noted in the previous chapter, factorial experiments have the important property of orthogonality, or balance. Each level of factor A appears the same number of times with each level of factor B. This means that in the analysis, the factors are independent of each other, so that when we are estimating their effects, we can use averages, which increases the precision of the estimates. Orthogonality makes it easier for us to understand what is going on in the experiment, and it also makes the analysis much simpler. This important property holds for the 2^n factorial and will also hold for the fractions that we will use. There are nonorthogonal fractions that are useful (see, for example, the book by John, 1971), but they are beyond the scope of this book.

17.2. EXPERIMENTING ON THE VERTICES OF A CUBE

The two experiments that were mentioned in the previous section come from vastly different fields, but they are mathematically equivalent. Each can be reduced to experimenting on the vertices of a cube with coordinates ± 1. In the chemical experiment, we can let the three factors be

A (temperature): $\quad x_1 = \dfrac{\text{temp.} - 250}{50}$,

B (pressure): $\quad x_2 = \dfrac{\text{pres.} - 525}{25}$,

C (flow rate): $\quad x_3 = \dfrac{\text{rate} - 10}{2}$.

In the potato experiment, we can let A, B, and C correspond to N, P, and K, respectively, with $x_i = $ (amount $- 2)/2$.

This formulation of the experiment was introduced by the scientists at ICI. Many agronomists use levels 0 and 1 rather than ± 1; Taguchi denotes the levels by 1 and 2. In any case, we can talk about the high and low levels of the factors. If a factor is qualitative, such as a choice between sodium hydroxide and potassium hydroxide as a catalyst, we can arbitrarily call one of them the low level of the factor "catalyst" and the other the high level. With only two factors, we are, as we will see shortly, experimenting at the vertices of a unit square. With more than three factors, we are experimenting at the vertices of an n-dimensional hypercube.

Each vertex of the cube represents a set of operating conditions, sometimes called a treatment combination. Yates gave a useful notation for denoting the various treatment combinations. In the three-factor experiment, the one point with all three factors at their low levels is denoted by (1). The point with A at its high level and the other two factors at their low levels is denoted by a. The point with A and B at high levels and C at its low level is denoted by ab. The point with all three factors at their high levels is abc. The experiment has eight points: (1), a, b, ab, c, ac, bc, and abc. The extension to more than three factors is obvious.

The points are usually written in standard order: first (1) and a, then, multiplying by b to include the second factor, b, ab. To add C, we multiply (1), a, b, and ab by c, getting c, ac, bc, and abc. In a 2^4 factorial, we add eight more points: d, ad, bd, abd, cd, acd, bcd, and $abcd$.

The notation serves double duty. The letter b denotes both the point with high B and low A and low C, and also the observation (or average of the observations) made under those conditions.

17.3. AN EXAMPLE OF A 2^2 EXPERIMENT

Suppose $(1) = 36$, $a = 48$, $b = 52$, and $ab = 54$. We can present the data at the vertices of a square as follows:

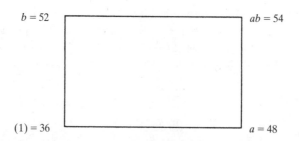

The (main) effect of the factor A is the average of the runs at high A minus the average of the runs at low A:

THE DESIGN MATRIX

$$A = \frac{48 + 54}{2} - \frac{36 + 52}{2} = 51.0 - 44.0 = 7.0.$$

The main effect of B is similarly defined:

$$B = \frac{52 + 54}{2} - \frac{36 + 48}{2} = 11.0.$$

Since there are two levels for each factor, each of these effects has 1 d.f.

The AB interaction is a measure of the failure of the simple additive model. Is the A effect the same at low B as it is at high B? Do we get the same improvement in yield with a ten degree increase in temperature at high pressure as we do at low pressure?

At high B, the effect of A is $ab - b = 54 - 52 = 2$; at low B, the effect of A is $a - (1) = 48 - 36 = 12$. The difference is $2 - 12 = -10$; the interaction is $(2 - 12)/2 = -5$.

We could just as well have compared the effect of changing the level of B at high A and low A, and obtained

$$(ab - a) - [b - (1)] = 54 - 48 - 52 + 36 = -10.$$

The interaction can be displayed graphically in an interaction plot, figure 17.3.1. If there were no interaction, the two lines on the plot would be parallel.

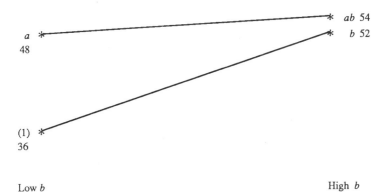

Figure 17.3.1. Interaction plot.

17.4. THE DESIGN MATRIX

The design matrix, **X**, is an array that shows the levels of the factors at the points of the design. In this example, **X** has two columns corresponding to A and B and four rows corresponding to (1), a, b, and ab. We usually write + and − instead of +1 and −1.

$$\mathbf{X} = \begin{matrix} A & B \\ - & - \\ + & - \\ + & + \\ + & + \end{matrix} \; ; \quad \mathbf{Y} = \begin{matrix} (1) \\ a \\ b \\ ab \end{matrix}$$

The estimated main effects can be obtained by multiplying the vector $\mathbf{Y}/2$ on the left by the transpose of $\mathbf{X}$:

$$A = \frac{\sum (x_1 = +1) - \sum (x_1 = -1)}{2} \quad \text{and}$$

$$B = \frac{\sum (x_2 = +1) - \sum (x_2 = -1)}{2} .$$

We note also that

$$AB = \frac{ab + (1) - a - b}{2} = \frac{\sum (x_1 x_2 = +1) - \sum (x_1 x_2 = -1)}{2},$$

so that AB corresponds to the product $x_1 x_2$. We can add another column to the design matrix to correspond to the interaction term.

$$\mathbf{X} = \begin{matrix} X_1 & X_2 & X_1 X_2 \\ - & - & + \\ + & - & - \\ - & + & - \\ + & + & + \end{matrix}$$

Notice that every pair of columns contains $(-1, -1)$, $(-1, +1)$, $(+1, -1)$, and $(+1, +1)$ exactly one time. This is orthogonality. The numerators of the fractions that give the estimates are called the A, B, and AB contrasts. In the general 2^n design, the estimates of the effects are obtained by dividing the appropriate contrast by $N/2$ or 2^{n-1}. An important consequence of the orthogonality of the design is that we are able to increase the precision of the estimates by using averages in this way.

17.5. THREE FACTORS

With three factors there are eight points,

$$(1), a, b, ab, c, ac, bc, abc ,$$

which can be represented as the vertices of a cube. There are seven effects

THREE FACTORS

of interest, each corresponding to one of the seven d.f. There are the three main effects, A, B, and C, and the three two-factor interactions: AB (x_1x_2), AC (x_1x_3), and BC (x_2x_3). Their estimates are obtained by dividing their contrasts by 4.

Example 17.5.1. This data comes from a chemical experiment. The three factors were A, temperature; B, pressure; and C, acid strength. The response, Y, is the coded yield of product.

The design matrix and the data are given in table 17.5.1. The A, B, and C contrasts are calculated by multiplying the **Y** vector by the appropriate column of the data matrix.

Table 17.5.1

	A (x_1)	B (x_2)	C (x_3)	Y
(1)	−	−	−	40
a	+	−	−	48
b	−	+	−	54
ab	+	+	−	56
c	−	−	+	44
ac	+	−	+	84
bc	−	+	+	42
abc	+	+	+	92

A	B	C
−40	−40	−40
+48	−48	−48
−54	+54	−54
+56	+56	−56
−44	−44	+44
+84	−84	+84
−42	+42	+42
+92	+92	+92
+100	+28	+64

The main effects are obtained by dividing these contrasts by 4. They are $A = 25$, $B = 7$, and $C = 16$. To calculate the contrasts for the interactions, we can make columns for x_1x_2, x_1x_3, and x_2x_3, and then proceed in the same way. This is shown in table 17.5.2.

The interactions are $AB = 4/4 = 1$, $AC = 20$, $BC = -4$, and $ABC = 4$. An explanation for the apparently large AC interaction can be found either by making an interaction plot or by making a 2×2 table of the totals for the four sets of levels of A and C, ignoring B.

Table 17.5.2

	AB (x_1x_2)	AC (x_1x_3)	BC (x_2x_3)	Y
(1)	+	+	+	40
a	−	−	+	48
b	−	+	−	54
ab	+	−	−	56
c	+	−	−	44
ac	−	+	−	84
bc	−	−	+	42
abc	+	+	+	92

	AB	AC	BC
	+40	+40	+40
	−48	−48	+48
	−54	+54	−54
	+56	−56	−56
	+44	−44	−44
	−84	+84	−84
	−42	−42	+42
	+92	+92	+92
	+4	+80	−16

		C	
A	Low	High	Total
High	104	176	280
Low	94	86	180
Total	198	262	460

The AC contrast is

$$(176 - 86) - (104 - 94) = 110 - 10 = 100.$$ □

17.6. THE REGRESSION MODEL

The calculations of section 17.5, although simple, can be tedious and error-prone when the number of factors increases. They can be made more easily on a PC by using a regression program to fit the following model:

$$y = \beta_0 + \beta_1 x_1 + \beta_2 x_2 + \beta_3 x_3 + \beta_{12} x_1 x_2 + \beta_{13} x_1 x_3 + \beta_{23} x_2 x_3 + e.$$
(17.6.1)

THE REGRESSION MODEL

The coefficients in the regression equation are multiplied by 2 to obtain the effects, e.g., $A = 2\beta_1$. This is illustrated in table 17.6.1, which is an edited regression printout from Minitab.

We could add another term, $\beta_{123} x_1 x_2 x_3$, to equation (17.6.1), corresponding to the three-factor interaction, ABC. That would make a model with eight terms, including the constant term, and eight points, which would force a perfect fit. We usually make the working assumption that interactions involving three or more factors are negligible a priori and omit them from the model. It is important to note that because of the orthogonality of the design, the estimates of the main effects and of the two-factor interactions will be the same whether we include or exclude the higher-order interactions. For the remainder of this book, we will restrict the word interaction to mean only two-factor interactions unless we make a specific statement to the contrary.

Table 17.6.1. Fitting the Regression Model for a 2^3 Factorial

```
MTB   > set x1 in c1
DATA > -1 +1 -1 +1 -1 +1 -1 +1
DATA > end
MTB   > set x2 in c2
DATA > -1 -1 +1 +1 -1 -1 +1 +1
DATA > end
MTB   > set x3 in c3
DATA > -1 -1 -1 -1 +1 +1 +1 +1
DATA > end
MTB   > let c4 = c1*c2  (This puts x1x2 in c4)
MTB   > let c5 = c1*c3  (This puts x1x3 in c5)
MTB   > let c6 = c2*c3  (This puts x2x3 in c6)
MTB   > set y in c10
DATA > 40 48 54 56 44 84 42 92
DATA > end
MTB   > name c1 'x1' c2 'x2' c3 'x3' c4 'x1x2' c5 'x1x3' c6 'x2x3'
MTB   > regress c10 6 c1-c6
```

The regression equation is

$$y = 57.5 + 12.5\ x1 + 3.50\ x2 + 8.00\ x3$$
$$+ 0.50\ x1x2 + 10.0\ x1x3 - 2.00\ x2x3$$

Analysis of Variance

SOURCE	DF	SS	MS	F	p
Regression	6	2694.00	449.00	14.03	0.202
Error	1	32.00	32.00		
TOTAL	7	2726.00			

Table 17.6.2

	A	B	AB	C	AC	BC	ABC
	1	2	3	4	5	6	7
(1)	−	−	+	−	+	+	−
a	+	−	−	−	−	+	+
b	−	+	−	−	+	−	+
ab	+	+	+	−	−	−	−
c	−	−	+	+	−	−	+
ac	+	−	−	+	+	−	−
bc	−	+	+	+	−	+	−
abc	+	+	+	+	+	+	+

The sum of squares for an effect is obtained by squaring the contrast and dividing by 2^n (or, equivalently, by squaring the effect and multiplying by 2^{n-2}). When we omit the higher-order interactions from the model, the total of their sums of squares can provide us with an error term. We mention this again in section 17.9.

With the model that includes all the interactions, including those with more than two factors, we can rewrite the design matrix to incorporate columns for the higher order terms, see table 17.6.2. Again, we notice the orthogonality. Take any pair of columns. They contain +1, +1 twice; +1, −1 twice; −1, +1 twice; and −1, −1 twice.

17.7. YATES' ALGORITHM

This is another way of calculating the contrasts that does not require a PC with a regression program. Yates (1937) presented an algorithm for calculating all the contrasts simultaneously. It is much faster than computing each contrast separately, as we did in section 17.5. In this section, we illustrate the algorithm by applying it to the 2^3 example of section 17.5; see table 17.7.1.

The first step in the algorithm is to write the observations in the standard order: (1), *a*, *b*, *ab*, *c*, *ac*, *bc*, and *abc*, and enter them in column 1 of the work sheet. We then go through repetitive cycles of adding and subtracting. Column 2 is made up of sums and differences of pairs of observations; the entries, in order, are the sums: $a + (1)$, $ab + b$, $ac + c$, and $abc + bc$, followed by the differences: $a - (1)$, $ab - b$, $ac - c$, and $abc - bc$. The differences are always taken in the same way—the second observation minus the first.

Column 3 is obtained from column 2 in the same way, and then column 4 from column 3. (In a 2^n factorial, the operation is continued for *n* cycles, through column $n + 1$). Column 4 contains the contrasts in the standard order, together with the grand total, which appears in the first place. In column 5, we divide every entry in column 3 by 4, except for the total,

AN EXAMPLE WITH FOUR FACTORS

Table 17.7.1. Yates' Algorithm for 2^3

	$1^\dagger$	2	3	$4^\dagger$	$5^\dagger$	$6^\dagger$	
(1)	40	88	198	460	57.5	26450	Mean
a	48	110	262	100	25	1250	A
b	54	128	10	28	7	98	B
ab	56	134	90	4	1	2	AB
c	44	8	22	64	16	512	C
ac	84	2	6	80	20	800	AC
bc	42	40	−6	−16	−4	32	BC
abc	92	50	10	16	4	32	ABC

† Column 1 contains the data, column 4 the contrasts, column 5 the estimated effects, and column 6 the sums of squares.

which is divided by 8; these are the effects. Column 6 is made by squaring the entries in column 4 and dividing them by 8; these are the sums of squares for the effects. In the general case, we should divide the contrasts by 2^{n-1} and their squares by 2^n. An engineer who plans to conduct several factorial experiments can make macros for 2^3, 2^4, and 2^5.

17.8. AN EXAMPLE WITH FOUR FACTORS

These data come from an experiment conducted by A. C. Church (1966) at the General Tire and Rubber Company. He was investigating the homogeneity of mixing a vinyl compound used in manufacturing tires. A sample of the compound is placed in a vat and mixed with heads driven by a motor. The response recorded is the torque of the motor when the heads are running at a constant speed. The four factors are A, the speed of the motor; B, the temperature of the vat; C, whether the sample is preheated (no = low, yes = high); and D, the weight of the sample. The output from applying Yates' algorithm is shown in table 17.8.1.

It is clear that the main effects of B and D are very much bigger than the other effects, and that those two factors are surely important. It is also clear that C is unimportant. What about A and AB? If we know σ, we can argue as follows.

With 16 points, an effect is the difference between two averages of eight points each. Its variance is $\sigma^2/8 + \sigma^2/8 = \sigma^2/4$. With 2^n points, the variance of an effect is $\sigma^2/2^{n-2}$. We can argue that an effect will not be significant unless the absolute value of its estimate is greater than twice its standard deviation:

$$|\text{effect}| > \frac{\sigma}{\sqrt{2^{n-4}}}$$

This is sometimes called the minimum hurdle.

Table 17.8.1. Yates' Algorithm for the Church Data

OBS	DATA	CONTRAST	EFFECT	SS	RANK	
(1)	1460	25560	3195.00	40832100	16	
a	1345	−790	−98.75	39006	12	A
b	990	−3510	−438.75	770006	14	B
ab	1075	900	112.50	50625	13	AB
c	1455	160	20.00	1600	2	C
ac	1330	−270	−33.75	4556	5	
bc	1000	−470	−58.75	13806	8	
abc	1065	300	37.50	5625	6	
d	2235	6120	765.00	2340900	15	D
ad	2070	−610	−76.25	23256	11	
bd	1795	−590	−73.75	21756	10	
abd	1730	120	15.00	900	1	
cd	2540	200	25.00	2500	3	
acd	2100	−210	−26.25	2756	4	
bcd	1700	−510	−63.75	16256	9	
abcd	1670	320	40.00	6400	7	

17.9. ESTIMATING THE VARIANCE

If σ is unknown, it can be estimated by taking duplicate observations at the points of the fraction or by taking several center points, i.e., points where each x_i is midway between −1 and +1. In the absence of such duplicates, we have a problem. A traditional approach is to decide that, as a working rule, interactions with three or more factors are negligible. Their sums of squares, therefore, must be due only to noise, and so we can pool them to give us a sum of squares for error. This is what happens when we use a regression model that has terms only for the main effects and the two-factor interactions.

In the example in the previous section, we would pool the sums of squares for ABC, ABD, ACD, BCD, and $ABCD$. This gives a pooled sum of squares $5625 + 900 + 2756 + 16{,}256 + 6400 = 31{,}937$, whence $s^2 = 6387$. The 5% value for $F(1, 5)$ is 6.61; one could then declare as significant any effect whose sum of squares exceeds $6.61 \times 6387 = 42218$. We would thus declare D, B, and AB to be significant, and come very near to declaring A to be significant, too.

The following alternative procedure is sometimes suggested. It is fraught with danger and should *not* be used. Instead of restricting themselves to using the higher interactions for the error term, engineers might set some arbitrary ad hoc level of smallness and pool all the effects that are less than that standard.

NORMAL PLOTS

In this example, the engineer might decide that 10,000 is a fairly low value for a sum of squares in this group and pool all sums of squares that are less than 10,000. That would give a pool consisting of C, AC, ABC, ABD, CD, ACD, and $ABCD$, totaling 24,337 with 7 d.f. and $s^2 = 3477$. The 5% value of $F(1,7)$ is 5.59 and we should declare any effect to be significant if its sum of squares exceeds $5.59 \times 3477 = 19436$. The list contains six effects: A, B, AB, D, AD, and BD!

The risk in this procedure is that we are using a biased estimate, s^2, of the variance in the denominator of the F statistics. The bias makes the estimate too low. We introduce it when we construct the pooled sum of squares by loading it with the smallest terms that we can find.

17.10. NORMAL PLOTS

The procedure that was described in the first part of section 17.9 was a standard procedure until 30 years ago, when it became apparent that it led engineers to declare too many false significant effects. (A similar phenomenon occurs in comparing all the treatments in the one-way analysis of variance; it led to the introduction of alternatives to Fisher's least significant difference; see section 16.4.) Since then statisticians have turned more toward graphical procedures, beginning with Cuthbert Daniel's (1959) half-normal plot, in which he plotted the absolute values of the effects. This has lately given way to normal plots.

The rationale of the normal plot is that the $2^n - 1$ contrasts for the various effects are orthogonal to one another and are linear combinations of

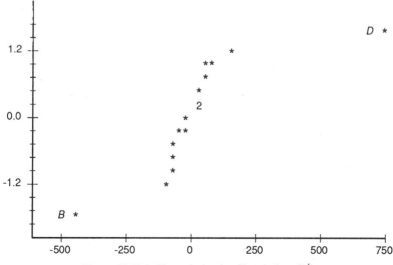

Figure 17.10.1. Normal plot for Church data (2^4).

normal variables. If the effects were all zero, we would have $2^n - 1$ independent normal random variables, each with variance $2^n\sigma^2$ and zero expectation. If we were to plot them on normal probability paper, the points ought to lie more or less on a straight line. If there are a few large effects, either positive or negative, the plotted points for the others will still lie on a line, but the points for the large effects will appear as outliers.

Figure 17.10.1 is a normal plot for the Church data. Normal scores for the 15 contrasts are calculated by computer, as in chapter five. The normal scores are then plotted against the actual contrasts. It is obvious that there are two extreme contrasts (two significant effects): D on the positive side in the upper right of the plot, and B on the negative side in the lower left. The other contrasts, including A and AB, lie on a straight line.

17.11. FRACTIONS

In the 2^3 experiment that we described earlier, we had eight observations and, hence, seven degrees of freedom. We used three of those degrees of freedom for the main effects of factors and the other four for interactions. Could we, instead, use those four degrees of freedom for extra factors, and thus accommodate seven factors in eight runs? The answer is yes, at a price.

Such designs are called fractional factorials. They were introduced by Finney (1945). The complete 2^7 factorial calls for $2 \times 2 \times 2 \times 2 \times 2 \times 2 \times 2 = 128$ runs and we will run only eight—a one-sixteenth fraction. The famous industrial statistician Cuthbert Daniel would call it a 2^{7-4} fraction, or write $2^7//8$—to denote a seven-factor experiment in eight runs.

17.12. THREE FACTORS IN FOUR RUNS

The complete two-factor design has four runs with three degrees of freedom. We added a third column to the design matrix to include the interaction. The three columns are orthogonal. The only other way to add a third column that is orthogonal to the first two is to use $-x_1x_2$ instead of $+x_1x_2$.

Suppose that we add a new factor, C, giving it the values in column 3 of the matrix. We get a $2^3//4$—a half replicate—consisting of the points c, a, b, and abc.

We could go ahead and fit a main-effects model for three factors:

$$y = \beta_0 + \beta_1 x_1 + \beta_2 x_2 + \beta_3 x_3 + e.$$

The estimate of β_3 would be half the C contrast:

$$\frac{abc + c - a - b}{2}.$$

What price have we paid by doing this? We have set $x_3 = x_1 x_2$. The main effect of C is completely confused with the AB interaction. The technical term for this is that C is *aliased* with AB; some say confounded. Since $x_3^2 = 1$ at each point, we can say that this fraction is defined by the relationship $x_1 x_2 x_3 = 1$. It follows that $x_1 = x_2 x_3$, and so A is aliased with BC. Similarly, B is aliased with AC. The cost of using the fraction in order to save time and money is that the main effects are aliased with two-factor interactions. It is a cost that we may or may not wish to pay. If we set $x_3 = -x_1 x_2$, or $x_1 x_2 x_3 = -1$, we obtain the complementary half replicate: (1), ab, ac, and bc. In either case, we are obtaining a reduction in the number of runs in our experiment at a price—we cannot distinguish between main effects and interactions. If we see a large A contrast, we must be prepared to assume that it is due to the main effect A and not to the interaction BC.

Two factors in a design are orthogonal if the products of their coordinates sum to zero. Thus, if $\Sigma x_1 x_2 = 0$, factors A and B are orthogonal. A design like this in which all the factors are orthogonal to each other, i.e., $\Sigma x_1 x_2 = \Sigma x_1 x_3 = \Sigma x_2 x_3 = 0$, is called an orthogonal array. Taguchi calls them lattices.

17.13. FRACTIONS WITH EIGHT RUNS

Suppose that we were to take the 2^3 factorial and add a fourth factor, D, by setting $x_4 = x_1 x_2 x_3$ or, equivalently, $x_1 x_2 x_3 x_4 = +1$. The fourth factor, D, is orthogonal to A, B, and C, and all four main effects are aliased with higher-order interactions that we are prepared to regard as negligible. In that sense, we are better off than we were in the 2^{3-1} case, where main effects were aliased with two-factor interactions. However, the two-factor interactions are aliased in pairs, AB with CD, AC with BD, and AD with BC. The design matrix for this half replicate is shown in table 17.13.1. Notice that the D column is identical with the ABC column in table 17.6.2.

Table 17.13.1

A	B	C	D	
−	−	−	−	(1)
+	−	−	+	ad
−	+	−	+	bd
+	+	−	−	ab
−	−	+	+	cd
+	−	+	−	ac
−	+	+	−	bc
+	+	+	+	abcd

17.14. SEVEN FACTORS IN EIGHT RUNS

We said earlier that we could obtain an orthogonal array for seven factors in eight runs. We take the design matrix for the complete 2^3 factorial and use all four interaction columns, AB, AC, BC, and ABC, for new factors. That procedure gives us the design in table 17.14.1.

We have equated D to ABC, E to AB, F to AC, and G to BC. Alternatively, we have set

$$x_1 x_2 x_5 = x_1 x_3 x_6 = x_2 x_3 x_7 = x_1 x_2 x_3 x_4 = 1 .$$

These products, which define the fraction, are called the defining contrasts. Taking all the products of these defining contrasts, we see that we actually have a fraction defined by

$$1 = x_1 x_2 x_3 x_4 = x_1 x_2 x_5 = x_3 x_4 x_5 = x_1 x_3 x_6 = x_2 x_4 x_6 = x_2 x_3 x_5 x_6$$
$$= x_1 x_4 x_5 x_6 = x_2 x_3 x_7 = x_1 x_4 x_7 = x_1 x_3 x_5 x_7 = x_2 x_4 x_5 x_7 = x_1 x_2 x_6 x_7$$
$$= x_3 x_4 x_6 x_7 = x_5 x_6 x_7 = x_1 x_2 x_3 x_4 x_5 x_6 x_7 ,$$

or, equivalently, by the relationship

$$I = ABCD = ABE = CDE = ACF = BDF = BCEF = ADEF = BCG$$
$$= ADG = ACEG = BDEG = ABFG = CDFG = EFG = ABCDEFG .$$

Multiplying through by x_1 and ignoring the terms with more than two factors, we see that

$$x_1 = x_2 x_5 = x_3 x_6 = x_4 x_7 ,$$

or $A = BE = CF = DG$. The main effect A is aliased with three interactions; we can write the four effects in an alias chain:

$$A + BE + CF + DG .$$

Table 17.14.1

	A	B	E	C	F	G	D
efg	−	−	+	−	+	+	−
adg	+	−	−	−	−	+	+
bdf	−	+	−	−	+	−	+
abe	+	+	+	−	−	−	−
cde	−	−	+	+	−	−	+
acf	+	−	−	+	+	−	−
bcg	−	+	−	+	−	+	−
abcdefg	+	+	+	+	+	+	+

There are six similar chains:

$$B + AE + CG + DF$$
$$C + AF + BG + DE$$
$$D + AG + BF + CE$$
$$E + AB + CD + FG$$
$$F + AC + BD + EG$$
$$G + AD + BC + EF$$

17.15. THE $L(8)$ LATTICE

If we multiply the columns for E, F, and G in table 17.14.1 by -1, we maintain orthogonality and get another fraction in the same family, which is shown in table 17.15.1. This time, we have $D = ABC$, $E = -AB$, $F = -AC$, and $G = -BC$. From a statistical point of view, one fraction is as good as the other. It is just that some like to have the base point (1) in their design. The alias chains are changed to $A - BE - CF - DG$, etc.

Taguchi changes the order of the columns and writes 1 and 2 where we have written $-$ and $+$. He calls the design his $L(8)$ lattice. Table 17.15.2 shows both the design matrix for this lattice and also the interaction table.

The engineer who uses this lattice to design, for example, an experiment with four factors, chooses four columns of the lattice for the factors. Suppose that column 1 is chosen for A and column 2 for B. The entry in the interaction table in row (1) and column 2 is 3; this says that the AB interaction appears in column 3, i.e., any factor assigned to column 3 will be aliased with the AB interaction, and so that column should not be used for C. Suppose column 4 is chosen for C. The AC interaction is in column 5, and the BC interaction is in column 6. That leaves column 7 free for the fourth factor, and one can confirm from the interaction table, that this

Table 17.15.1

A	B	E	C	F	G	D	
−	−	−	−	−	−	−	(1)
+	−	+	−	+	−	+	adef
−	+	+	−	−	+	+	bdeg
+	+	−	−	+	+	−	abfg
−	−	−	+	+	+	+	cdfg
+	−	+	+	−	+	−	aceg
−	+	+	+	+	−	−	bcef
+	+	−	+	−	−	+	abcd

Table 17.15.2. Seven Factors in 8 Runs

Design Matrix for L(8)

Col. No.	1	2	3	4	5	6	7
	1	1	1	1	1	1	1
	1	1	1	2	2	2	2
	1	2	2	1	1	2	2
	1	2	2	2	2	1	1
	2	1	2	1	2	1	2
	2	1	2	2	1	2	1
	2	2	1	1	2	2	1
	2	2	1	2	1	1	2

Interaction Table for L(8)

Col. No.	1	2	3	4	5	6	7
	(1)	3	2	5	4	7	6
		(2)	1	6	7	4	5
			(3)	7	6	5	4
				(4)	1	2	3
					(5)	3	2
						(6)	1
							(7)

assignment of factors to columns gives a design with main effects clear of interactions.

If the engineer goes through with that assignment of factors, the following design results:

Col. No.	1	2	4	7	
	A	B	C	D	
	1	1	1	1	(1)
	1	1	2	2	cd
	1	2	1	2	bd
	1	2	2	1	bc
	2	1	1	2	ad
	2	1	2	1	ac
	2	2	1	1	ab
	2	2	2	2	abcd

This is the design that we obtained before in table 17.13.1.

17.16. THE L(16) LATTICE

The method that we used in section 17.14, of adding extra factors by equating them to interactions in the basic factorial design, can be extended

SCREENING EXPERIMENTS

Table 17.16.1. Fifteen Factors in 16 Runs

Design Matrix for L(16)

Col. No.	1	2	3	4	5	6	7	8	9	10	11	12	13	14	15
	1	1	1	1	1	1	1	1	1	1	1	1	1	1	1
	1	1	1	1	1	1	1	2	2	2	2	2	2	2	2
	1	1	1	2	2	2	2	1	1	1	1	2	2	2	2
	1	1	1	2	2	2	2	2	2	2	2	1	1	1	1
	1	2	2	1	1	2	2	1	1	2	2	1	1	2	2
	1	2	2	1	1	2	2	2	2	1	1	2	2	1	1
	1	2	2	2	2	1	1	1	1	2	2	2	2	1	1
	1	2	2	2	2	1	1	2	2	1	1	1	1	2	2
	2	1	2	1	2	1	2	1	2	1	2	1	2	1	2
	2	1	2	1	2	1	2	2	1	2	1	2	1	2	1
	2	1	2	2	1	2	1	1	2	1	2	2	1	2	1
	2	1	2	2	1	2	1	2	1	2	1	1	2	1	2
	2	2	1	1	2	2	1	1	2	2	1	1	2	2	1
	2	2	1	1	2	2	1	2	1	1	2	2	1	1	2
	2	2	1	2	1	1	2	1	2	2	1	2	1	1	2
	2	2	1	2	1	1	2	2	1	1	2	1	2	2	1

Interaction Table for L(16)

Col. No.	1	2	3	4	5	6	7	8	9	10	11	12	13	14	15
	(1)	3	2	5	4	7	6	9	8	11	10	13	12	15	14
		(2)	1	6	7	4	5	10	11	8	9	14	15	12	13
			(3)	7	6	5	4	11	10	9	8	15	14	13	12
				(4)	1	2	3	12	13	14	15	8	9	10	11
					(5)	3	2	13	12	15	14	9	8	11	10
						(6)	1	14	15	12	13	10	11	8	9
							(7)	15	14	13	12	11	10	9	8
								(8)	1	2	3	4	5	6	7
									(9)	3	2	5	4	7	6
										(10)	1	6	7	4	5
											(11)	7	6	5	4
												(12)	1	2	3
													(13)	3	2
														(14)	1

to generate designs for 15 factors in 16 runs, 31 factors in 32 runs, and so on. The $L(16)$ lattice is shown in table 17.16.1.

17.17. SCREENING EXPERIMENTS

When all seven columns of the $L(8)$ lattice or all 15 columns of the $L(16)$ lattice are used for factors, the design is said to be saturated. In designs that are saturated, or nearly saturated, the main effects are all aliased with several interactions. As a consequence, they are used mainly for screening experiments.

In a screening experiment, the engineer considers a relatively large number of factors. The engineer assumes that only a few of them will be important, and wants to spot those important factors to be investigated more closely in a subsequent experiment. The engineer argues that the few important factors will stick up through the undergrowth and be detected.

Example 17.17.1. (An example of a screening experiment.) An $L(8)$ lattice is used in a screening experiment for seven factors. The order of the columns and the assignment of factors to columns is slightly different from the design in table 17.15.1. It is obtained from the 2^3 factorial in A, B, and C by setting $D = -AB$, $E = -AC$, $F = -BC$, and $G = ABC$. The lattice and the data are shown in table 17.17.1.

To calculate the effects, we have three choices. If we know what defining contrasts were used to obtain the fraction, we can carry out Yates' algorithm. In this case, we should carry out the algorithm for 2^3 in ABC and then equate D to $-AB$, etc.

	1	2	3	4	5	6 Eff.	7 SS	
(1)	17.4	43.6	84.5	173.7				
$a(deg)$	26.2	40.9	89.2	11.5	2.875	16.53	A
$b(dfg)$	22.7	43.8	4.3	-1.1	-0.275	0.15	B
$ab(ef)$	18.2	45.4	7.2	-16.9	-4.225	35.70	AB
$c(efg)$	19.2	8.8	-2.7	4.7	1.175	2.76	C
$ac(df)$	24.6	-4.5	1.6	2.9	0.725	1.05	AC
$bc(de)$	21.8	5.4	-13.3	4.3	1.075	2.31	BC
$abc(g)$	23.6	1.8	-3.6	9.7	2.425	11.76	ABC

The D effect is $-AB = +4.225$, the E effect is $-AC = -0.725$, the F effect is $-BC = -1.075$, and the G effect is $+ABC = 2.425$. The sums of squares are not affected by the changes of sign.

Table 17.17.1. The Data

	A	B	C	D	E	F	G	Y
(1)	$-$	$-$	$-$	$-$	$-$	$-$	$-$	17.4
$adeg$	$+$	$-$	$-$	$+$	$+$	$-$	$+$	26.2
$bdfg$	$-$	$+$	$-$	$+$	$-$	$+$	$+$	22.7
$abef$	$+$	$+$	$-$	$-$	$+$	$+$	$-$	18.2
$cefg$	$-$	$-$	$+$	$-$	$+$	$+$	$+$	19.2
$acdf$	$+$	$-$	$+$	$+$	$-$	$+$	$-$	24.6
$bcde$	$-$	$+$	$+$	$+$	$+$	$-$	$-$	21.8
$abcg$	$+$	$+$	$+$	$-$	$-$	$-$	$+$	23.6

FOLDOVER DESIGNS   303

Table 17.17.2. Calculating the Effects

A	B	C	D	E	F	G
−17.4	−17.4	−17.4	−17.4	−17.4	−17.4	−17.4
+26.2	−26.2	−26.2	+26.2	+26.2	−26.2	+26.2
−22.7	+22.7	−22.7	+22.7	−22.7	+22.7	+22.7
+18.2	+18.2	−18.2	−18.2	+18.2	+18.2	−18.2
−19.2	−19.2	+19.2	−19.2	+19.2	+19.2	+19.2
+24.6	−24.6	+24.6	+24.6	−24.6	+24.6	−24.6
−21.8	+21.8	+21.8	+21.8	+21.8	−21.8	−21.8
+23.6	+23.6	+23.6	−23.6	−23.6	−23.6	+23.6
+11.5	−1.1	+4.7	+16.9	−2.9	−4.3	+9.7
Dividing by 4:						
+2.88	−0.28	+1.18	+4.22	−0.72	−1.08	+2.42
A			D			G

If the defining contrasts are not known, the engineer can use the method of section 17.5 or regression, as in section 17.6. Table 17.17.2 shows the calculations for the method of section 17.5. We see that A, D, and G stand out as candidates for further investigation. □

17.18. FOLDOVER DESIGNS

The array in Table 17.15.1 can also be obtained from table 17.14.1 in the following way. Take each point in the first design and change the levels of every factor. Then *def* becomes *abcd*, and so on. This procedure is known as folding over.

The combined fractions form a 16-point design. The three-factor defining contrasts, such as $x_1 x_2 x_5$, drop out, as does $x_1 x_2 x_3 x_4 x_5 x_6 x_7$, and the 16 points satisfy the defining relations

$$1 = x_1 x_2 x_3 x_4 = x_2 x_3 x_5 x_6 = x_1 x_4 x_5 x_6 = x_1 x_3 x_5 x_7$$
$$= x_2 x_4 x_5 x_7 = x_1 x_2 x_6 x_7 = x_3 x_4 x_6 x_7 \, ,$$

or

$$I = ABCD = BCEF = ADEF = ACEG = BDEG = ABFG = CDFG \, .$$

In this larger fraction, the main effects are aliased only with higher-order interactions and are "clear" of two-factor interactions.

Such fractions are said to have resolution IV, because their defining contrasts have at least four letters. The design in table 17.13.1 is also a resolution IV design. The design in table 17.14.1 has resolution III because some of its defining contrasts have only three letters. In resolution III designs, some main effects are aliased with two-factor interactions. Whenever a resolution III fraction is folded over, the two parts combine to form a resolution IV fraction.

17.19. RESOLUTION V FRACTIONS

There are also fractions with resolution V, in which we can estimate all the main effects and all the two-factor interactions clear of one another. A useful example is the half replicate of the 2^5 factorial defined by

$$x_1 x_2 x_3 x_4 x_5 = -1.$$

It consists of 16 points:

$$(1), ae, be, ab, ce, ac, bc, abce,$$
$$de, ad, bd, abde, cd, acde, bcde, abcd.$$

Each main effect is aliased only with a four-factor interaction; each two-factor interaction is aliased only with a three-factor interaction.

This fraction can be obtained from the $L(16)$ lattice by assigning A to column 1, B to column 2, C to column 4, D to column 8, and E to column 15.

17.20. FRACTIONS WITH 12 POINTS

During World War II, two British statisticians who were engaged in war work, R. J. Plackett and J. P. Burman (1946), developed an orthogonal

Table 17.20.1. Eleven Factors in 12 runs (Plackett and Burman, 1946)

A	B	C	D	E	F	G	H	J	K	L	
1	1	1	1	1	1	1	1	1	1	1	(1)
2	2	1	2	2	2	1	1	1	2	1	abdefk
1	2	2	1	2	2	2	1	1	1	2	bcefgl
2	1	2	2	1	2	2	2	1	1	1	acdfgh
1	2	1	2	2	1	2	2	2	1	1	bdeghj
1	1	2	1	2	2	1	2	2	2	1	cefhjk
1	1	1	2	1	2	2	1	2	2	2	dfgjkl
2	1	1	1	2	1	2	2	1	2	2	aeghkl
2	2	1	1	1	2	1	2	2	1	2	abfhjl
2	2	2	1	1	1	2	1	2	2	1	abcgjk
1	2	2	2	1	1	1	2	1	2	2	bcdhkl
2	1	2	2	2	1	1	1	2	1	2	acdejl

EXERCISES

array for 11 factors in 12 runs; see table 17.20.1. The derivation of this array is quite different from the derivation of the other fractions that we have discussed, and what we have said about aliasing does not apply to it. These 12-point arrays should not be used when there are interactions.

EXERCISES

These exercises involve 2^{n-k} factorials. Note that in some of them you will find that you reach different decisions about which factors are significant depending upon whether you rely upon normal plots or regression!

17.1. These are the data from a 2^3 factorial experiment.

(1), 20.0 ;   a, 23.4 ;   b, 24.5 ;   ab, 27.2 ;
c, 20.6 ;   ac, 26.6 ;   bc, 26.0 ;   abc, 29.6

Calculate the A, B, and C contrasts. Which effects are important?

17.2. Calculate the contrasts for the main effects in this 2^4 factorial. Which effects are important?

(1), 43.2 ;   a, 46.8 ;   b, 41.4 ;   ab, 43.2 ;   c, 43.2 ;
ac, 47.6 ;   bc, 42.4 ;   abc, 49.2 ;   d, 40.0 ;   ad, 46.6 ;
bd, 40.8 ;   abd, 46.6 ;   cd, 49.0 ;   acd, 48.4 ;
bcd, 44.8 ;   abcd, 47.0

17.3. The following data come from a 2^{7-4} fraction. Match each of the factors with a column in the $L(8)$ array. Calculate the contrasts. Do any of the factors seem to be important?

(1), 68.7 ;   adeg, 67.1 ;   bdfg, 59.0 ;   abef, 58.2 ;
cefg, 56.7 ;   acdf, 54.3 ;   bcde, 59.2 ;   abcg, 64.4

17.4. This fractional factorial for seven factors in 16 runs was used in investigating a stitch bond process. The response measured was the percentage of defective stitches. The fraction was constructed by taking a full factorial in A, B, C, and E, and setting $E = -BD$, $F = -CD$, and $G = -ABCD$. Which columns in the $L(16)$ array correspond to the seven factors?

(1)	defg	cfg	cde	beg	bdf	bcef	bcdg
95	42	92	9	90	65	15	3

ag	adef	acf	acdeg	abe	abdfg	abcefg	abcd
94	4	32	5	95	24	6	3

Calculate the contrasts and identify the important factors. Which interactions are aliased with A? with E?

17.5. An engineer makes a 16-point fraction for seven factors by setting $E = -ABCD$, $F = ACD$, and $G = ABD$. Which points are in the fraction and which interactions are aliased with A? Which columns of the $L(16)$ lattice give you this fraction?

17.6. An engineer carried out a 2^3 factorial and obtained the following set of data:

(1), 47.5; a, 46.3; b, 44.4; ab, 51.7;

c, 46.2; ac, 46.8; bc, 46.2; abc, 56.6

The experiment is repeated later and provided these data.

(1), 48.1; a, 47.9; b, 44.5; ab, 52.6;

c, 56.6; ac, 47.2; bc, 47.0; abc, 56.6

Was there a significant difference between the two runnings? Which of the factors A, B, C is significant?

17.7. An engineer wishes to look at five factors in an $L(8)$ array, assuming that the only interactions that might exist are AB and AC. Find an assignment of factors to columns that achieves this aim. Which columns correspond to the interactions? Show that the fraction

(1), ae, bde, abd, cde, acd, bc, $abce$

meets the requirements. If you want to estimate AB and CD instead of AB and AC with an $L(8)$ array can you do it?

17.8. Find a 16-point fraction of a 2^6 experiment (find an assignment of the six factors to the columns of the $L(16)$ lattice) that will enable you to estimate A and the five interactions that contain A whether or not the other interactions are zero. Show that you can also estimate the other five main effects if the other interactions are zero.

17.9. Suppose that a fifth factor had been added to the example in table 17.8.1, so that the fraction was the half replicate defined by $x_1 x_2 x_3 x_4 x_5 = -1$ and the experimental points became

(1), ae, be, ab, ce, ac, bc, abce,
de, ad, bd, abde, cd, acde, bcde, abcd

with the same data. Which lines in the Yates table now correspond to the main effect of E and to the interactions that contain E? Notice that you now have a saturated fraction. Each degree of freedom corresponds to either a main effect or an interaction. You can make a normal plot in the usual way. Will the plot that we made in the text be changed? How could you handle the analysis using regression? What could you use for an error term?

17.10. McBeth (1989) investigates the effect of three temperatures on the thickness of the film of resist that is put on a wafer. The wafer is first placed on a cold plate for 30 seconds to stabilize its temperature; C is the temperature of the cold plate. Then the photoresist material is applied; B is the temperature of the photoresist. A is the ambient temperature. He measured two responses: Y_1 is the average thickness of the films on four wafers at each set of conditions; Y_2 is a measure of the uniformity of the film. He used two different photoresist materials, KTI 820 (low) and KTI 825 (high); they constitute factor D. The levels of A were 20.4°C and 23.5°C. He does not mention the levels of the other factors.

(1)	256	30.8	d	389	36.6
a	767	55.1	ad	765	65.6
b	325	11.0	bd	412	15.6
ab	805	44.1	abd	857	42.9
c	389	13.8	cd	457	12.7
ac	737	50.4	acd	738	54.9
bc	440	14.0	bcd	540	16.6
abc	942	32.0	abcd	880	33.9

Answer the following questions for each of the two responses. Which are the important factors? Is there any significant difference between the two materials?

17.11. It is clear in exercise 17.10 that the uniformity is greater at the higher level of the ambient temperature, A. Suppose that we decide to operate at that level of A. Consider the smaller factorial that consists of the eight points at high A. Are any of the other factors significant? Are any of your results shown in interactions?

17.12. These data are taken from an example that is quoted by Cornell (1981). Fish patties were made at the University of Florida from a fish called mullet. Three factors were involved. First the patties were deep fried (factor A) for 25 or 40 seconds. Then they were cooked in an oven at either of two temperatures (factor B), 375 or 425 degrees, for a given time (factor C), 25 or 40 minutes. The response recorded was a measure of the hardness of the texture of the patty. The data are

(1), 1.84;   a, 2.86;   b, 3.01;   ab, 4.13;
c, 1.65;   ac, 2.32;   bc, 3.04;   abc, 4.13.

Which are the important factors and interactions? Explain.

17.13. A petroleum engineer investigated dynamic control of the production of synthesis gas from underground coal gasification and determined an accurate rate equation for new catalysts that are resistant to sulfur poisoning by developing a simple model using five factors. Y is the coded rate constant as determined by a complex chemical equation. The factors are A, carbon monoxide; B, steam; C, carbon dioxide; D, hydrogen; and E, the temperature, each at two levels; the design used was a 2^5 factorial. Fit a model to the data using $Z = \ln(Y)$ as the response.

(1), 29;   a, 61;   b, 62;   ab, 140;   c, 15;   ac, 55;
bc, 56;   abc, 120;   d, 32;   ad, 64;   bd, 59;
abd, 140;   cd, 29;   acd, 62;   bcd, 67;   abcd, 140;
e, 16;   ae, 32;   be, 36;   abe, 73;   ce, 17;
ace, 36;   bce, 39;   abce, 81;   de, 11;   ade, 23;
bde, 26;   abde, 53;   cde, 13;   acde, 28;
bcde, 30;   abcde, 63.

17.14. Charles Hendrix of Union Carbide gives the following example of a fraction for seven factors in 16 runs. Show that it is a foldover design. What interactions are aliased with the main effects of the factors? Show that A and F are significant factors. The term involving AF is also significant. Which other interactions are aliased with AF? Would you be content to attribute all that sum of squares to AF, and not to the others? Make up a two-way table to explain that interaction.

(1), 67.6;   adg, 77.7;   bdf, 66.4;   abfg, 47.8;
cdfg, 61.9;   acf, 41.2;   bcg, 68.7;   abcd, 66.7;
efg, 68.4;   adef, 42.6;   bdeg, 59.0;   abe, 81.0;
cde, 78.6;   aceg, 86.4;   bcef, 65.0;   abcdefg, 38.7.

EXERCISES

17.15. Otto Dykstra (1959) gave an example of partial duplication of a factorial experiment.

$$(1), 18.9; \quad ab, 24.2; \quad ac, 25.3; \quad bc, 23.1;$$
$$a, 25.3; \quad b, 24.2; \quad c, 20.3; \quad abc, 29.8;$$
$$(1), 21.2; \quad ab, 26.6; \quad ac, 27.6; \quad bc, 25.1.$$

We cannot use Yates' algorithm for this problem. We can use multiple regression, letting X_1 be the coordinate of A, and so on. The values of the predictor variables are

X_1	X_2	X_3
-1	-1	-1
$+1$	$+1$	-1
$+1$	-1	$+1$
-1	$+1$	$+1$
$+1$	-1	-1
-1	$+1$	-1
-1	-1	$+1$
$+1$	$+1$	$+1$
-1	-1	-1
$+1$	$+1$	-1
$+1$	-1	$+1$
-1	$+1$	$+1$

We can add the interactions in the usual way by letting $X_4 = X_1 X_2$, $X_5 = X_1 X_3$, $X_6 = X_2 X_3$. Make two fits, first with only the terms for main effects, and then including the interactions. Are the interactions orthogonal to the main effects as they were in the $L(8)$ array? Can you estimate the main effect of A by subtracting the average at low A from the average at high A?

17.16. In the data table in exercise 17.15 the responses in the third row (the duplicated portion) are higher than the responses in the first row. Suppose that the last four points had actually been run a week later using a different batch of raw material. Include raw material as a fourth factor, D, which takes the values: $-1, -1, -1, -1, -1, -1, -1, -1, +1, +1, +1, +1$; repeat your analysis. This design is a three-quarter replicate (John, 1971).

We can estimate from the 12 points all four main effects and all six interactions. Notice that AB, AC, and BC are estimated from only the first eight points; A, B, and C are estimated from the last eight points; AD, BD, and CD are estimated from the first and third rows of the data table. D is estimated by the average of the two estimates from row 1 + row 3, and from row 2 + row 3.

17.17. A 2^4 experiment has the following data, in standard order:

| 138 | 145 | 149 | 157 | 154 | 166 | 159 | 158 |
| 124 | 153 | 151 | 148 | 158 | 150 | 152 | 150 |

One of the observations seems to be a bad observation. Which is it? What would you suggest as a value to replace it?

17.18. In a simulation study an engineer measures the delay time of a device. Short times are desirable. The engineer begins with seven factors in eight runs using an $L(8)$ lattice:

(1), 475;   *adfg*, 702;   *befg*, 483;   *abde*, 522;

cdef, 484;   *aceg*, 517;   *bcdg*, 399;   *abcf*, 568;

and then folds the lattice over and makes eight more runs

degh, 562;   *aefh*, 521;   *bdfh*, 506;   *abgh*, 452;

cfgh, 490;   *acdh*, 586;   *bceh*, 389;   *abcdefgh*, 561.

The engineer is interested in finding the best case and the worst case, which are the sets of conditions that minimize and maximize the delay. What are those conditions? What would you estimate the delays to be in the two cases?

17.19. The following data come from an experiment using the Plackett and Burman (1946) $L(12)$ lattice in table 17.20.1. Calculate the contrasts for the 11 factors and decide which are important.

(1), 5.02;   *abdefk*, 5.87;   *bcefgl*, 7.08;   *acdfgh*, 7.00;

bdeghj, 4.44;   *cefhjk*, 6.50;   *dfgjkl*, 5.48;   *aeghkl*, 4.82;

abfhjl, 5.48;   *abcgjk*, 5.55;   *bcdhkl*, 6.33;   *acdejl*, 5.55.

CHAPTER EIGHTEEN

Design of Experiments: Factorial Experiments at Several Levels

18.1. FACTORS WITH THREE LEVELS

The 2^n factorial designs are flexible and powerful, but because they have only two levels for each factor, they cannot handle quadratic models. The model for the 2^n contains terms in x_i, $x_h x_i$, $x_h x_i x_j$, etc., but not in x_i^2. The latter cannot be included because each factor appears at only two levels, and $x_i^2 = +1$ at every point. In order to incorporate quadratic terms in a model or to estimate operating conditions that optimize a response, we need at least three levels of each factor. Therefore, we turn to the 3^n factorials, n factors each at three levels.

There is, again, a difference of opinion about notation. Some statisticians denote the three levels by $-$, 0, and $+$. Others denote them by 1, 2, and 3. Yates (1937) used 0, 1, and 2; we will see an argument in favor of this approach when we consider fractionation.

18.2. TWO FACTORS

The two-factor experiment is an ordinary two-way layout with two factors, each at three levels. The analysis of variance table gives a subdivision into three sums of squares: the main effects of A and B, each with two degrees of freedom, and the residual and/or $A \times B$ interaction with four d.f.

We discussed in chapter sixteen how the sum of squares for each factor can be divided into components with single degrees of freedom. In the case of factors at three levels, we can break the sum of squares into linear and quadratic components. Suppose, for example, that we denote the sum of the observations at $A = 0$, 1, and 2 by T_0, T_1, and T_2, respectively, and that there are N points in the design, which may be a complete factorial or a fraction. The linear and quadratic contrasts for A are given by

$$\text{Lin } A = T_2 - T_0 \quad \text{and} \quad \text{Quad } A = 2T_1 - T_0 - T_2$$

and their sums of squares by

$$\text{Lin } A \quad \text{and} \quad \frac{3(\text{Lin } A)^2}{2N}; \quad \text{Quad } A \quad \text{and} \quad \frac{(\text{Quad } A)^2}{2N}.$$

18.3. FRACTIONS: FOUR FACTORS IN NINE POINTS

The 3^{4-2} factorial is a one-ninth replicate of the complete design. We use the notation $3^4//9$ to denote that it is a fraction for four factors, each at three levels, in nine runs. The third and fourth factors are introduced by setting $x_3 = x_1 + x_2$ and $x_4 = 2x_1 + x_2$. The sums are then reduced mod(3), which means that the actual numbers are replaced by the remainder when they are divided by 3; for example, when $x_1 = x_2 = 2$, $x_3 = 2 + 2 = 4$, which becomes 1. The nine points in the fraction are shown in table 18.3.1.

Example 18.3.1. This example is taken from an experiment to optimize a nitride etch process on a single plasma etcher (Yin and Jillie, 1987). There are four factors: A is the power applied to the cathode, 275, 300, and 325 watts; B is the pressure in the reaction chamber, 450, 500, and 550 mTorr; C is the gap between the anode and cathode, 0.8, 1.0, and 1.2 cm; and D is the flow rate of the etching gas, C_2F_6, 125, 160, and 200 sccm. The engineers considered three responses: etch rate in Å/min, uniformity, and selectivity. We look at the first response, etch rate; the other two are left as exercises for the reader. The data are given in table 18.3.2.

With all the degrees of freedom used for the factors, there is no error term and F tests are not available. We can compute the averages and the sums of squares for the individual factors in the usual way. They are as

Table 18.3.1

A	B	C	D
0	0	0	0
0	1	1	1
0	2	2	2
1	0	1	2
1	1	2	0
1	2	0	1
2	0	2	1
2	1	0	2
2	2	1	0

THREE FACTORS IN 27 RUNS

Table 18.3.2

A	B	C	D	Y_1	Y_2	Y_3
0	0	0	0	1075	2.7	1.63
0	1	1	1	633	4.9	1.37
0	2	2	2	406	4.6	1.10
1	0	1	2	860	3.4	1.58
1	1	2	0	561	4.6	1.26
1	2	0	1	868	4.6	1.65
2	0	2	1	669	5.0	1.42
2	1	0	2	1138	2.9	1.69
2	2	1	0	749	5.6	1.54

follows:

	Averages					
Factor Levels	0	1	2	SS	Lin	Quad
A	704.7	763.0	852.0	33031	32561	470
B	868.0	777.3	674.3	56336	56260	76
C	1027.0	747.3	545.3	351020	348004	3016
D	795.0	723.3	801.3	11260	60	11200

We have no way to tell which sums of squares are significant, but it is clear that the most important factor influencing the etch rate is C, the gap. Each sum of squares has been divided into its linear and quadratic components. For factors A, B, and C, there is no appreciable quadratic term. For D, the quadratic term dominates; there is obviously a minimum at level 1. The optimum set of conditions, if we are only interested in a high etch rate, is to run at the lowest level of B and C and the highest level of A and D. That point, 2, 0, 0, 2, was not included in the fraction. Another run should be made under those conditions to confirm that conclusion. The authors did that and obtained an etch rate of 1293 Å/min. It should be emphasized that we have had to assume that there are no interactions. □

18.4. THREE FACTORS IN 27 RUNS

We can treat this complete factorial by following the procedure described for a three-factor analysis of variance in section 16.13. The following example illustrates a chemical engineering application. When we have the complete factorial, we can consider the interactions.

Example 18.4.1 (Synthesis of tartaric acid by the hydroxylation of maleic acid). We start with maleic acid and treat it with hydrogen peroxide. The

Table 18.4.1

Temp (°C)	60			70			80		
H_2O_2 (%)									
WO_3 (%)	5	15	25	5	15	25	5	15	25
0.1	26.0	71.5	43.0	64.0	83.5	79.5	59.5	69.5	77.5
0.5	37.0	75.5	50.5	66.5	84.5	79.0	57.5	67.0	72.5
1.0	65.5	78.5	55.5	68.0	84.5	77.0	54.5	62.0	72.0

three factors are

  (i) the temperature,
  (ii) the concentration of hydrogen peroxide,
  (iii) the level of catalyst, which is tungstic acid, WO_3.

The data are in table 18.4.1.
The analysis of variance table is table 18.4.2.
  The sums of squares for the three main effects are

$$S_a = 1871.2, \qquad S_b = 1787.0, \qquad S_c = 107.6.$$

The totals at the various levels of each factor, together with the contrasts and the sums of squares for the linear and quadratic components, are as follows:

	A	B	C
$x = 0$	503.0	498.5	574.0
$x = 1$	686.5	676.5	590.0
$x = 2$	592.0	606.5	617.5
Lin	89.0	108.0	43.5
Quad	278.0	248.0	−11.5
SS Lin	440.06	648.00	105.12
SS Quad	1431.18	1138.96	2.45

Table 18.4.2. Analysis of Variance

SOURCE	DF	SS	MS	F
A	2	1871.2	935.62	36.79
B	2	1787.0	893.48	35.13
C	2	107.6	53.79	2.11
AB	4	889.4	222.34	8.74
AC	4	539.3	134.81	5.30
BC	4	157.4	39.34	1.55
Error	8	203.5	25.43	
TOTAL	26	5555.2	213.66	

The linear and quadratic contrasts for a factor were introduced in section 16.7. For a factor with three levels, they are the following: linear: $L =$ (sum at $x = 2$) − (sum at $x = 0$) $= (-1, 0, 1)$; Quad: $Q =$ (sum at $x = 0$) − 2(sum at $x = 1$) + (sum at $x = 2$) $= (-1, +2, -1)$. Since there are nine observations at each level, the sums of squares are $L^2/(2 \times 9)$ and $Q^2/(6 \times 9)$.

There are clearly quadratic trends in A and B; in both cases, we should find, if we fitted a parabola, that the maximum response occurs somewhere between $x = 0$ and $x = 2$.

We notice that the F test for the main effect of C is not significant at $\alpha = 5\%$, and yet there is a significant AC interaction. It seems at first that, on the average, there is no difference between the levels of tungstic acid. The totals for the three levels are

$$0.1\%, 574.0; \quad 0.5\%, 590.0; \quad 1.0\%, 617.5.$$

The spread in the totals for C is less than the spread for A or for B. However, the linear component alone has an F value of $105.12/25.43 = 4.13$, which, although not significant at $\alpha = 5\%$, exceeds the 10% value, which is $F^*(1, 8) = 3.46$. There is thus some evidence for a linear trend, but nothing to indicate any curvature in the response.

Table 18.4.3, totals for the levels of A and C, shows the interaction. At 60 and 70 degrees, higher levels of acid are better. At 80 degrees, the lowest level of acid is best.

Table 18.4.3. Temperature vs. Tungstic Acid Totals

WO$_3$ (%)	Temperature			
	60	70	80	Sum
0.1	140.5	227.0	206.5	574.0
0.5	163.0	230.0	197.0	590.0
1.0	199.5	229.5	188.5	617.5
Sum	503.0	686.5	592.0	1781.5
Lin C	59.0	2.5	−18.0	43.5
Quad C	−14.0	3.5	−1.0	−11.5

Another way to say this is to note that the linear contrast for C is positive at low temperature and negative at high temperature. How can we evaluate the significance of this switch in sign? One way is to divide the four d.f. for the AC interaction into single d.f.: $A_L C_L$, $A_L C_Q$, $A_Q C_L$, and $A_Q C_Q$. The Lin A Lin C contrast represents a change in the linear trend in Lin C as we change levels of A. It is the difference between the Lin C at high A and Lin C at low A. In this case,

$$A_L C_L = -18.0 - 59.0 = -77.0.$$

Since there are three observations at each combination of levels of A and C, the corresponding sum of squares is

$$\frac{(77.0)^2}{3 \times 2 \times 2} = 494.08.$$

The Quad A Lin C contrast is obtained by applying $(-1, 2, -1)$ to the Lin C terms, and is

$$-59.0 + 2(2.5) - (-18.0) = -36.0.$$

Its sum of squares is $(36.0)^2/(3 \times 2 \times 6) = 36.0$.

The Lin A Quad C and Quad A Quad C terms are obtained in similar fashion from the quadratic contrasts at the three levels of C. Their values are the following:

Lin A Quad $C = (-1.0) - (-14.0) = 13.0$; SS $= (13.0)^2/(3 \times 2 \times 6) = 4.7$;
Quad A Quad $C = -(-14.0) + 2(3.5) - (-1.0) = 22.0$, with SS
484.0/$(3 \times 6 \times 6) = 4.5$.

The sums of squares for the four components add up to the sum of squares for AC interaction, 539.3. The lion's share is accounted for by the linear by linear component. □

18.5. FOUR FACTORS IN 27 RUNS

We can add a fourth factor to the complete 3^3 factorial and still have a design that has main effects clear of interactions, which is analogous to a resolution IV fraction. We can, for example, set

$$x_4 = x_1 + x_2 + x_3,$$

reduced (modulo 3), which was done in the following example.

Table 18.5.1

A	B	C	D	Y	A	B	C	D	Y	A	B	C	D	Y
0	0	0	0	5.63	1	0	0	1	6.24	2	0	0	2	6.84
0	0	1	1	6.17	1	0	1	2	7.01	2	0	1	0	6.18
0	0	2	2	6.85	1	0	2	0	6.23	2	0	2	1	6.64
0	1	0	1	5.42	1	1	0	2	5.81	2	1	0	0	5.45
0	1	1	2	5.84	1	1	1	0	5.29	2	1	1	1	5.82
0	1	2	0	5.92	1	1	2	1	5.72	2	1	2	2	6.46
0	2	0	2	5.09	1	2	0	0	4.69	2	2	0	1	5.44
0	2	1	0	4.98	1	2	1	1	5.31	2	2	1	2	5.83
0	2	2	1	5.27	1	2	2	2	5.75	2	2	2	0	5.50

FOUR FACTORS IN 27 RUNS

Table 18.5.2. Analysis of Variance

SOURCE	DF	SS	Lin	Quad
A	2	0.525	0.497	0.028
B	2	5.567	5.478	0.089
C	2	0.773	0.773	0.000
D	2	1.779	1.748	0.031
Error	18	0.514	$s^2 = 0.0286$	
TOTAL	26	9.158		

Example 18.5.1. In this experiment, the response was the propagation delay in nanoseconds in an integrated-circuit device. The data are shown in table 18.5.1. The analysis of variance table is table 18.5.2.

$F^*(1, 18) = 4.41$, and so a contrast will be significant if its sum of squares exceeds $4.41 s^2 = 0.114$. None of the quadratic terms is significant; all of the linear terms are.

The averages for the various levels of the factors are as follows:

	0	1	2
A	5.686	5.783	6.018
B	6.421	5.748	5.318
C	5.623	5.826	6.038
D	5.541	5.781	6.164

To minimize the propagation delay, the best conditions are the lowest level of A, C, and D and the highest level of B.

It is not necessary to have a computer program that handles four-factor analyses of variance for this problem. Because the factors are orthogonal, and we are only interested in main effects, we can make two passes on a two-factor ANOVA program. On the first pass, we ignore C and D and do a two-way ANOVA for A and B. This gives us the sums of squares for A and for B and the total; on the second pass, we ignore A and B and obtain the sums of squares for C and D. The sum of squares for error is then obtained by subtracting the sums of squares for A, B, C, and D from the total.

The first pass gives the following table:

SOURCE	DF	SS	MS
A	2	0.525	0.262
B	2	5.567	2.783
interaction	4	0.170	0.043
error	18	2.896	0.161
TOTAL	26	9.158	

The interaction sum of squares is spurious, and so is the error term, but the sums of squares for A and B are what are needed together with the total. □

18.6. THIRTEEN FACTORS IN 27 RUNS

Earlier, we obtained a 2^{7-4} fraction by starting with a 2^3 factorial and adding the four extra factors by equating them to interactions. A similar procedure enables us to derive an orthogonal array for 13 factors, with three levels each, in 27 runs. This array is shown in table 18.6.1. The actual derivation of the array, which goes back to Fisher, is more complicated than it was in the two-level case and will not be given here. The design used in example 18.5.1 can be obtained from this lattice by assigning A to column 1, B to column 2, C to column 5, and D to column 9. The interactions in the three-level

Table 18.6.1. The $L(27)$ Lattice

1	2	3	4	5	6	7	8	9	10	11	12	13
0	0	0	0	0	0	0	0	0	0	0	0	0
0	0	0	0	1	1	1	1	1	1	1	1	1
0	0	0	0	2	2	2	2	2	2	2	2	2
0	1	1	1	0	0	0	1	1	1	2	2	2
0	1	1	1	1	1	1	2	2	2	0	0	0
0	1	1	1	2	2	2	0	0	0	1	1	1
0	2	2	2	0	0	0	2	2	2	1	1	1
0	2	2	2	1	1	1	0	0	0	2	2	2
0	2	2	2	2	2	2	1	1	1	0	0	0
1	0	1	2	0	1	2	0	1	2	0	1	2
1	0	1	2	1	2	0	1	2	0	1	2	0
1	0	1	2	2	0	1	2	0	1	2	0	1
1	1	2	0	0	1	2	1	2	0	2	0	1
1	1	2	0	1	2	0	2	0	1	0	1	2
1	1	2	0	2	0	1	0	1	2	1	2	0
1	2	0	1	0	1	2	2	0	1	1	2	0
1	2	0	1	1	2	0	0	1	2	2	0	1
1	2	0	1	2	0	1	1	2	0	0	1	2
2	0	2	1	0	2	1	0	2	1	0	2	1
2	0	2	1	1	0	2	1	0	2	1	0	2
2	0	2	1	2	1	0	2	1	0	2	1	0
2	1	0	2	0	2	1	1	0	2	2	1	0
2	1	0	2	1	0	2	2	1	0	0	2	1
2	1	0	2	2	1	0	0	2	1	1	0	2
2	2	1	0	0	2	1	2	1	0	1	0	2
2	2	1	0	1	0	2	0	2	1	2	1	0
2	2	1	0	2	1	0	1	0	2	0	2	1

18.7. THE $L(18)$ LATTICE

This is a lattice for $3^7//18$. The eighteen points fall into triples. We can add an eighth factor, H, at two levels by running the first three of the triples at the low level of H and the other three at the high level of H. See table 18.7.1. We can say nothing about the interactions, and this lattice can be used for main effects only. An example of the use of this lattice is given in the next chapter.

Table 18.7.1. The $L(18)$ Lattice

A	B	C	D	E	F	G	H
0	0	0	0	0	0	0	0
0	1	1	1	1	1	1	0
0	2	2	2	2	2	2	0
1	0	1	2	1	2	0	0
1	1	2	0	2	0	1	0
1	2	0	1	0	1	2	0
2	0	2	1	1	0	2	0
2	1	0	2	2	1	0	0
2	2	1	0	0	2	1	0
0	0	2	2	0	1	1	1
0	1	0	0	1	2	2	1
0	2	1	1	2	0	0	1
1	0	0	1	2	2	1	1
1	1	1	2	0	0	2	1
1	2	2	0	1	1	0	1
2	0	1	0	2	1	2	1
2	1	2	1	0	2	0	1
2	2	0	2	1	0	1	1

18.8. THE $L(36)$ LATTICE

The last of the standard lattices that we present is the $L(36)$ lattice, which is given in table 18.8.1. The first 12 columns of the lattice form an orthogonal array for 3^{12}. The 36 points in that array form 12 groups of three points each. We can add a thirteenth factor with three levels to the experiment by assigning four groups to level 0 of the new factor, four to level 1, and four to level 2. In the table, we have also added two more factors at two levels, and thus obtained an orthogonal array for $2^3 3^{13}//36$.

Table 18.8.1. $2^2 3^{13}//36$

0	0	0	0	0	0	0	0	0	0	0	0	0	0	0
1	1	1	1	1	1	1	1	1	1	1	1	0	0	0
2	2	2	2	2	2	2	2	2	2	2	2	0	0	0
0	0	0	0	1	1	1	1	2	2	2	2	0	0	1
1	1	1	1	2	2	2	2	0	0	0	0	0	0	1
2	2	2	2	0	0	0	0	1	1	1	1	0	0	1
0	0	1	2	0	1	2	2	0	1	1	2	0	1	0
1	1	2	0	1	2	0	0	1	2	2	0	0	1	0
2	2	0	1	2	0	1	1	2	0	0	1	0	1	0
0	0	2	1	0	2	1	2	1	0	2	1	0	1	1
1	1	0	2	1	0	2	0	2	1	0	2	0	1	1
2	2	1	0	2	1	0	1	0	2	1	0	0	1	1
0	1	2	0	2	1	0	2	2	1	0	1	1	0	0
1	2	0	1	0	2	1	0	0	2	1	2	1	0	0
2	0	1	2	1	0	2	1	1	0	2	0	1	0	0
0	1	2	1	0	0	2	1	2	2	1	0	1	0	1
1	2	0	2	1	1	0	2	0	0	2	1	1	0	1
2	0	1	0	2	2	1	0	1	1	0	2	1	0	1
0	1	0	2	2	2	0	1	1	0	1	2	1	1	0
1	2	1	0	0	0	1	2	2	1	2	0	1	1	0
2	0	2	1	1	1	2	0	0	2	0	1	1	1	0
0	1	1	2	2	0	1	0	0	2	2	1	1	1	1
1	2	2	0	0	1	2	1	1	0	0	2	1	1	1
2	0	0	1	1	2	0	2	2	1	1	0	1	1	1
0	2	1	0	1	2	2	0	2	0	1	1	2	0	0
1	0	2	1	2	0	0	1	0	1	2	2	2	0	0
2	1	0	2	0	1	1	2	1	2	0	0	2	0	0
0	2	1	1	1	0	0	2	1	2	0	2	2	0	1
1	0	2	2	2	1	1	0	2	0	1	0	2	0	1
2	1	0	0	0	2	2	1	0	1	2	1	2	0	1
0	2	2	2	1	2	1	1	0	1	0	0	2	1	0
1	0	0	0	2	0	2	2	1	2	1	1	2	1	0
2	1	1	1	0	1	0	0	2	0	2	2	2	1	0
0	2	0	1	2	1	2	0	1	1	2	0	2	1	1
1	0	1	2	0	2	0	1	2	2	0	1	2	1	1
2	1	2	0	1	0	1	2	0	0	1	2	2	1	1

We could also start with the first 12 columns and add a Plackett and Burman design. One row of the $L(12)$ lattice is added to each group of three points. The resulting array is an orthogonal design for $2^{11}3^{12}//36$. It is shown in table 18.8.2.

Table 18.8.2. $2^{11}3^{12}//36$

0	0	0	0	0	0	0	0	0	0	0	0	0	0	0	0	0	0	0	0	0	0	0
1	1	1	1	1	1	1	1	1	1	1	1	0	0	0	0	0	0	0	0	0	0	0
2	2	2	2	2	2	2	2	2	2	2	2	0	0	0	0	0	0	0	0	0	0	0
0	0	0	0	1	1	1	1	2	2	2	2	1	1	0	1	1	1	0	0	0	1	0
1	1	1	1	2	2	2	2	0	0	0	0	1	1	0	1	1	1	0	0	0	1	0
2	2	2	2	0	0	0	0	1	1	1	1	1	1	0	1	1	1	0	0	0	1	0
0	0	1	2	0	1	2	2	0	1	1	2	0	1	1	0	1	1	1	0	0	0	1
1	1	2	0	1	2	0	0	1	2	2	0	0	1	1	0	1	1	1	0	0	0	1
2	2	0	1	2	0	1	1	2	0	0	1	0	1	1	0	1	1	1	0	0	0	1
0	0	2	1	0	2	1	2	1	0	2	1	1	0	1	1	0	1	1	1	0	0	0
1	1	0	2	1	0	2	0	2	1	0	2	1	0	1	1	0	1	1	1	0	0	0
2	2	1	0	2	1	0	1	0	2	1	0	1	0	1	1	0	1	1	1	0	0	0
0	1	2	0	2	1	0	2	2	1	0	1	0	1	0	1	1	0	1	1	1	0	0
1	2	0	1	0	2	1	0	0	2	1	2	0	1	0	1	1	0	1	1	1	0	0
2	0	1	2	1	0	2	1	1	0	2	0	0	1	0	1	1	0	1	1	1	0	0
0	1	2	1	0	0	2	1	2	2	1	0	0	0	1	0	1	1	0	1	1	1	0
1	2	0	2	1	1	0	2	0	0	2	1	0	0	1	0	1	1	0	1	1	1	0
2	0	1	0	2	2	1	0	1	1	0	2	0	0	1	0	1	1	0	1	1	1	0
0	1	0	2	2	2	0	1	1	0	1	2	0	0	0	1	0	1	1	0	1	1	1
1	2	1	0	0	0	1	2	2	1	2	0	0	0	0	1	0	1	1	0	1	1	1
2	0	2	1	1	1	2	0	0	2	0	1	0	0	0	1	0	1	1	0	1	1	1
0	1	1	2	2	0	1	0	0	2	2	1	1	0	0	0	1	0	1	1	0	1	1
1	2	2	0	0	1	2	1	1	0	0	2	1	0	0	0	1	0	1	1	0	1	1
2	0	0	1	1	2	0	2	2	1	1	0	1	0	0	0	1	0	1	1	0	1	1
0	2	1	0	1	2	2	0	2	0	1	1	1	1	0	0	0	1	0	1	1	0	1
1	0	2	1	2	0	0	1	0	1	2	2	1	1	0	0	0	1	0	1	1	0	1
2	1	0	2	0	1	1	2	1	2	0	0	1	1	0	0	0	1	0	1	1	0	1
0	2	1	1	1	0	0	2	1	2	0	2	1	1	1	0	0	0	1	0	1	1	0
1	0	2	2	2	1	1	0	2	0	1	0	1	1	1	0	0	0	1	0	1	1	0
2	1	0	0	0	2	2	1	0	1	2	1	1	1	1	0	0	0	1	0	1	1	0
0	2	2	2	1	2	1	1	0	1	0	0	0	1	1	1	0	0	0	1	0	1	1
1	0	0	0	2	0	2	2	1	2	1	1	0	1	1	1	0	0	0	1	0	1	1
2	1	1	1	0	1	0	0	2	0	2	2	0	1	1	1	0	0	0	1	0	1	1
0	2	0	1	2	1	2	0	1	1	2	0	1	0	1	1	1	0	0	0	1	0	1
1	0	1	2	0	2	0	1	2	2	0	1	1	0	1	1	1	0	0	0	1	0	1
2	1	2	0	1	0	1	2	0	0	1	2	1	0	1	1	1	0	0	0	1	0	1

18.9. A NONORTHOGONAL FRACTION

Throughout the last three chapters of this book, we are focusing on fractional factorials that are orthogonal arrays. In this section, we give an example of a nonorthogonal fraction, a Box–Behnken design. We can only touch on the area represented by this example, which is called response surface methodology. The Box–Behnken designs are, in essence, 3^n factorials without the vertices of the cube and with several center points. The most commonly used designs handle three or four factors.

The levels of the factors are taken as -1, 0, $+1$, and the engineer fits a complete quadratic model:

$$y = \beta_0 + \beta_1 x_1 + \beta_2 x_2 + \beta_3 x_3 + \beta_{11} x_1^2 + \beta_{22} x_2^2 + \beta_{33} x_3^2 + \beta_{12} x_1 x_2 + \beta_{13} x_1 x_3 + \beta_{23} x_2 x_3 .$$

The reader who remembers analytic geometry will recall that this equation is a three-dimensional generalization of conic sections (ellipses, parabolas, and hyperbolas). The contours for given values of y can be a set of concentric quadratic surfaces. The center of the system is the set of conditions that maximizes, or minimizes, the response, y.

Example 18.9.1. This example comes from an etching study. The three factors are A, the ratio of gases used; B, pulse; and C, power. Several responses are observed; we will look at oxide uniformity. The design and the observed responses are shown in table 18.9.1.

Table 18.9.1

Ratio	Pulse	Power	Oxide Uniformity
0	+1	+1	21.2
−1	0	−1	12.5
+1	0	−1	13.9
0	−1	−1	12.1
0	0	0	25.0
0	+1	−1	22.0
+1	+1	0	25.5
+1	0	+1	34.4
−1	−1	0	38.1
0	0	0	37.4
−1	+1	0	18.5
+1	−1	0	28.0
−1	0	+1	43.0
0	−1	+1	45.6
0	0	0	28.0

The initial regression fit of the model to the data is

$$y = 30.1 - 1.29 x_1 - 4.57 x_2 + 10.5 x_3 - 0.94 x_1^2 - 1.67 x_2^2 \\ - 3.24 x_3^2 + 4.27 x_1 x_2 - 2.50 x_1 x_3 - 8.57 x_2 x_3 .$$

The only significant terms at $\alpha = 10\%$ are x_2, x_3, and $x_2 x_3$. By fitting the reduced model, the regression equation is

$$y = 27.01 - 4.58x_2 + 10.46x_3 - 8.58x_2x_3.$$

For this set of data, the gas ratio is not important, and the model reduces to a quadratic in x_2 and x_3, which is a hyperbola, or a saddle surface. The model clearly indicates interaction between the two remaining factors. The predicted responses at the four "corners," all of which are outside the area in which the observations were taken, are as follows:

x_1	x_2	Estimate
-1	-1	12.55
-1	$+1$	50.63
$+1$	-1	20.55
$+1$	$+1$	24.31 .

□

18.10. LATIN SQUARES

A Latin Square is an arrangement of p letters in a $p \times p$ square in such a way that each letter appears exactly once in each row and each column. It is an orthogonal array for three factors, each at p levels. The rows represent the levels of the first factor, the columns represent the levels of the second factor, and the letters the levels of the third factor. For $p = 3$, we have a square

$$\begin{array}{ccc} a & b & c \\ b & c & a \\ c & a & b \end{array}$$

This is a fraction for 3^3 in nine points. We denote the factors by P, Q, and R instead of our usual A, B, and C because we are using a, b, and c as the letters in the square. The points in the first row are at level 0 of P; the points in the first column are at level 0 of Q; and so on. Thus, the nine points are

0 0 0, 0 1 1, 0 2 2, 1 0 1, 1 1 2, 1 2 0,
2 0 2, 2 1 0, 2 2 1.

They are the points of the $L(9)$ array, table 18.3.1, with the last factor omitted.

18.11. GRAECO-LATIN SQUARES

Suppose that we take two Latin squares of side p, and, following tradition, use Latin letters for the first square and Greek letters for the second. The two squares are said to be orthogonal if, when we put one on top of the other, each Latin letter appears exactly once with each Greek letter. For example, with $p = 3$, we can take the two squares

$$\begin{array}{ccc} a & b & c \\ b & c & a \\ c & a & b \end{array} \quad \text{and} \quad \begin{array}{ccc} \alpha & \beta & \gamma \\ \gamma & \alpha & \beta \\ \beta & \gamma & \alpha \end{array}$$

Superimposing the squares, we get the Graeco-Latin square,

$$\begin{array}{ccc} a\alpha & b\beta & c\gamma \\ b\gamma & c\alpha & a\beta \\ c\beta & a\gamma & b\alpha \end{array}$$

We can now make an orthogonal array for four factors by letting the rows correspond to the levels of the first factor, the columns to the second, the Latin letters to the third, and the Greek letters to the fourth. The square that we have just shown provides a fraction for 3^4 in nine points:

$$\begin{array}{llll} 0\ 0\ 0\ 0, & 0\ 1\ 1\ 1, & 0\ 2\ 2\ 2, \\ 1\ 0\ 1\ 2, & 1\ 1\ 2\ 0, & 1\ 2\ 0\ 1, \\ 2\ 0\ 2\ 1, & 2\ 1\ 0\ 2, & 2\ 2\ 1\ 0, \end{array}$$

which is the lattice of table 18.3.1.

18.12. HYPER-GRAECO-LATIN SQUARES

If we can find three or more squares that are orthogonal to each other, we can superimpose them to form a hyper-Graeco-Latin square. This, in turn, provides an orthogonal array for five or more factors, each at p levels. We will give an example of a square of side four that can be used for a 4^5 factorial in 16 runs, and of an array for 5^6 in 25 runs. Algebraists have proved that one can always find a set of $p - 1$ mutually orthogonal squares when p is a prime number (2, 3, 5, 7, 11, . . .) or a power of a prime number ($4 = 2^2$, $8 = 2^3$, . . .). On the other hand, one cannot even find two orthogonal Latin squares of side six, and so no Graeco-Latin square of side six exists.

18.13. FIVE FACTORS AT FOUR LEVELS IN 16 RUNS

Table 18.13.1 shows four mutually orthogonal Latin squares of side four. Table 18.13.2 is the orthogonal array for five factors, each at four levels, in 16 runs that we derive from the three squares.

Table 18.13.1. Orthogonal Squares with $p = 4$

a	b	c	d	α	β	γ	δ	A	B	C	D
b	a	d	c	γ	δ	α	β	D	C	B	A
c	d	a	b	δ	γ	β	α	B	A	D	C
d	c	b	a	β	α	δ	γ	C	D	A	B

Table 18.13.2. 4^5 in 16 Runs

A	B	C	D	E		A	B	C	D	E		A	B	C	D	E		A	B	C	D	E
0	0	0	0	0		0	1	1	1	1		0	2	2	2	2		0	3	3	3	3
1	0	1	2	3		1	1	0	3	2		1	2	3	0	1		1	3	2	1	0
2	0	2	3	1		2	1	3	2	0		2	2	0	1	3		2	3	1	0	2
3	0	3	1	2		3	1	2	0	3		3	2	1	3	0		3	3	0	2	1

The data in the example are shown in table 18.13.3. The observations are written in the same order as table 18.13.2. The analysis of variance table is table 18.13.4.

Table 18.13.3. Data

20.97	22.15	22.10	24.58
22.80	26.43	21.98	23.05
23.87	22.74	23.54	23.13
21.27	22.09	23.16	21.10

Table 18.13.4. Analysis of Variance

SOURCE	DF	SS	MS
A	3	7.11	2.37
B	3	2.68	0.89
C	3	0.28	0.09
D	3	15.86	5.29
E	3	3.08	1.03

It is obvious that D is a major factor and that A is also worthy of attention. The price of accommodating five factors is that all 15 d.f. are used for the factors, and there is no error term for F tests. For those reasons, some statisticians suggest that this design should be used with only three factors so that there remain six d.f. for the error term in the analysis of variance. This is a matter of choice. Those who use all five factors argue that at the screening stage, the purpose of the experiment is merely to detect the one or two factors that clearly stand out as important, so that they may be the targets of further investigation, and that no formal test of significance is necessary at this stage.

18.14. FOUR-LEVEL FACTORS AND TWO-LEVEL FACTORS

A factor with four levels can be accommodated in a 2^n factorial. There are three d.f. for the four-level factor; they are associated with two of the

Table 18.14.1. Contrasts for a Four-Level Factor

	1	2	3	4
I	−1	+1	−1	+1
II	−1	−1	+1	+1
III	+1	−1	−1	+1

two-level factors and their interaction. We demonstrate the rationale in the next paragraph and follow that by two examples.

To be consistent with the notation used in the $L(8)$ lattice in table 17.2.2, we write the levels of the four-level factor by 1, 2, 3, and 4. We can follow chapter sixteen and associate the three degrees of freedom with orthogonal contrasts, as shown in table 18.14.1.

If we replace 1, 2, 3, and 4 by (1) b, a, ab, respectively, contrast I becomes the B contrast, II becomes the A contrast, and III the AB contrast.

Example 18.14.1. The first example is an orthogonal main-effects array for five factors with four at two levels each and the fifth at four levels. We take the $L(8)$ array and assign A and B to columns 1 and 2; their interaction is in column 3. The other columns are available for the two-level factors. The combination 1 1 in the first two columns of $L(8)$ (and, consequently, 1 in column 3) becomes level 0; 2 1 becomes 1; 1 2 becomes 2; and 2 2 becomes 3. The array is shown in table 18.14.2 with the four-level factor in the first column.

We cannot extend this idea to handle two factors at four levels in eight runs, but we can handle five four-level factors in 16 runs, as we saw in the previous section. □

Example 18.14.2. If, in the $L(16)$ lattice, we assign A, B, C, and D to columns 1, 2, 4, and 8, respectively, the remaining columns will represent:

3, AB;   5, AC;   6, BC;   7, ABC;   9, AD;   10, BD;   11, ABD;
12, CD;   13, ACD;   14, BCD;   15, $ABCD$.

Table 18.14.2. $2^4 4^1$ in Eight Runs

1	1	1	1	1
1	1	1	1	1
1	2	2	2	2
2	1	1	2	2
2	2	2	1	1
3	1	2	1	2
3	2	1	2	1
4	1	2	2	1
4	2	1	1	2

FOUR-LEVEL FACTORS AND TWO-LEVEL FACTORS 327

Table 18.14.3. 4^5 in the $L(16)$ Array

P	Q	R	S	T
1	1	1	1	1
1	2	2	2	2
1	3	3	3	3
1	4	4	4	4
2	1	2	3	4
2	2	1	4	3
2	3	4	1	2
2	4	3	2	1
3	1	3	4	2
3	2	4	3	1
3	3	1	2	4
3	4	2	1	3
4	1	4	2	3
4	2	3	1	4
4	3	2	4	1
4	4	1	3	2

We assign the five factors at four levels to the columns in the following way:

P   1, 2, 3 (A, B, AB)
Q   4, 8, 12 (C, D, CD)
R   5, 10, 15 (AC, BD, $ABCD$)
S   7, 9, 14 (ABC, AD, BCD)
T   1, 11, 13 (BC, ABD, ACD)

The array is shown in table 18.14.3. It is left to the reader to fill in the details of the derivation. Changing the levels to 0, 1, 2, and 3 gives the array in table 18.13.2. □

Table 18.15.1

1.	4	2	1	3	1	2	6.	3	4	1	4	2	3	11.	5	1	1	2	3	1
2.	5	4	2	5	1	5	7.	2	1	2	1	2	2	12.	1	3	2	3	3	3
3.	1	1	3	4	1	4	8.	4	3	3	2	2	5	13.	3	5	3	5	3	2
4.	3	3	4	1	1	1	9.	5	5	4	3	2	4	14.	2	2	4	4	3	5
5.	2	5	5	2	1	3	10.	1	2	5	5	2	1	15.	4	4	5	1	3	4
16.	2	3	1	5	4	4	21.	1	5	1	1	5	5							
17.	4	5	2	4	4	1	22.	3	2	2	2	5	4							
18.	5	2	3	1	4	3	23.	2	4	3	3	5	1							
19.	1	4	4	2	4	2	24.	4	1	4	5	5	3							
20.	3	1	5	3	4	5	25.	5	3	5	4	5	2							

18.15. FACTORS AT FIVE LEVELS

The method of section 18.13 can also be used to construct an orthogonal array for six factors, each at five levels, in 25 runs. Such an array is shown in table 18.15.1. An application appears in the exercises.

EXERCISES

18.1. In example 18.3.1, we analyzed the response etch rate in the experiment by Yin and Jillie. Repeat the analysis for the other two responses, uniformity and selectivity.

18.2. In a paper, T. Oikawa and T. Oka (1987) report the use of a $3 \times 3 \times 3$ factorial to investigate differences between two methods of estimating membrane stress. The first column gives the approximate value of the stress; the second column gives the value from a more detailed analysis. The authors were interested in the ratio of Col. 2/Col. 1. The levels of the three factors, A, B, and C, are given in the last three columns. Carry out the analysis of variance.

191.8	204.5	1	1	1
230.5	232.6	1	1	2
269.6	278.0	1	1	3
159.3	169.5	1	2	1
175.5	185.0	1	2	2
186.3	201.3	1	2	3
154.8	158.7	1	3	1
167.6	171.8	1	3	2
154.8	187.1	1	3	3
227.5	246.5	2	1	1
295.1	294.5	2	1	2
369.0	349.7	2	1	3
174.6	198.1	2	2	1
231.5	235.8	2	2	2
278.3	267.3	2	2	3
153.7	176.7	2	3	1
213.3	216.6	2	3	2
236.0	235.8	2	3	3
208.6	250.6	3	1	1
293.6	304.1	3	1	2
369.8	361.5	3	1	3
170.5	201.1	3	2	1
234.0	245.7	3	2	2
286.0	281.8	3	2	3
152.4	177.1	3	3	1
206.1	217.9	3	3	2
251.9	250.2	3	3	3

EXERCISES

18.3. This is an example of a Box–Benkhen design with three factors and three center points for a total of 15. It is adapted from an experiment carried out in the development of semiconductors.

	X_1	X_2	X_3	Y		X_1	X_2	X_3	Y
1.	0	+1	+1	36	6.	0	+1	−1	7
2.	−1	0	−1	12	7.	+1	+1	0	20
3.	+1	0	−1	14	8.	+1	0	+1	68
4.	0	−1	−1	21	9.	−1	−1	0	35
5.	0	0	0	28	10.	0	0	0	26

	X_1	X_2	X_3	Y
11.	−1	+1	0	16
12.	+1	−1	0	48
13.	−1	0	+1	43
14.	0	−1	+1	68
15.	0	0	0	30

The values taken by the X variables are coded values for the levels taken by the factors. Thus, the first factor is the power in Mtorr at levels 10, 15, 20; $X_1 = $ (power $- 10)/10$.

Show that X_1, X_2, and X_3 are uncorrelated. This an important point about the design. Fit the model

$$y = \beta_0 + \beta_1 x_1 + \beta_2 x_2 + \beta_3 x_3.$$

There are substantial t-values for each of the predictors and a large value of R^2, which might lead you to conclude that the model gave a very good fit to the data. Print out the observed values, predicted values, and residuals. There are a few residuals that seem to be large—but that could, perhaps, be due to a high level of noise (note the value of s).

18.4. (Continuation of exercise 18.3.) Fit the complete quadratic model

$$y = \beta_0 + \beta_1 x_1 + \beta_2 x_2 + \beta_3 x_3 + \beta_{11} x_1^2 + \beta_{22} x_2^2 + \beta_{33} x_3^2 \\ + \beta_{12} x_1 x_2 + \beta_{13} x_1 x_3 + \beta_{23} x_2 x_3$$

to the data in exercise 18.3.

This is overkill; there are several terms whose coefficients have small t-values. Drop one of the two terms with the smallest t-value and rerun the regression. Continue this procedure until you have a model in which every coefficient has a significant t-value.

18.5. (Continuation of exercise 18.4). Fit the model

$$y = \beta_0 + \beta_1 x_1 + \beta_2 x_2 + \beta_3 x_3 + \beta_{33} x_3^2 + \beta_{13} x_1 x_3 + \beta_{23} x_2 x_3$$

to the data of exercise 18.3 and compare the fit to the results of exercise 18.3. Notice that the estimated standard deviation, s, has dropped to less than half its earlier value. Compare the predicted values and the residuals in the two fits. Notice that there is a marked reduction in the size of the residuals in the second fit, particularly for the two points with the highest responses ($y = 68$).

18.6. Work out the details of the derivation of the lattice array for 4^5 in table 18.14.3.

18.7. The following data set comes from a simulation study of an output transformerless (OTL) push-pull circuit reported by Wu et al. (1987). The response, Y, is the midpoint voltage of the system. There are five factors, each at five levels. They are related to components of the circuit. The array that was used is the first five columns of the array in table 18.15.1. The observations in the order that is given in the table are as follows:

```
0.066  2.153  7.704  0.471  1.326  0.326  3.685  0.085  2.917
8.005  1.164  6.177  0.423  4.189  0.260  3.560  0.248  1.464
5.391  1.522  3.150  1.085  2.850  0.076  1.254 .
```

Which are the important factors? Which levels of those factors give the largest voltage?

18.8. Wu et al. (1987) also report the data for an $L(9)$ array for three factors. The experiment involves a heat exchanger. The response is the outlet temperature minus 360 degrees.

A	B	C	Y	A	B	C	Y
1	1	2	54.90	1	2	1	58.18
1	3	3	125.64	2	1	1	67.03
2	2	3	81.71	2	3	2	85.25
3	1	3	19.78	3	2	2	19.82
3	3	1	14.97				

Which are the important factors? Do they exhibit linear or quadratic trends? Are these conclusions dominated by the single large response, 125.64? Would your conclusions be changed if you used as the response $\ln(Y)$ or $\sqrt{Y}$? If the objective was to minimize Y, would the engineer be justified in recommending that A should be at level 3 and that the levels of the other factors are immaterial?

CHAPTER NINETEEN

Taguchi and Orthogonal Arrays

19.1. INTRODUCTION

In the traditional strategy of experimentation in engineering, agriculture, and other sciences, attention has usually been focused on a response such as the yield. The agronomist wants to compare different varieties of wheat and different levels of fertilizer with the idea of optimizing the yield: can we develop a hybrid that will yield more bushels per acre, or can we improve our husbandry to coax a higher yield from our land? They have had enormous success in achieving their aims.

The chemical engineer wants to know which levels of pressure and temperature will optimize the amount of product coming out of the reactor. This is another area in which statistics has played a major role. Sequences of well-designed experiments lead the engineer to the operating conditions that optimize the yield and also map the region around that maximum so that the operator can adjust for such departures from optimum as the gradual loss of potency of a catalyst. This is the topic of response surface methods, which were developed by Box and others in the 1950s and 1960s. Many engineers are now using these methods, especially the series of experimental designs introduced by Box and his student D. H. Behnken. An example of these designs was given in the previous chapter.

In the last decade, our attention has been attracted to a third field—manufacturing engineering—in which the objective is not to squeeze the extra bushel or the extra octane number, but to produce thousands and thousands of widgets that are as close to some specifications as possible. This is an area that was not of major concern in the West until recently. The Japanese, however, recognized the problem after World War II and, under the guidance of Deming, worked hard at solving it. One of the leaders in the field is Japanese engineer and statistician G. Taguchi. He has been mentioned several times in previous chapters. Much of this chapter is based on his contributions.

In earlier chapters, we painted a picture of the old days when the engineer had a process rolling along smoothly in a state of statistical control;

between the engineer and the factory gate was a squad of inspectors who, in theory, spotted all the bad widgets and allowed only good ones to leave the plant. Two things must be said at this stage. The first is to remind the reader that even though a process can be rolling along smoothly under control, i.e., within its natural limits, those limits may bear little relationship to the actual specifications. The second is that the inspectors can spot most of the bad batches and turn them back for rework or for scrap. By their diligence, they can reduce the percentage of substandard widgets that leave the plant, but they cannot directly improve quality. If you are producing 20% defective widgets, the inspectors cannot alter that figure. All they can do is decrease the fraction of that 20% that goes out to the customer. *You cannot inspect quality into a process.*

So, what can you do? You can begin by rejecting the old simplistic notion that to be outside specification is bad, but to be within specification, even by only a whisker, is satisfactory. Replace it by the new ideas of process capability that were discussed in section 9.15 in connection with the criteria C_p and C_{pk}. One of the major contributions of the Japanese approach is to make the engineer think in terms of aiming for a target value, μ_0, and work to reduce the mean square error, $E(y - \mu_0)^2$. One might think of two golfers playing iron shots from the fairway. One aims for the green: she either meets the specification of being on the green, in which case she is ready to go into the next stage of production, putting, or she misses the green and is out of specification, doing rework in the rough. The other player is not content with the criterion of being either on or off the green. She takes the hole itself for her target and aims for the flag with the criterion of getting as close to the hole as possible; as her game improves, her mean square error becomes smaller, and she works continually to reduce it even further.

The mean square error is the sum of two parts: the square of the bias, which is the difference between the actual process mean, μ, and the target value, μ_0, and the variance. We want to control both of them; we would like to refine our manufacturing process to have zero bias and the smallest possible variance. Although we cannot improve quality by inspection, we can, and must, build quality by designing it into the production process. Too often the word design is used in a restricted sense to mean the work of a group of specialist engineers, sometimes far away physically and intellectually from the shop floor, who design new items and perhaps build a single prototype model. Taguchi calls this the system design phase. We are more concerned with the next step, which is to take their prototype and make it manufacturable. When we talk about designing quality into the process, we mean carrying out research-and-development experiments to find the best operating conditions for manufacturing not just one item but thousands. This is called off-line quality control, or robust design.

Statisticians use the term robust to mean insensitive to departures from set conditions. The sample median can be preferred to the sample mean as an estimator of $E(Y)$ because it is robust against extreme observations (outliers). Wilcoxon's nonparametric rank test can be preferred to the two-sample t-test because it is insensitive to departures from the assumption of a normal distribution. Taguchi has led the way in making engineers think of carrying out experiments in the manufacturing process to make the product robust to variation both in the parts that are used in the manufacture and in the environmental conditions in which the item is to be employed.

The Taguchi approach to robust design makes the engineer focus on reducing the variability of the process. He advocates a systematic approach to achieve that aim.

19.2. TERMINOLOGY

The generic observation has different labels in different fields. The agronomist usually talks about the yield of a plot. The engineer talks about y as the response. In the present context, the variable that is being observed is often called a performance characteristic. The two most commonly used performance characteristics for a sample of data are the sample mean, $\bar{y}$, for location, and the sample variance, s^2, for dispersion. We may also use transformations of them to measure the dispersion, such as $\ln(s^2)$, and some other statistics, called signal-to-noise ratios, which are advocated by Taguchi.

19.3. THE THREE STAGES OF PROCESS DESIGN

Taguchi suggests that there are three stages of design for a process: system design, which we mentioned earlier, parameter design, and tolerance design.

We discuss parameter design in the next sections. The process involves several factors, which we call the design factors. In the parameter design stage, we conduct experiments to determine the "best" nominal settings (or design parameters) for these factors.

In the final stage, tolerance design, the engineer finds the tolerances around the nominal settings that were determined by the parameter design. We may learn that some of the parts used in the process require very close tolerances, while others are not so demanding. Therefore we may save cost by using in the second group cheaper parts than we had originally projected.

19.4. A STRATEGY FOR PARAMETER DESIGN

Suppose that we are manufacturing an electric circuit and that we are interested in the output of a certain stage of the circuit. One of the variables that controls the output is the input ratio at a certain point in the circuit.

Figure 19.4.1 shows a plot of output against the ratio. The target value, $y_0 = 55.5$, corresponds to a ratio, $x_0 = 13$. As the ratio varies about the value x_0, the output varies about y_0; the slope of the curve is quite steep at that point. If the ratio is increased to $x = 20$, the output is increased to $y = 60$, but the sensitivity of y to variation in the ratio is decreased because of the nonlinearity of the response curve. Perhaps we can find another part in the circuit, such as a capacitor, that we can vary to adjust the output back to y_0 without losing the advantage of lower sensitivity to small changes in the ratio that we discovered at $x = 20$.

This suggests a design strategy for parameter design experiments. We will perform a factorial experiment to investigate the influence of these factors. From it, we will learn that changing the levels of some factors affects the process mean, but not the variance; changing the levels of some other factors affects the variance. We should use the latter group to reduce the variance as much as we can, and then use the others to adjust the performance characteristic to its target value. As part of this strategy, Taguchi makes extensive use of orthogonal arrays.

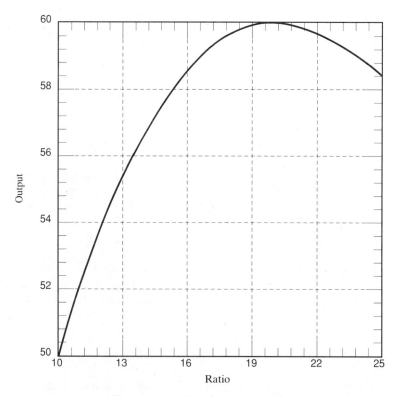

Figure 19.4.1. Plot of output vs. ratio.

19.5. PARAMETER DESIGN EXPERIMENTS

The purpose of the parameter design experiment is to determine which of the factors are important in the manufacturing process and to find the optimum set of working conditions, i.e., the best levels of those factors, or parameters. We start with a simple example that illustrates the basic principles.

Suppose that we have three factors in our process and that we decide to look at two levels of each. We run the complete factorial (all eight points) and we make four runs at each of the eight sets of conditions. The data are shown in table 19.5.1.

We can analyze the data in two ways. First, we take as the response the means of the sets of five observations. That tells us which factors have a major influence on the location of the process, e.g., the average output voltage. We find that A and B are the important factors. Increasing A decreases $\bar{y}$; increasing B increases $\bar{y}$. The effect of changing the level of C, though positive, is modest.

In the second analysis, we take as the performance characteristic the estimated variance at each of the eight sets of conditions. We learn that changing the levels of A or B has little effect on the variance; on the other hand, changing the level of C produces a significant change in the variance, which is greater at the high level of C. Many statisticians prefer to use the natural logarithm of s^2, rather than s^2 itself, because it is more nearly normally distributed. In this example, we obtain the same result with either choice.

The combined analyses suggest that we should run the process at the low level of C to cut down on the variability, and then adjust the levels of A and B to get the desired value of the process mean.

19.6. SIGNAL-TO-NOISE RATIOS

In the example of the previous section, we used the sample variance as the measure of variability, or dispersion, and $\bar{y}$ as the measure of the process

Table 19.5.1

A	B	C				
-1	-1	-1	60.5	61.7	60.5	60.8
$+1$	-1	-1	47.0	46.3	46.7	47.2
-1	$+1$	-1	92.1	91.0	92.0	91.6
$+1$	$+1$	-1	71.0	71.7	71.1	70.0
-1	-1	$+1$	65.2	66.8	64.3	65.2
$+1$	-1	$+1$	49.5	50.6	49.5	50.5
-1	$+1$	$+1$	91.2	90.5	91.5	88.7
$+1$	$+1$	$+1$	76.0	76.0	78.3	76.4

mean. Taguchi advocates several other measures, which he calls signal-to-noise ratios. It has been said that he has developed over 70 different such signal-to-noise measures!

The three most commonly used of Taguchi's signal-to-noise ratios are these:

(i) $$SN_S = -10 \log_{10}\left(\frac{\sum y_i^2}{n}\right),$$

where n is the number of points involved, when the objective is to make y as small as possible;

(ii) $$SN_L = -10 \log_{10}\left(\frac{\sum (1/y_i)^2}{n}\right),$$

when the objective is to make y as large as possible;

(iii) $$SN_N = 10 \log_{10}\left[\left(\frac{\bar{y}}{s}\right)^2 - \frac{1}{n}\right],$$

when y is to attain a specific target value.

They are given the slogans "smaller is better," "larger is better," and "on target is best."

The fact that the ratios are expressed in terms of $\log_{10}$ rather than natural logarithms is not important. In all cases, the objective of the engineer is to optimize the process by choosing the set of working conditions—the levels of the experimental factors—that makes the signal-to-noise ratio as large as possible. That is why some of them are prefaced with minus signs.

The use of the signal-to-noise ratios has been a matter of dispute. It can be argued that the first of them, SN_S, is a transformation of the mean square error when the target value is zero. Pushing that a bit further, the second is related to the mean square error for $1/y$ when its target value is zero.

The last one, however, is more difficult to justify. It is clearly a transformation of the well-known coefficient of variation, which is the ratio of the standard deviation of the response to its mean. For a set of data

$$\text{c.v.} = \frac{s}{\bar{y}};$$

it is often expressed as a percentage. It can be shown that this ratio is mathematically equivalent to the logarithm of the variance of $\log(\bar{y})$. Why we should be considering the variance of $\log(\bar{y})$ rather than looking at $\bar{y}$ or $\log(\bar{y})$ is not clear. It may well be more appropriate in some cases, but not as a general procedure.

Furthermore, SN_N contains no mention of the target value, which makes it hard to justify as a measure of meeting that target. Maximizing the quantity $\bar{y}^2/s^2$ places a premium on points with large values of $\bar{y}$ and small values of s. Certainly, we want small values of s, but we do not necessarily want to have $\bar{y}$ as large as possible. In a letter to the editor of *Quality Progress*, J. S. Hunter (1987) has presented a strongly written and convincing argument against using this criterion, and it has also been the subject of articles by Box and others.

19.7. INNER AND OUTER ARRAYS

The example in section 19.5 was simple in two ways. There were only three design factors, each at two levels, and so we were able to run the complete factorial experiment. We obtained our values of the variance by making four observations at each of the eight vertices of the cube. With more factors and more levels, the experimental design is more complicated.

We consider two types of factor: the design factors that we have already discussed and noise factors. The first step in planning the experiment is to select the design factors and to choose their levels. With three design factors at two levels each, the total number of possibilities is only eight. However, with six design factors, each at three levels, the total number of possibilities is 729, which exceeds most budgets. Therefore, we choose a fractional factorial—one of the orthogonal arrays, perhaps the $L(18)$ lattice for six factors in 18 runs. This lattice for the design factors is called the inner array.

The design factors are factors that we can control—more or less. They are the knobs on the black box that is our process. We can decide that factor A will be a certain resistor with three levels: 80 ohms, 100 ohms, and 120 ohms. There are other factors that affect the performance of the final product that we either cannot, or do not choose to, control. Examples of these are the temperature in the work place, the humidity, and the pollen count. These are the noise factors, representing environmental noise. We could also treat as a noise factor the resistance of the component that was mentioned earlier by perturbing it by a few ohms above and below its nominal values. That is the sort of thing that might happen if we used cheaper parts.

The outer array is another orthogonal array, this time in the noise factors. The overall experiment consists of running the outer array once at each point (vertex) of the inner array. The analysis proceeds in two steps. Suppose that there are N vertices in the inner array and n observations in each outer array, giving a total of Nn observations in the complete experiment. This can turn out to be an awful lot of observations! At each vertex, we calculate two performance characteristics of the outer array: the average observation, $\bar{y}_i$, and the overall variance of the n points, s_i^2. We can also calculate one or more signal-to-noise ratios.

Now we have the inner array with two performance characteristics observed at each vertex, and we make an analysis for each, just as in section 19.5. If nature is kind, we will find some factors, which we can call the signal factors, that affect $\bar{y}_i$, but do not alter the dispersion, s_i^2, and some other factors that control the dispersion, but do not affect the location. If that is so, we will put the dispersion factors at the levels that minimize the variability, and use the adjustment factors to get $\bar{y}$ back on target.

19.8. AN EXAMPLE OF A PARAMETER DESIGN EXPERIMENT

We consider an example in which there are five factors that are possible control factors. It is decided that these factors be considered at three levels each and that the inner array consist of the $L(18)$ lattice described in section 18.7.1. The factors are assigned to the first five columns of the lattice. The outer arrays consist of nine points each. Table 19.8.1 shows the design of the inner array and the summary statistics, $\bar{y}$, s^2, and SN_N, for the outer arrays carried out at each of the 18 vertices.

We begin with an analysis of the data for the response $\bar{y}$. Because the array is orthogonal, we can use analysis of variance. The easiest way is to make three runs on a two-way analysis of variance program, following the pattern of example 18.5.1—one with factors A and B, the second with C and D, and the third with E and one of the others. We note that the linear contrasts for A and both the linear and quadratic contrasts for B are

Table 19.8.1. The $L(18)$ Lattice for the Inner Array

A	B	C	D	E	$\bar{y}$	s^2	SN_N
0	0	0	0	0	39.08	10.1767	21.7598
0	1	1	1	1	41.82	7.8668	23.4675
0	2	2	2	2	39.77	10.1985	21.9026
1	0	1	2	1	42.15	9.1639	22.8727
1	1	2	0	2	46.82	9.5696	23.5976
1	2	0	1	0	43.05	12.1668	21.8245
2	0	2	1	1	46.28	11.3950	22.7382
2	1	0	2	2	46.80	9.3032	23.7165
2	2	1	0	0	45.67	7.8275	24.2546
0	0	2	2	0	39.30	7.8870	22.9163
0	1	0	0	1	42.65	10.6673	22.3150
0	2	1	1	2	41.37	8.6670	22.9526
1	0	0	1	2	39.91	9.3533	22.3092
1	1	1	2	0	45.21	10.8302	22.7558
1	2	2	0	1	45.51	8.4237	23.9051
2	0	1	0	2	43.47	8.8222	23.3058
2	1	2	1	0	46.07	9.6859	23.4048
2	2	0	2	1	46.67	11.3564	22.8258

Table 19.8.2. Averages for $\bar{y}$

	0	1	2
A	40.7	43.8	45.8
B	41.7	44.9	43.7
C	43.5	43.8	42.9
D	43.8	43.5	42.9
E	42.7	43.5	44.1

significant at the 5% level, and that the linear component of E would be significant if we took $\alpha = 10\%$. This tells us that A, B, and, possibly, E are signal factors.

It is not always really necessary to carry out the analysis of variance procedure. We already saw that in the more obvious cases, one can come to the same conclusions from looking at a table of the averages at the various levels, table 19.8.2. The reason for this is that the use of the factorial experiment, or its orthogonal fractions, enables us to estimate effects by averages.

It is clear from an examination of table 19.8.2 that factors A and B, and, to a lesser extent, E have a big influence upon $\bar{y}$.

Analyzing the variance response, s^2, shows that factor D and, to a certain extent, the linear component of E are important for controlling the variability. The averages are shown in table 19.8.3. It is clear that one should try to operate at the middle level of D. The table of averages of $\ln(s^2)$, Table 19.8.4, points in the same direction.

A reasonable set of recommendations would be to run the process at the middle level of D, and to adjust the level of $\bar{y}$ by altering A and B. If the target value for μ is about 43.0, we should certainly move to the middle level of E because that brings us near to the target with a further reduction in variability.

If the objective is to maximize $\bar{y}$, the decision process is easier. One chooses the "best" values for each of the factors A and B, namely, $A = 2$ and $B = 1$, together with $D = E = 1$, and leaves the choice of the level of C to other considerations.

The use of a signal-to-noise ratio like SN_N makes life simpler for the engineer because there is only one criterion to be considered, but that

Table 19.8.3. Averages for s^2

	0	1	2
A	9.24	9.92	9.73
B	9.47	9.65	9.77
C	9.32	9.70	9.78
D	11.10	8.71	9.09
E	9.89	9.16	9.85

Table 19.8.4. Averages of $\ln(s^2)$

	0	1	2
A	2.219	2.287	2.266
B	2.241	2.262	2.266
C	2.227	2.263	2.279
D	2.405	2.163	2.202
E	2.286	2.202	2.282

simplification may not always be realistic. The averages for the ratio SN_N are given in table 19.8.5.

We could decide that the optimum conditions for our process are those that maximize SN_N for each factor. In this example, they are $A = 2$ and $B = C = D = E = 1$.

19.9. THE CONFIRMATION EXPERIMENT

When all this has been said and done, the result is a set of operating conditions that appear to be the best. They are the best in the sense that if we run the process at those conditions, we will, we hope, manufacture product that is close to the target with a minimum of variability. But there may be a pitfall.

In deciding which of the factors should be varied to minimize the variance and which should be varied to adjust the response, we relied on an orthogonal array—an orthogonal main-effects plan. In the analysis, we assumed that there were no two-factor interactions. But we could have been wrong!

This leads us to the last stage of the experimental plan for parameter design—the confirmation experiment. We should run the outer array at the set of values for the factors in the inner array that we are going to recommend.

The results could be disastrous, in which case we have to stop and think (and be thankful that we did not make a rash recommendation that would stand as a memorial to us). We need some more data. Perhaps we need to expand our inner array to make it a resolution IV fraction, or a resolution V

Table 19.8.5. Averages for SN_N

	0	1	2
A	22.55	22.88	23.37
B	22.65	23.21	22.94
C	22.09	22.98	22.74
D	22.37	23.37	23.06
E	22.66	23.18	22.96

FINALE   341

fraction. This can become costly in terms of both money and time, and calls for some serious thinking and statistical planning.

The results could confirm our earlier conclusions. That happened in the example discussed in section 18.3. We have our optimum conditions, relatively speaking, and the outer array at the confirmation run provides some data about the area around that set of conditions where we will be working. We also have a basis for the next stage of the investigation—tolerance analysis.

19.10. TOLERANCE DESIGN

Having settled on the parameter values of our process, we turn to tolerances. Taguchi recommends that the parameter design experiment be done with low-price components that have, by the nature of their prices, some variability, and not with precision parts in a white coat laboratory setting. If the outer array on our confirmation experiment has shown a satisfactory level of variability, we can go into production with the low-cost parts and a corresponding savings in production costs. If the variability is too high, we are going to have to use better parts in some places. But which parts need to be improved?

A tolerance design experiment is an experiment to find out where the variability occurs and where the adjustments should be made. We run a second outer array centered on the optimum parameter conditions. This time the noise factors are noise in the parts themselves. If factor A is a resistor at a nominal level of 100 ohms, it becomes, in the noise array, a factor with levels, say, 100 ± 2. When the array has been run, we find how much of the variability is associated with each factor, and, hence, where we have to make our further improvements. We may find that A is critical and that the best way to reduce variability is to use very precise, and more costly, resistors for that item; on the other hand, we may find that "wiggling" the level of A does not contribute much to the variability of the final product. Not only will improving the precision of A not help us much, but we might just as well continue to use low-cost parts for A, and save a lot of money.

19.11. FINALE

In the last three chapters, we have given a short introduction to the design of experiments. In this chapter, we have focused on the approach of Taguchi, which has its strengths and its weaknesses.

Until about 60 years ago, experimenters were taught to change one variable at a time. This classical approach to experimentation has two drawbacks: inefficiency and the inability to detect interactions. With three factors at two levels and four runs, we have two choices.

The classical one-at-a-time design consists of four points: (1), a, b, and c; the engineer runs the base point, and then investigates A by keeping B and C fixed and varying A alone. The estimate of the A effect is $a - (1)$ with variance $2\sigma^2$. The 2^{3-1} half replicate also has four points, (1), ab, ac, and bc, but now the A effect is estimated by $[ab + ac - bc - (1)]/2$ with variance σ^2; instead of one point at high A minus one point at low A, we have the average of two at high A and two at low A. Furthermore, the estimates of A, B, and C in the factorial experiment are uncorrelated; the estimates in the one-at-a-time experiment are correlated.

The halving of the variance, or doubling of efficiency, and the statistical independence of the estimates were achieved by considering, and varying, several factors simultaneously. Multifactor experimentation, as we know it, was developed at Rothamsted Agricultural Station by Fisher and Yates in the 1930s; it followed and built upon Fisher's earlier introduction of the procedure for the analysis of variance. Fractional factorials were developed by Finney, also at Rothamsted, during World War II. The strength of these designs lies in their efficiency and their ability to detect interactions. The efficiency comes from the balance. Because the designs are orthogonal, we are able to compare averages—the *average* of all the points at high A minus the *average* of all the points at low A. This property is sometimes called the hidden replication property of factorial designs.

The ideas of multifactor experiments took a while to move from agriculture to engineering. At the end of World War II, scientists at the laboratories of Imperial Chemicals Industries (ICI) in Great Britain, particularly George Box, were using them in chemical engineering research. The methods that they were using were gathered in a book (*Design and Analysis of Industrial Experiments*, edited by O. L. Davies) first published in 1954. For a long time, it was the standard book from which most industrial statisticians learned about design of experiments. It is still very good reading for both engineers and statisticians.

The growth in the use of these ideas in engineering in the United States was slow. The main reason has been that, unlike the Japanese, American engineering education has not included much emphasis on statistical methodology. It is still difficult to find room for statistics courses in most engineering curricula. Now that practicing engineers are aware, particularly in the semiconductor industry, of the importance of statistics, and of the gains that Japanese have made through intensive use of statistical methods, American engineers are learning about designed experiments through programs arranged by their companies, mainly in-plant short courses.

Many of them are hearing about these things for the first time in the framework of the Taguchi methodology. Unfortunately, some of the more zealous disciples of Taguchi get overheated and proclaim that all the ideas of multifactor experimentation are Japanese, and were invented by Taguchi himself. An article in one of the major U.S. newspapers conveyed the absurd allegation that the western statisticians are still advocating one-at-a-time experiments and are unaware of the principles of multifactor experimentation, which it claimed were developed outside the West. Some rabid disciples also go so far as to deny the existence of interactions, arguing that they do not exist in the real world. Such blind partisanship by zealots is regrettable. It can too easily get out of hand and result in engineers blindly following the instructions in the handbook while their brains and engineering judgment are left hanging on the hook outside the door.

No matter what language you speak, the important thing about well-designed experiments is that they are efficient and simple. Because they are orthogonal factorial experiments, and use orthogonal arrays, the calculations are straightforward. But there is more than that. Because of the orthogonality, the results are easy to present and easy to understand.

In the last three chapters, we have outlined a systematic approach to experimentation. But some caveats must be mentioned again. There really are interactions, and you need to watch out for them. That is one of the reasons for the confirmation experiment in section 19.9. Another caveat concerns variability. One of the important uses of the analysis of variance is to obtain a valid estimate of the error. Fisher had to discourage some of his agronomist colleagues from going off half-cocked and claiming that effects were significant when they were not. It is tempting to load up an $L(32)$ lattice with 31 factors, but remember: not only will the interactions (if any) be aliased all over the place, but you will have no good handle on the variability. A policy of picking the winner among the means of three levels of a factor may get you into trouble if there is no significant difference.

Setting up a multifactor experiment in an orthogonal array is a straightforward matter. If you do it properly, you can hardly avoid learning something! Good luck!

ORTHOGONALITY = BALANCE = SIMPLICITY = EFFICIENCY

EXERCISES

19.1. Carry out the details of the analysis of the data set in table 19.5.1, including the calculation of all three signal-to-noise ratios.

19.2. This data set comes from Karlin (1987). He is reviewing a software package and he includes this example of a design with five factors in an $L(8)$ lattice and five observations at each point. It is an example for a Taguchi parameter design.

−1	−1	−1	+1	+1	4.5	9.0	0.5	5.0	3.5
+1	−1	−1	−1	−1	9.5	8.0	3.5	7.0	4.5

−1	+1	−1	−1	+1	0.5	4.0	1.5	6.0	7.0
+1	+1	−1	+1	−1	11.5	9.5	6.6	17.5	9.5
−1	−1	+1	+1	−1	7.0	8.5	19.5	15.5	16.0
+1	−1	+1	−1	+1	2.5	5.0	1.0	7.0	4.5
−1	+1	+1	−1	−1	9.0	13.5	0.5	5.5	7.0
+1	+1	+1	+1	+1	10.5	4.5	4.0	1.5	2.5

Suppose that you were to use the signal-to-noise ratio $(SN)_L$. Which would you regard as the important factors?

19.3. Pigniatello and Ramberg (1985) reported an experiment on the heat treatment of leaf springs on trucks. The engineers considered five factors in the manufacturing process: B, high heat temperature; C, heating time; D, transfer time; E, hold-down time; and O, quench oil temperature. The design of the experiment was an $L(8)$ array in factors B, C, D, and E, which was repeated at both levels of O. There were three runs at each set of conditions. The response, Y, is the free height of the spring. Its target value is eight inches. They were prepared to accept that BC, BD, and CD were the only interactions of interest. The data follow:

B	C	BC	D	BD	CD	E	High O			Low O		
1	1	2	1	2	2	1	7.78	7.78	7.81	7.50	7.25	7.12
2	1	1	1	1	2	2	8.15	8.18	7.88	7.88	7.88	7.44
1	2	1	1	2	1	2	7.50	7.56	7.50	7.50	7.56	7.50
2	2	2	1	1	1	1	7.59	7.56	7.75	7.63	7.75	7.56
1	1	2	2	1	1	2	7.94	8.00	7.88	7.32	7.44	7.44
2	1	1	2	2	1	1	7.69	8.09	8.06	7.56	7.69	7.62
1	2	1	2	1	2	1	7.56	7.62	7.44	7.18	7.18	7.25
2	2	2	2	2	2	2	7.56	7.81	7.69	7.81	7.50	7.59

They took as their signal-to-noise ratio

$$Z = 10 \log_{10}\left(\frac{\bar{y}^2}{s^2}\right).$$

What levels of the factors would you recommend? Which factors would you adjust to minimize the variance? Having done that, which factors would you adjust to move the response to $y = 8.00$?

19.4. Kackar and Shoemaker (1986) used designed experiments to improve an integrated-circuit fabrication process. They investigated eight factors in an $L(16)$ array. For each run, they measured the thickness of the epitaxial layer at five places on each of 14 wafers (70 measurements in all). The factors were: A, susceptor rotation method; B, code of wafers; C, deposition temperature; D, deposition time; E, arsenic gas flow rate; F HCl etch temp.; G, HCl flow rate; and H, nozzle position. The orthogonal array follows together with the values of $\bar{y}$ and s^2 at each point. Which factors should you control to minimize the variance? Which factors should you then adjust to obtain a mean thickness of 14.5 μm?

A	B	C	D	E	F	G	H	$\bar{y}$	log (s^2)
1	1	1	2	1	1	1	1	14.821	−0.4425
1	1	1	2	2	2	2	2	14.888	−1.1989
1	1	2	1	1	1	2	2	14.037	−1.4307
1	1	2	1	2	2	1	1	13.880	−0.6505
1	2	1	1	1	2	1	2	14.165	−1.4230
1	2	1	1	2	1	2	1	13.860	−0.4969
1	2	2	2	1	2	2	1	14.757	−0.3267
1	2	2	2	2	1	1	2	14.921	−0.6270
2	1	1	1	1	2	2	1	13.972	−0.3467
2	1	1	1	2	1	1	2	14.032	−0.8563
2	1	2	2	1	2	1	2	14.843	−0.4369
2	1	2	2	2	1	2	1	14.415	−0.3131
2	2	1	2	1	1	2	2	14.878	−0.6154
2	2	1	2	2	2	1	1	14.932	−0.2292
2	2	2	1	1	1	1	1	13.907	−0.1190
2	2	2	1	2	2	2	2	13.914	−0.8625

19.5. Phadke (1986) used an $L(32)$ lattice in an experiment to compare the lives of router bits. He had eight factors at two levels each and two factors at four levels. He assigned the four-level factors to three columns of the lattice; see chapter 18. The data follow. Y is the number of cycles to failure of the bit. If the bit failed during the eighteenth cycle, for example, Y was recorded as 17.5; 14 of the 32 bits failed during the first cycle. Which were the important factors? Given that large values of Y are desirable, at what levels of the four-level factors should you run your process? Does it matter?

A	B	C	D	E	F	G	H	I	Y
1	1	1	1	1	1	1	1	1	3.5
1	1	1	2	2	2	2	1	1	0.5
1	1	1	3	4	1	2	2	1	0.5
1	1	1	4	3	2	1	2	1	17.5
1	2	2	3	1	2	2	1	1	0.5
1	2	2	4	2	1	1	1	1	2.5
1	2	2	1	4	2	1	2	1	0.5
1	2	2	2	3	1	2	2	1	0.5
2	1	2	4	1	1	2	2	1	17.5

2	1	2	3	2	2	1	2	1	2.5
2	1	2	2	4	1	1	1	1	0.5
2	1	2	1	3	2	2	1	1	3.5
2	2	1	2	1	2	1	2	1	0.5
2	2	1	1	2	1	2	2	1	2.5
2	2	1	4	4	2	2	1	1	0.5
2	2	1	3	3	1	1	1	1	3.5
1	1	1	1	1	1	1	1	2	17.5
1	1	1	2	2	2	2	1	2	0.5
1	1	1	3	4	1	2	2	2	0.5
1	1	1	4	3	2	1	2	2	17.5
1	2	2	3	1	2	2	1	2	0.5
1	2	2	4	2	1	1	1	2	17.5
1	2	2	1	4	2	1	2	2	14.5
1	2	2	2	3	1	2	2	2	0.5
2	1	2	4	1	1	2	2	2	17.5
2	1	2	3	2	2	1	2	2	3.5
2	1	2	2	4	1	1	1	2	17.5
2	1	2	1	3	2	2	1	2	3.5
2	2	1	2	1	2	1	2	2	0.5
2	2	1	1	2	1	2	2	2	3.5
2	2	1	4	4	2	2	1	2	0.5
2	2	1	3	3	1	1	1	2	17.5

References

Box, G. E. P., Hunter, J. S., and Hunter, W. G. (1978), *Statistics for Experimenters*, Wiley, New York.

Church, Alonzo C., Jr. (1966), "Analysis of data when the response is a curve," *Technometrics*, vol. 8, 229–246.

Cornell, J. A. (1981), *Experiments with Mixtures*, Wiley, New York.

Daniel, C. (1959), "Use of half-normal plots in interpreting factorial two-level experiments," *Technometrics*, vol. 1, 311–342.

Daniel, C., and Wood, F. (1980), *Fitting Equations to Data: Computer Analysis of Multifactor Data*, 2nd edition, Wiley, New York.

Davies, O. L., (Ed.) (1956), *Design and Analysis of Industrial Experiments*, 2nd edition, Hafner, New York.

Dodge, H. F. (1943), "A sampling inspection plan for continuous production," *Annals of Mathematical Statistics*, vol. 14, 264–279.

Dodge, H. F. (1955a), "Chain sampling inspection plans," *Industrial Quality Control*, vol. 11, no. 4, 10–13.

Dodge, H. F. (1955b), "Skip-lot sampling plans," *Industrial Quality Control*, vol. 11, no. 5, 3–5.

Dodge, H. F. (1977), Nine of his papers were collected in a memorial issue of *Journal of Quality Technology*, vol. 9, 93–157.

Dodge, H. F., and Romig, H. G. (1959), *Sampling Inspection Tables*, 2nd edition, Wiley, New York.

Dodge, H. F., and Torrey, M. N. (1951), "Additional continuous sampling inspection plans," *Industrial Quality Control*, vol. 7, no. 5, 7–12.

Draper, N. J., and Smith, H. (1981), *Applied Regression Analysis*, 2nd edition, Wiley, New York.

Duncan, A. J. (1986), *Quality Control and Industrial Statistics*, 5th edition, Irwin, Homewood, Illinois.

Dykstra, Otto (1959), "Partial duplication of factorial experiments." *Technometrics*, vol. 1, 63–70.

Finney, D. J. (1945), "The fractional replication of factorial arrangements," *Annals of Eugenics*, vol. 12, 291–301.

Hunter, J. S. (1986), "The exponentially weighted moving average," *Journal of Quality Technology*, vol. 18, no. 4, 203–210.

Hunter, J. S. (1987), "Letter to the editor," *Quality Progress*, vol. 20, May, 7–8.

Ishikawa, K. (1976), *Guide to Quality Control*, Asian Productivity Organization, UNIPUB, New York.

John, Peter W. M. (1971), *Statistical Design and Analysis of Experiments*, Macmillan, New York.

Kackar, R. N., and Shoemaker, A. C. (1986), "Robust Design: A cost-effective method for improving manufacturing processes," *A. T. & T. Technical Journal*, vol. 65, issue 2, 39–50.

Karlin, E. W. (1987), "Software on review," *Quality Progress*, vol. 20, January, 54–57.

Lucas, J. M. (1973), "A modified V-mask control scheme," *Technometrics*, vol. 15, 833–847.

Lucas, J. M. (1976), "The design and use of cumulative sum quality control schemes," *Journal of Quality Technology*, vol. 8, 1–12.

Lucas, J. M. (1982), "Combined Shewhart-cusum quality control schemes," *Journal of Quality Technology*, vol 14, 51–59.

MacBeth, Gordon (1989), "Check thermal effects in photoresist coating process," *Semiconductor International*, May 1989, 130–134.

Montgomery, D. C. (1985), *Introduction to Statistical Quality Control*, Wiley, New York.

Nelson, L. S. (1983a), "Transformations for attribute data," *Journal of Quality Technology*, vol. 15, 55–56.

Nelson, L. S. (1983b), "The deceptiveness of moving averages," *Journal of Quality Technology*, vol. 15, 99.

Oikawa, T., and Oka, T. (1987), "New techniques for approximating the stress in pad-type nozzles attached to a spherical shell," *Transactions of the American Society of Mechanical Engineers*, May, 188–192.

Ott, E. R., and Snee, R. D. (1973), "Identifying useful differences in a multiple-head machine," *Journal of Quality Technology*, vol. 5, 47–57.

Page, E. S. (1954), "Continuous inspection schemes," *Biometrika*, vol. 41, 100–115.

Page, E. S. (1961), "Cumulative sum control charts," *Technometrics*, vol. 3, 1–9.

Phadke, M. S. (1985), "Design optimization case studies," *A. T. & T. Technical Journal*, vol 65, issue 2, 51–68.

Pigniatello, J. J., and Ramberg, J. S. (1985), "Discussion of a paper by R. N. Kackar," *Journal of Quality Technology*, vol. 17, 198–206.

Plackett, R. L., and Burman, J. P. (1946), "The design of optimum multifactorial experiments," *Biometrika*, vol. 33, 305–325.

Roberts, S. W. (1959), "Control chart tests based on geometric moving averages," *Technometrics*, vol. 1, 239–250.

Scheffé, H. (1958), "Experiments with mixtures," *Journal of the Royal Statistical Society*, series B, vol. 20, 344–360.

Schilling, E. G., and Nelson, P. R. (1976), "The effect of nonnormality on the control limits of $\bar{x}$ charts," *Journal of Quality Technology*, vol. 8, 183–188.

REFERENCES

Schilling, E. G. (1982), *Acceptance Sampling in Quality Control*, Dekker, New York.

Schilling, E. G. (1984), "An overview of acceptance control," *Quality Progress*, vol. 17, 22–25.

Shewhart, W. A. (1931), *Economic Control of Quality of Manufactured Product*, Van Nostrand, New York.

Snee, R. D. (1981), "Developing blending models for gasoline and other mixtures," *Technometrics*, vol. 23, 119–130.

Tukey, J. W. (1959), "A quick compact two sample test to Duckworth's specifications," *Technometrics*, vol. 1, 31–48.

Tukey, J. W., (1977), *Exploratory Data Analysis*, Addison-Wesley, Reading, Massachusetts.

Wald, A. (1947), *Sequential Analysis*, Wiley, New York.

Wu, C. F. J., Mao, S. S., and Ma, F. S. (1987), "An investigation of OA-based methods for parameter design optimization," Report No. 24, Center for Quality and Productivity Improvement, University of Wisconsin, Madison.

Yates, F. (1937), "The design and analysis of factorial experiments," Imperial Bureau of Soil Science, Harpenden, England.

Yin, Gerald Z., and Jillie, Don W. (1987), "Orthogonal design for process optimization and its application in plasma etching," *Solid State Technology*, May, 127–132.

Tables

Table I. The Normal Distribution*†

z	.00	.01	.02	.03	.04	.05	.06	.07	.08	.09
.0	.5000	.5040	.5080	.5120	.5160	.5199	.5239	.5279	.5319	.5359
.1	.5398	.5438	.5478	.5517	.5557	.5596	.5636	.5675	.5714	.5753
.2	.5793	.5832	.5871	.5910	.5948	.5987	.6026	.6064	.6103	.6141
.3	.6179	.6217	.6255	.6293	.6331	.6368	.6406	.6443	.6480	.6517
.4	.6554	.6591	.6628	.6664	.6700	.6736	.6772	.6808	.6844	.6879
.5	.6915	.6950	.6985	.7019	.7054	.7088	.7123	.7157	.7190	.7224
.6	.7257	.7291	.7324	.7357	.7389	.7422	.7454	.7486	.7517	.7549
.7	.7580	.7611	.7642	.7673	.7703	.7734	.7764	.7794	.7823	.7852
.8	.7881	.7910	.7939	.7967	.7995	.8023	.8051	.8078	.8106	.8133
.9	.8159	.8186	.8212	.8238	.8264	.8289	.8315	.8340	.8365	.8389
1.0	.8413	.8438	.8461	.8485	.8508	.8531	.8554	.8577	.8599	.8621
1.1	.8643	.8665	.8686	.8708	.8729	.8749	.8770	.8790	.8810	.8830
1.2	.8849	.8869	.8888	.8907	.8925	.8944	.8962	.8980	.8997	.90147
1.3	.90320	.90490	.90658	.90824	.90988	.91149	.91309	.91466	.91621	.91774
1.4	.91924	.92073	.92220	.92364	.92507	.92647	.92785	.92922	.93056	.93189
1.5	.93319	.93448	.93574	.93699	.93822	.93943	.94062	.94179	.94295	.94408
1.6	.94520	.94630	.94738	.94845	.94950	.95053	.95154	.95254	.95352	.95449
1.7	.95543	.95637	.95728	.95818	.95907	.95994	.96080	.96164	.96246	.96327
1.8	.96407	.96485	.96562	.96638	.96712	.96784	.96856	.96926	.96995	.97062
1.9	.97128	.97193	.97257	.97320	.97381	.97441	.97500	.97558	.97615	.97670

Table I (continued)

z	.00	.01	.02	.03	.04	.05	.06	.07	.08	.09
2.0	.97725	.97778	.97831	.97882	.97932	.97982	.98030	.98077	.98124	.98169
2.1	.98214	.98257	.98300	.98341	.98382	.98422	.98461	.98500	.98537	.98574
2.2	.98610	.98645	.98679	.98713	.98745	.98778	.98809	.98840	.98870	.98899
2.3	.98928	.98956	.98983	.9^{2}0097	.9^{2}0358	.9^{2}0613	.9^{2}0863	.9^{2}1106	.9^{2}1344	.9^{2}1576
2.4	.9^{2}1802	.9^{2}2024	.9^{2}2240	.9^{2}2451	.9^{2}2656	.9^{2}2857	.9^{2}3053	.9^{2}3244	.9^{2}3431	.9^{2}3613
2.5	.9^{2}3790	.9^{2}3963	.9^{2}4132	.9^{2}4297	.9^{2}4457	.9^{2}4614	.9^{2}4766	.9^{2}4915	.9^{2}5060	.9^{2}5201
2.6	.9^{2}5339	.9^{2}5473	.9^{2}5604	.9^{2}5731	.9^{2}5855	.9^{2}5975	.9^{2}6093	.9^{2}6207	.9^{2}6319	.9^{2}6427
2.7	.9^{2}6533	.9^{2}6636	.9^{2}6736	.9^{2}6833	.9^{2}6928	.9^{2}7020	.9^{2}7110	.9^{2}7197	.9^{2}7282	.9^{2}7365
2.8	.9^{2}7445	.9^{2}7523	.9^{2}7599	.9^{2}7673	.9^{2}7744	.9^{2}7814	.9^{2}7882	.9^{2}7948	.9^{2}8012	.9^{2}8074
2.9	.9^{2}8134	.9^{2}8193	.9^{2}8250	.9^{2}8305	.9^{2}8359	.9^{2}8411	.9^{2}8462	.9^{2}8511	.9^{2}8559	.9^{2}8605
3.0	.9^{2}8650	.9^{2}8694	.9^{2}8736	.9^{2}8777	.9^{2}8817	.9^{2}8856	.9^{2}8893	.9^{2}8930	.9^{2}8965	.9^{2}8999
3.1	.9^{3}0324	.9^{3}0646	.9^{3}0957	.9^{3}1260	.9^{3}1553	.9^{3}1836	.9^{3}2112	.9^{3}2378	.9^{3}2636	.9^{3}2886
3.2	.9^{3}3129	.9^{3}3363	.9^{3}3590	.9^{3}3810	.9^{3}4024	.9^{3}4230	.9^{3}4429	.9^{3}4623	.9^{3}4810	.9^{3}4991
3.3	.9^{3}5166	.9^{3}5335	.9^{3}5499	.9^{3}5658	.9^{3}5811	.9^{3}5959	.9^{3}6103	.9^{3}6242	.9^{3}6376	.9^{3}6505
3.4	.9^{3}6631	.9^{3}6752	.9^{3}6869	.9^{3}6982	.9^{3}7091	.9^{3}7197	.9^{3}7299	.9^{3}7398	.9^{3}7493	.9^{3}7585
3.5	.9^{3}7674	.9^{3}7759	.9^{3}7842	.9^{3}7922	.9^{3}7999	.9^{3}8074	.9^{3}8146	.9^{3}8215	.9^{3}8282	.9^{3}8347
3.6	.9^{3}8409	.9^{3}8469	.9^{3}8527	.9^{3}8583	.9^{3}8637	.9^{3}8689	.9^{3}8739	.9^{3}8787	.9^{3}8834	.9^{3}8879
3.7	.9^{3}8922	.9^{3}8964	.9^{4}0039	.9^{4}0426	.9^{4}0799	.9^{4}1158	.9^{4}1504	.9^{4}1838	.9^{4}2159	.9^{4}2468
3.8	.9^{4}2765	.9^{4}3052	.9^{4}3327	.9^{4}3593	.9^{4}3848	.9^{4}4094	.9^{4}4331	.9^{4}4558	.9^{4}4777	.9^{4}4988
3.9	.9^{4}5190	.9^{4}5385	.9^{4}5573	.9^{4}5753	.9^{4}5926	.9^{4}6092	.9^{4}6253	.9^{4}6406	.9^{4}6554	.9^{4}6696
4.0	.9^{4}6833	.9^{4}6964	.9^{4}7090	.9^{4}7211	.9^{4}7327	.9^{4}7439	.9^{4}7546	.9^{4}7649	.9^{4}7748	.9^{4}7843

* $\Phi(z) = \dfrac{1}{\sqrt{2\pi}} \int_{-\infty}^{z} e^{-x^2/2}\,dx$

† From *Statistical Tables and Formulas*, by A. Hald, John Wiley and Sons, New York (1952); reproduced by permission of Professor A. Hald and the publishers.

TABLES

Table II. The t-Distribution*,†,‡

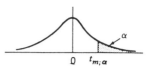

m \ α	.25	.1	.05	.025	.01	.005
1	1.000	3.078	6.314	12.706	31.821	63.657
2	.816	1.886	2.920	4.303	6.965	9.925
3	.765	1.638	2.353	3.182	4.541	5.841
4	.741	1.533	2.132	2.776	3.747	4.604
5	.727	1.476	2.015	2.571	3.365	4.032
6	.718	1.440	1.943	2.447	3.143	3.707
7	.711	1.415	1.895	2.365	2.998	3.499
8	.706	1.397	1.860	2.306	2.896	3.355
9	.703	1.383	1.833	2.262	2.821	3.250
10	.700	1.372	1.812	2.228	2.764	3.169
11	.697	1.363	1.796	2.201	2.718	3.106
12	.695	1.356	1.782	2.179	2.681	3.055
13	.694	1.350	1.771	2.160	2.650	3.012
14	.692	1.345	1.761	2.145	2.624	2.977
15	.691	1.341	1.753	2.131	2.602	2.947
16	.690	1.337	1.746	2.120	2.583	2.921
17	.689	1.333	1.740	2.110	2.567	2.898
18	.688	1.330	1.734	2.101	2.552	2.878
19	.688	1.328	1.729	2.093	2.539	2.861
20	.687	1.325	1.725	2.086	2.528	2.845
21	.686	1.323	1.721	2.080	2.518	2.831
22	.686	1.321	1.717	2.074	2.508	2.819
23	.685	1.319	1.714	2.069	2.500	2.807
24	.685	1.318	1.711	2.064	2.492	2.797
25	.684	1.316	1.708	2.060	2.485	2.787
26	.684	1.315	1.706	2.056	2.479	2.779
27	.684	1.314	1.703	2.052	2.473	2.771
28	.683	1.313	1.701	2.048	2.467	2.763
29	.683	1.311	1.699	2.045	2.462	2.756
30	.683	1.310	1.697	2.042	2.457	2.750
40	.681	1.303	1.684	2.021	2.423	2.704
60	.679	1.296	1.671	2.000	2.390	2.660
120	.677	1.289	1.658	1.980	2.358	2.617
∞	.674	1.282	1.645	1.960	2.326	2.576

* That is, values of $t_{m;\alpha}$, where m equals degrees of freedom and

$$\int_{t_{m;\alpha}}^{\infty} \frac{\Gamma[(m+1)/2]}{\sqrt{\pi m}\,\Gamma(m/2)} \left(1 + \frac{t^2}{m}\right)^{-(m+1)/2} dt = \alpha.$$

† From *Biometrika Tables for Statisticians*, Vol. 1 (2nd edition) Cambridge University Press (1958); edited by E. S. Pearson and H. O. Hartley; reproduced by permission of the publishers.

‡ Where necessary, interpolation should be carried out using the reciprocals of the degrees of freedom, and for this the function $120/m$ is convenient.

Table III. The Chi-Square Distribution*,†

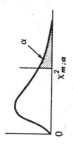

$\chi^2_{m;\alpha}$

α / m	.995	.990	.975	.950	.050	.025	.010	.005
1	392704 × 10⁻¹⁰	157088 × 10⁻⁹	982069 × 10⁻⁹	393214 × 10⁻⁸	3.84146	5.02389	6.63490	7.87944
2	.0100251	.0201007	.0506356	.102587	5.99147	7.37776	9.21034	10.5966
3	.0717212	.114832	.215795	.351846	7.81473	9.34840	11.3449	12.8381
4	.206990	.297110	.484419	.710721	9.48773	11.1433	13.2767	14.8602
5	.411740	.554300	.831211	1.145476	11.0705	12.8325	15.0863	16.7496
6	.675727	.872085	1.237347	1.63539	12.5916	14.4494	16.8119	18.5476
7	.989265	1.239043	1.68987	2.16735	14.0671	16.0128	18.4753	20.2777
8	1.344419	1.646482	2.17973	2.73264	15.5073	17.5346	20.0902	21.9550
9	1.734926	2.087912	2.70039	3.32511	16.9190	19.0228	21.6660	23.5893
10	2.15585	2.55821	3.24697	3.94030	18.3070	20.4831	23.2093	25.1882
11	2.60321	3.05347	3.81575	4.57481	19.6751	21.9200	24.7250	26.7569
12	3.07382	3.57056	4.40379	5.22603	21.0261	23.3367	26.2170	28.2995
13	3.56503	4.10691	5.00874	5.89186	22.3621	24.7356	27.6883	29.8194
14	4.07468	4.66043	5.62872	6.57063	23.6848	26.1190	29.1413	31.3193
15	4.60094	5.22935	6.26214	7.26094	24.9958	27.4884	30.5779	32.8013
16	5.14224	5.81221	6.90766	7.96164	26.2962	28.8454	31.9999	34.2672
17	5.69724	6.40776	7.56418	8.67176	27.5871	30.1910	33.4087	35.7185
18	6.26481	7.01491	8.23075	9.39046	28.8693	31.5264	34.8053	37.1564
19	6.84398	7.63273	8.90655	10.1170	30.1435	32.8523	36.1908	38.5822

Table III (continued)

α \ m	.995	.990	.975	.950	.050	.025	.010	.005
20	7.43386	8.26040	9.59083	10.8508	31.4104	34.1696	37.5662	39.9968
21	8.03366	8.89720	10.28293	11.5913	32.6705	35.4789	38.9321	41.4010
22	8.64272	9.54249	10.9823	12.3380	33.9244	36.7807	40.2894	42.7956
23	9.26042	10.19567	11.6885	13.0905	35.1725	38.0757	41.6384	44.1813
24	9.88623	10.8564	12.4011	13.8484	36.4151	39.3641	42.9798	45.5585
25	10.5197	11.5240	13.1197	14.6114	37.6525	40.6465	44.3141	46.9278
26	11.1603	12.1981	13.8439	15.3791	38.8852	41.9232	45.6417	48.2899
27	11.8076	12.8786	14.5733	16.1513	40.1133	43.1944	46.9630	49.6449
28	12.4613	13.5648	15.3079	16.9279	41.3372	44.4607	48.2782	50.9933
29	13.1211	14.2565	16.0471	17.7083	42.5569	45.7222	49.5879	52.3356
30	13.7867	14.9535	16.7908	18.4926	43.7729	46.9792	50.8922	53.6720
40	20.7065	22.1643	24.4331	26.5093	55.7585	59.3417	63.6907	66.7659
50	27.9907	29.7067	32.3574	34.7642	67.5048	71.4202	76.1539	79.4900
60	35.5346	37.4848	40.4817	43.1879	79.0819	83.2976	88.3794	91.9517
70	43.2752	45.4418	48.7576	51.7393	90.5312	95.0231	100.425	104.215
80	51.1720	53.5400	57.1532	60.3915	101.879	106.629	112.329	116.321
90	59.1963	61.7541	65.6466	69.1260	113.145	118.136	124.116	128.299
100	67.3276	70.0648	74.2219	77.9295	124.342	129.561	135.807	140.169

* That is, values of $\chi^2_{m;\,\alpha}$, where m represents degrees of freedom and

$$\int_{\chi^2_{m,\,\alpha}}^{\infty} \frac{1}{2\Gamma(m/2)} \left(\frac{\chi^2}{2}\right)^{(m/2)-1} e^{-\chi^2/2}\, d\chi^2 = \alpha.$$

For $m < 100$, linear interpolation is adequate. For $m > 100$, $\sqrt{2\chi^2_m}$ is approximately normally distributed with mean $\sqrt{2m-1}$ and unit variance, so that percentage points may be obtained from Table II.

† From *Biometrika Tables for Statisticians*, Vol. 1 (2nd edition), Cambridge University Press (1958); edited by E. S. Pearson and H. O. Hartley; reproduced by permission of the publishers.

Table IV. The F Distribution*,†

$\alpha = .10$

m_2 \ m_1	1	2	3	4	5	6	7	8	9
1	39.864	49.500	53.593	55.833	57.241	58.204	58.906	59.439	59.858
2	8.5263	9.0000	9.1618	9.2434	9.2926	9.3255	9.3491	9.3668	9.3805
3	5.5383	5.4624	5.3908	5.3427	5.3092	5.2847	5.2662	5.2517	5.2400
4	4.5448	4.3246	4.1908	4.1073	4.0506	4.0098	3.9790	3.9549	3.9357
5	4.0604	3.7797	3.6195	3.5202	3.4530	3.4045	3.3679	3.3393	3.3163
6	3.7760	3.4633	3.2888	3.1808	3.1075	3.0546	3.0145	2.9830	2.9577
7	3.5894	3.2574	3.0741	2.9605	2.8833	2.8274	2.7849	2.7516	2.7247
8	3.4579	3.1131	2.9238	2.8064	2.7265	2.6683	2.6241	2.5893	2.5612
9	3.3603	3.0065	2.8129	2.6927	2.6106	2.5509	2.5053	2.4694	2.4403
10	3.2850	2.9245	2.7277	2.6053	2.5216	2.4606	2.4140	2.3772	2.3473
11	3.2252	2.8595	2.6602	2.5362	2.4512	2.3891	2.3416	2.3040	2.2735
12	3.1765	2.8068	2.6055	2.4801	2.3940	2.3310	2.2828	2.2446	2.2135
13	3.1362	2.7632	2.5603	2.4337	2.3467	2.2830	2.2341	2.1953	2.1638
14	3.1022	2.7265	2.5222	2.3947	2.3069	2.2426	2.1931	2.1539	2.1220
15	3.0732	2.6952	2.4898	2.3614	2.2730	2.2081	2.1582	2.1185	2.0862
16	3.0481	2.6682	2.4618	2.3327	2.2438	2.1783	2.1280	2.0880	2.0553
17	3.0262	2.6446	2.4374	2.3077	2.2183	2.1524	2.1017	2.0613	2.0284
18	3.0070	2.6239	2.4160	2.2858	2.1958	2.1296	2.0785	2.0379	2.0047
19	2.9899	2.6056	2.3970	2.2663	2.1760	2.1094	2.0580	2.0171	1.9836
20	2.9747	2.5893	2.3801	2.2489	2.1582	2.0913	2.0397	1.9985	1.9649
21	2.9609	2.5746	2.3649	2.2333	2.1423	2.0751	2.0232	1.9819	1.9480
22	2.9486	2.5613	2.3512	2.2193	2.1279	2.0605	2.0084	1.9668	1.9327
23	2.9374	2.5493	2.3387	2.2065	2.1149	2.0472	1.9949	1.9531	1.9189
24	2.9271	2.5383	2.3274	2.1949	2.1030	2.0351	1.9826	1.9407	1.9063
25	2.9177	2.5283	2.3170	2.1843	2.0922	2.0241	1.9714	1.9292	1.8947
26	2.9091	2.5191	2.3075	2.1745	2.0822	2.0139	1.9610	1.9188	1.8841
27	2.9012	2.5106	2.2987	2.1655	2.0730	2.0045	1.9515	1.9091	1.8743
28	2.8939	2.5028	2.2906	2.1571	2.0645	1.9959	1.9427	1.9001	1.8652
29	2.8871	2.4955	2.2831	2.1494	2.0566	1.9878	1.9345	1.8918	1.8560
30	2.8807	2.4887	2.2761	2.1422	2.0492	1.9803	1.9269	1.8841	1.8498
40	2.8354	2.4404	2.2261	2.0909	1.9968	1.9269	1.8725	1.8289	1.7929
60	2.7914	2.3932	2.1774	2.0410	1.9457	1.8747	1.8194	1.7748	1.7380
120	2.7478	2.3473	2.1300	1.9923	1.8959	1.8238	1.7675	1.7220	1.6843
∞	2.7055	2.3026	2.0838	1.9449	1.8473	1.7741	1.7167	1.6702	1.6315

Table IV (continued)

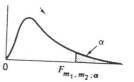

$\alpha = .10$

10	12	15	20	24	30	40	60	120	∞
60.195	60.705	61.220	61.740	62.002	62.265	62.529	62.794	63.061	63.328
9.3916	9.4081	9.4247	9.4413	9.4496	9.4579	9.4663	9.4746	9.4829	9.4913
5.2304	5.2156	5.2003	5.1845	5.1764	5.1681	5.1597	5.1512	5.1425	5.1337
3.9199	3.8955	3.8689	3.8443	3.8310	3.8174	3.8036	3.7896	3.7753	3.7607
3.2974	3.2682	3.2380	3.2067	3.1905	3.1741	3.1573	3.1402	3.1228	3.1050
2.9369	2.9047	2.8712	2.8363	2.8183	2.8000	2.7812	2.7620	2.7423	2.7222
2.7025	2.6681	2.6322	2.5947	2.5753	2.5555	2.5351	2.5142	2.4928	2.4708
2.5380	2.5020	2.4642	2.4246	2.4041	2.3830	2.3614	2.3391	2.3162	2.2926
2.4163	2.3789	2.3396	2.2983	2.2768	2.2547	2.2320	2.2085	2.1843	2.1592
2.3226	2.2841	2.2435	2.2007	2.1784	2.1554	2.1317	2.1072	2.0818	2.0554
2.2482	2.2087	2.1671	2.1230	2.1000	2.0762	2.0516	2.0261	1.9997	1.9721
2.1878	2.1474	2.1049	2.0597	2.0360	2.0115	1.9861	1.9597	1.9323	1.9036
2.1376	2.0966	2.0532	2.0070	1.9827	1.9576	1.9315	1.9043	1.8759	1.8462
2.0954	2.0537	2.0095	1.9625	1.9377	1.9119	1.8852	1.8572	1.8280	1.7973
2.0593	2.0171	1.9722	1.9243	1.8990	1.8728	1.8454	1.8168	1.7867	1.7551
2.0281	1.9854	1.9399	1.8913	1.8656	1.8388	1.8108	1.7816	1.7507	1.7182
2.0009	1.9577	1.9117	1.8624	1.8362	1.8090	1.7805	1.7506	1.7191	1.6856
1.9770	1.9333	1.8868	1.8368	1.8103	1.7827	1.7537	1.7232	1.6910	1.6567
1.9557	1.9117	1.8647	1.8142	1.7873	1.7592	1.7298	1.6988	1.6659	1.6308
1.9367	1.8924	1.8449	1.7938	1.7667	1.7382	1.7083	1.6768	1.6433	1.6074
1.9197	1.8750	1.8272	1.7756	1.7481	1.7193	1.6890	1.6569	1.6228	1.5862
1.9043	1.8593	1.8111	1.7590	1.7312	1.7021	1.6714	1.6389	1.6042	1.5668
1.8903	1.8450	1.7964	1.7439	1.7159	1.6864	1.6554	1.6224	1.5871	1.5490
1.8775	1.8319	1.7831	1.7302	1.7019	1.6721	1.6407	1.6073	1.5715	1.5327
1.8658	1.8200	1.7708	1.7175	1.6890	1.6589	1.6272	1.5934	1.5570	1.5176
1.8550	1.8090	1.7596	1.7059	1.6771	1.6468	1.6147	1.5805	1.5437	1.5036
1.8451	1.7989	1.7492	1.6951	1.6662	1.6356	1.6032	1.5686	1.5313	1.4906
1.8359	1.7895	1.7395	1.6852	1.6560	1.6252	1.5925	1.5575	1.5198	1.4784
1.8274	1.7808	1.7306	1.6759	1.6465	1.6155	1.5825	1.5472	1.5090	1.4670
1.8195	1.7727	1.7223	1.6673	1.6377	1.6065	1.5732	1.5376	1.4989	1.4564
1.7627	1.7146	1.6624	1.6052	1.5741	1.5411	1.5056	1.4672	1.4248	1.3769
1.7070	1.6574	1.6034	1.5435	1.5107	1.4755	1.4373	1.3952	1.3476	1.2915
1.6524	1.6012	1.5450	1.4821	1.4472	1.4094	1.3676	1.3203	1.2646	1.1926
1.5987	1.5458	1.4871	1.4206	1.3832	1.3419	1.2951	1.2400	1.1686	1.0000

Table IV (*continued*)

$\alpha = .05$

m_2 \ m_1	1	2	3	4	5	6	7	8	9
1	161.45	199.50	215.71	224.58	230.16	233.99	236.77	238.88	240.54
2	18.513	19.000	19.164	19.247	19.296	19.330	19.353	19.371	19.385
3	10.128	9.5521	9.2766	9.1172	9.0135	8.9406	8.8868	8.8452	8.8123
4	7.7086	6.9443	6.5914	6.3883	6.2560	6.1631	6.0942	6.0410	5.9988
5	6.6079	5.7861	5.4095	5.1922	5.0503	4.9503	4.8759	4.8183	4.7725
6	5.9874	5.1433	4.7571	4.5337	4.3874	4.2839	4.2066	4.1468	4.0990
7	5.5914	4.7374	4.3468	4.1203	3.9715	3.8660	3.7870	3.7257	3.6767
8	5.3177	4.4590	4.0662	3.8378	3.6875	3.5806	3.5005	3.4381	3.3881
9	5.1174	4.2565	3.8626	3.6331	3.4817	3.3738	3.2927	3.2296	3.1789
10	4.9646	4.1028	3.7083	3.4780	3.3258	3.2172	3.1355	3.0717	3.0204
11	4.8443	3.9823	3.5874	3.3567	3.2039	3.0946	3.0123	2.9480	2.8962
12	4.7472	3.8853	3.4903	3.2592	3.1059	2.9961	2.9134	2.8486	2.7964
13	4.6672	3.8056	3.4105	3.1791	3.0254	2.9153	2.8321	2.7669	2.7144
14	4.6001	3.7389	3.3439	3.1122	2.9582	2.8477	2.7642	2.6987	2.6458
15	4.5431	3.6823	3.2874	3.0556	2.9013	2.7905	2.7066	2.6408	2.5876
16	4.4940	3.6337	3.2389	3.0069	2.8524	2.7413	2.6572	2.5911	2.5377
17	4.4513	3.5915	3.1968	2.9647	2.8100	2.6987	2.6143	2.5480	2.4943
18	4.4139	3.5546	3.1599	2.9277	2.7729	2.6613	2.5767	2.5102	2.4563
19	4.3808	3.5219	3.1274	2.8951	2.7401	2.6283	2.5435	2.4768	2.4227
20	4.3513	3.4928	3.0984	2.8661	2.7109	2.5990	2.5140	2.4471	2.3928
21	4.3248	3.4668	3.0725	2.8401	2.6848	2.5727	2.4876	2.4205	2.3661
22	4.3009	3.4434	3.0491	2.8167	2.6613	2.5491	2.4638	2.3965	2.3419
23	4.2793	3.4221	3.0280	2.7955	2.6400	2.5277	2.4422	2.3748	2.3201
24	4.2597	3.4028	3.0088	2.7763	2.6207	2.5082	2.4226	2.3551	2.3002
25	4.2417	3.3852	2.9912	2.7587	2.6030	2.4904	2.4047	2.3371	2.2821
26	4.2252	3.3690	2.9751	2.7426	2.5868	2.4741	2.3883	2.3205	2.2655
27	4.2100	3.3541	2.9604	2.7278	2.5719	2.4591	2.3732	2.3053	2.2501
28	4.1960	3.3404	2.9467	2.7141	2.5581	2.4453	2.3593	2.2913	2.2360
29	4.1830	3.3277	2.9340	2.7014	2.5454	2.4324	2.3463	2.2782	2.2229
30	4.1709	3.3158	2.9223	2.6896	2.5336	2.4205	2.3343	2.2662	2.2107
40	4.0848	3.2317	2.8387	2.6060	2.4495	2.3359	2.2490	2.1802	2.1240
60	4.0012	3.1504	2.7581	2.5252	2.3683	2.2540	2.1665	2.0970	2.0401
120	3.9201	3.0718	2.6802	2.4472	2.2900	2.1750	2.0867	2.0164	1.9588
∞	3.8415	2.9957	2.6049	2.3719	2.2141	2.0986	2.0096	1.9384	1.8799

Table IV (*continued*)

$\alpha = .05$

10	12	15	20	24	30	40	60	120	∞
241.88	243.91	245.95	248.01	249.05	250.09	251.14	252.20	253.25	254.32
19.396	19.413	19.429	19.446	19.454	19.462	19.471	19.479	19.487	19.496
8.7855	8.7446	8.7029	8.6602	8.6385	8.6166	8.5944	8.5720	8.5494	8.5265
5.9644	5.9117	5.8578	5.8025	5.7744	5.7459	5.7170	5.6878	5.6581	5.6281
4.7351	4.6777	4.6188	4.5581	4.5272	4.4957	4.4638	4.4314	4.3984	4.3650
4.0600	3.9999	3.9381	3.8742	3.8415	3.8082	3.7743	3.7398	3.7047	3.6688
3.6365	3.5747	3.5108	3.4445	3.4105	3.3758	3.3404	3.3043	3.2674	3.2298
3.3472	3.2840	3.2184	3.1503	3.1152	3.0794	3.0428	3.0053	2.9669	2.9276
3.1373	3.0729	3.0061	2.9365	2.9005	2.8637	2.8259	2.7872	2.7475	2.7067
2.9782	2.9130	2.8450	2.7740	2.7372	2.6996	2.6609	2.6211	2.5801	2.5379
2.8536	2.7876	2.7186	2.6464	2.6090	2.5705	2.5309	2.4901	2.4480	2.4045
2.7534	2.6866	2.6169	2.5436	2.5055	2.4663	2.4259	2.3842	2.3410	2.2962
2.6710	2.6037	2.5331	2.4589	2.4202	2.3803	2.3392	2.2966	2.2524	2.2064
2.6021	2.5342	2.4630	2.3879	2.3487	2.3082	2.2664	2.2230	2.1778	2.1307
2.5437	2.4753	2.4035	2.3275	2.2878	2.2468	2.2043	2.1601	2.1141	2.0658
2.4935	2.4247	2.3522	2.2756	2.2354	2.1938	2.1507	2.1058	2.0589	2.0096
2.4499	2.3807	2.3077	2.2304	2.1898	2.1477	2.1040	2.0584	2.0107	1.9604
2.4117	2.3421	2.2686	2.1906	2.1497	2.1071	2.0629	2.0166	1.9681	1.9168
2.3779	2.3080	2.2341	2.1555	2.1141	2.0712	2.0264	1.9796	1.9302	1.8780
2.3479	2.2776	2.2033	2.1242	2.0825	2.0391	1.9938	1.9464	1.8963	1.8432
2.3210	2.2504	2.1757	2.0960	2.0540	2.0102	1.9645	1.9165	1.8657	1.8117
2.2967	2.2258	2.1508	2.0707	2.0283	1.9842	1.9380	1.8895	1.8380	1.7831
2.2747	2.2036	2.1282	2.0476	2.0050	1.9605	1.9139	1.8649	1.8128	1.7570
2.2547	2.1834	2.1077	2.0267	1.9838	1.9390	1.8920	1.8424	1.7897	1.7331
2.2365	2.1649	2.0889	2.0075	1.9643	1.9192	1.8718	1.8217	1.7684	1.7110
2.2197	2.1479	2.0716	1.9898	1.9464	1.9010	1.8533	1.8027	1.7488	1.6906
2.2043	2.1323	2.0558	1.9736	1.9299	1.8842	1.8361	1.7851	1.7307	1.6717
2.1900	2.1179	2.0411	1.9586	1.9147	1.8687	1.8203	1.7689	1.7138	1.6541
2.1768	2.1045	2.0275	1.9446	1.9005	1.8543	1.8055	1.7537	1.6981	1.6377
2.1646	2.0921	2.0148	1.9317	1.8874	1.8409	1.7918	1.7396	1.6835	1.6223
2.0772	2.0035	1.9245	1.8389	1.7929	1.7444	1.6928	1.6373	1.5766	1.5089
1.9926	1.9174	1.8364	1.7480	1.7001	1.6491	1.5943	1.5343	1.4673	1.3893
1.9105	1.8337	1.7505	1.6587	1.6084	1.5543	1.4952	1.4290	1.3519	1.2539
1.8307	1.7522	1.6664	1.5705	1.5173	1.4591	1.3940	1.3180	1.2214	1.0000

Table IV (*continued*)

$\alpha = .025$

m_2 \ m_1	1	2	3	4	5	6	7	8	9
1	647.79	799.50	864.16	899.58	921.85	937.11	948.22	956.66	963.28
2	38.506	39.000	39.165	39.248	39.298	39.331	39.355	39.373	39.387
3	17.443	16.044	15.439	15.101	14.885	14.735	14.624	14.540	14.473
4	12.218	10.649	9.9792	9.6045	9.3645	9.1973	9.0741	8.9796	8.9047
5	10.007	8.4336	7.7636	7.3879	7.1464	6.9777	6.8531	6.7572	6.6810
6	8.8131	7.2598	6.5988	6.2272	5.9876	5.8197	5.6955	5.5996	5.5234
7	8.0727	6.5415	5.8898	5.5226	5.2852	5.1186	4.9949	4.8994	4.8232
8	7.5709	6.0595	5.4160	5.0526	4.8173	4.6517	4.5286	4.4332	4.3572
9	7.2093	5.7147	5.0781	4.7181	4.4844	4.3197	4.1971	4.1020	4.0260
10	6.9367	5.4564	4.8256	4.4683	4.2361	4.0721	3.9498	3.8549	3.7790
11	6.7241	5.2559	4.6300	4.2751	4.0440	3.8807	3.7586	3.6638	3.5879
12	6.5538	5.0959	4.4742	4.1212	3.8911	3.7283	3.6065	3.5118	3.4358
13	6.4143	4.9653	4.3472	3.9959	3.7667	3.6043	3.4827	3.3880	3.3120
14	6.2979	4.8567	4.2417	3.8919	3.6634	3.5014	3.3799	3.2853	3.2093
15	6.1995	4.7650	4.1528	3.8043	3.5764	3.4147	3.2934	3.1987	3.1227
16	6.1151	4.6867	4.0768	3.7294	3.5021	3.3406	3.2194	3.1248	3.0488
17	6.0420	4.6189	4.0112	3.6648	3.4379	3.2767	3.1556	3.0610	2.9849
18	5.9781	4.5597	3.9539	3.6083	3.3820	3.2209	3.0999	3.0053	2.9291
19	5.9216	4.5075	3.9034	3.5587	3.3327	3.1718	3.0509	2.9563	2.8800
20	5.8715	4.4613	3.8587	3.5147	3.2891	3.1283	3.0074	2.9128	2.8365
21	5.8266	4.4199	3.8188	3.4754	3.2501	3.0895	2.9686	2.8740	2.7977
22	5.7863	4.3828	3.7829	3.4401	3.2151	3.0546	2.9338	2.8392	2.7628
23	5.7498	4.3492	3.7505	3.4083	3.1835	3.0232	2.9024	2.8077	2.7313
24	5.7167	4.3187	3.7211	3.3794	3.1548	2.9946	2.8738	2.7791	2.7027
25	5.6864	4.2909	3.6943	3.3530	3.1287	2.9685	2.8478	2.7531	2.6766
26	5.6586	4.2655	3.6697	3.3289	3.1048	2.9447	2.8240	2.7293	2.6528
27	5.6331	4.2421	3.6472	3.3067	3.0828	2.9228	2.8021	2.7074	2.6309
28	5.6096	4.2205	3.6264	3.2863	3.0625	2.9027	2.7820	2.6872	2.6106
29	5.5878	4.2006	3.6072	3.2674	3.0438	2.8840	2.7633	2.6686	2.5919
30	5.5675	4.1821	3.5894	3.2499	3.0265	2.8667	2.7460	2.6513	2.5746
40	5.4239	4.0510	3.4633	3.1261	2.9037	2.7444	2.6238	2.5289	2.4519
60	5.2857	3.9253	3.3425	3.0077	2.7863	2.6274	2.5068	2.4117	2.3344
120	5.1524	3.8046	3.2270	2.8943	2.6740	2.5154	2.3948	2.2994	2.2217
∞	5.0239	3.6889	3.1161	2.7858	2.5665	2.4082	2.2875	2.1918	2.1136

Table IV (*continued*)

$\alpha = .025$

10	12	15	20	24	30	40	60	120	∞
968.63	976.71	984.87	993.10	997.25	1001.4	1005.6	1009.8	1014.0	1018.3
39.398	39.415	39.431	39.448	39.456	39.456	39.473	39.481	39.490	39.498
14.419	14.337	14.253	14.167	14.124	14.081	14.037	13.992	13.947	13.902
8.8439	8.7512	8.6565	8.5599	8.5109	8.4613	8.4111	8.3604	8.3092	8.2573
6.6192	6.5246	6.4277	6.3285	6.2780	6.2269	6.1751	6.1225	6.0693	6.0153
5.4613	5.3662	5.2687	5.1684	5.1172	5.0652	5.0125	5.9589	4.9045	4.8491
4.7611	4.6658	4.5678	4.4667	4.4150	4.3624	4.3089	4.2544	4.1989	4.1423
4.2951	4.1997	4.1012	3.9995	3.9472	3.8940	3.8398	3.7844	3.7279	3.6702
3.9639	3.8682	3.7694	3.6669	3.6142	3.5604	3.5055	3.4493	3.3918	3.3329
3.7168	3.6209	3.5217	3.4186	3.3654	3.3110	3.2554	3.1984	3.1399	3.0798
3.5257	3.4296	3.3299	3.2261	3.1725	3.1176	3.0613	3.0035	2.9441	2.8828
3.3736	3.2773	3.1772	3.0728	3.0187	2.9633	2.9063	2.8478	2.7874	2.7249
3.2497	3.1532	3.0527	2.9477	2.8932	2.8373	2.7797	2.7204	2.6590	2.5955
3.1469	3.0501	2.9493	2.8437	2.7888	2.7324	2.6742	2.6142	2.5519	2.4872
3.0602	2.9633	2.8621	2.7559	2.7006	2.6437	2.5850	2.5242	2.4611	2.3953
2.9862	2.8890	2.7875	2.6808	2.6252	2.5678	2.5085	2.4471	2.3831	2.3163
2.9222	2.8249	2.7230	2.6158	2.5598	2.5021	2.4422	2.3801	2.3153	2.2474
2.8664	2.7689	2.6667	2.5590	2.5027	2.4445	2.3842	2.3214	2.2558	2.1869
2.8173	2.7196	2.6171	2.5089	2.4523	2.3937	2.3329	2.2695	2.2032	2.1333
2.7737	2.6758	2.5731	2.4645	2.4076	2.3486	2.2873	2.2234	2.1562	2.0853
2.7348	2.6368	2.5338	2.4247	2.3675	2.3082	2.2465	2.1819	2.1141	2.0422
2.6998	2.6017	2.4984	2.3890	2.3315	2.2718	2.2097	2.1446	2.0760	2.0032
2.6682	2.5699	2.4665	2.3567	2.2989	2.2389	2.1763	2.1107	2.0415	1.9677
2.6396	2.5412	2.4374	2.3273	2.2693	2.2090	2.1460	2.0799	2.0099	1.9353
2.6135	2.5149	2.4110	2.3005	2.2422	2.1816	2.1183	2.0517	1.9811	1.9055
2.5895	2.4909	2.3867	2.2759	2.2174	2.1565	2.0928	2.0257	1.9545	1.8781
2.5676	2.4688	2.3644	2.2533	2.1946	2.1334	2.0693	2.0018	1.9299	1.8527
2.5473	2.4484	2.3438	2.2324	2.1735	2.1121	2.0477	1.9796	1.9072	1.8291
2.5286	2.4295	2.3248	2.2131	2.1540	2.0923	2.0276	1.9591	1.8861	1.8072
2.5112	2.4120	2.3072	2.1952	2.1359	2.0739	2.0089	1.9400	1.8664	1.7867
2.3882	2.2882	2.1819	2.0677	2.0069	1.9429	1.8752	1.8028	1.7242	1.6371
2.2702	2.1692	2.0613	1.9445	1.8817	1.8152	1.7440	1.6668	1.5810	1.4822
2.1570	2.0548	1.9450	1.8249	1.7597	1.6899	1.6141	1.5299	1.4327	1.3104
2.0483	1.9447	1.8326	1.7085	1.6402	1.5660	1.4835	1.3883	1.2684	1.0000

Table IV (continued)

$\alpha = .01$

m_1 \ m_2	1	2	3	4	5	6	7	8	9
1	4052.2	4999.5	5403.3	5624.6	5763.7	5859.0	5928.3	5981.6	6022.5
2	98.503	99.000	99.166	99.249	99.299	99.332	99.356	99.374	99.388
3	34.116	30.817	29.457	28.710	28.237	27.911	27.672	27.489	27.345
4	21.198	18.000	16.694	15.977	15.522	15.207	14.976	14.799	14.659
5	16.258	13.274	12.060	11.392	10.967	10.672	10.456	10.289	10.158
6	13.745	10.925	9.7795	9.1483	8.7459	8.4661	8.2600	8.1016	7.9761
7	12.246	9.5466	8.4513	7.8467	7.4604	7.1914	6.9928	6.8401	6.7188
8	11.259	8.6491	7.5910	7.0060	6.6318	6.3707	6.1776	6.0289	5.9106
9	10.561	8.0215	6.9919	6.4221	6.0569	5.8018	5.6129	5.4671	5.3511
10	10.044	7.5594	6.5523	5.9943	5.6363	5.3858	5.2001	5.0567	4.9424
11	9.6460	7.2057	6.2167	5.6683	5.3160	5.0692	4.8861	4.7445	4.6315
12	9.3302	6.9266	5.9526	5.4119	5.0643	4.8206	4.6395	4.4994	4.3875
13	9.0738	6.7010	5.7394	5.2053	4.8616	4.6204	4.4410	4.3021	4.1911
14	8.8616	6.5149	5.5639	5.0354	4.6950	4.4558	4.2779	4.1399	4.0297
15	8.6831	6.3589	5.4170	4.8932	4.5556	4.3183	4.1415	4.0045	3.8948
16	8.5310	6.2262	5.2922	4.7726	4.4374	4.2016	4.0259	3.8896	3.7804
17	8.3997	6.1121	5.1850	4.6690	4.3359	4.1015	3.9267	3.7910	3.6822
18	8.2854	6.0129	5.0919	4.5790	4.2479	4.0146	3.8406	3.7054	3.5971
19	8.1850	5.9259	5.0103	4.5003	4.1708	3.9386	3.7653	3.6305	3.5225
20	8.0960	5.8489	4.9382	4.4307	4.1027	3.8714	3.6987	3.5644	3.4567
21	8.0166	5.7804	4.8740	4.3688	4.0421	3.8117	3.6396	3.5056	3.9381
22	7.9454	5.7190	4.8166	4.3134	3.9880	3.7583	3.5867	3.4530	3.3458
23	7.8811	5.6637	4.7649	4.2635	3.9392	3.7102	3.5390	3.4057	3.2986
24	7.8229	5.6136	4.7181	4.2184	3.8951	3.6667	3.4959	3.3629	3.2560
25	7.7698	5.5680	4.6755	4.1774	3.8550	3.6272	3.4568	3.3239	3.2172
26	7.7213	5.5263	4.6366	4.1400	3.8183	3.5911	3.4210	3.2884	3.1818
27	7.6767	5.4881	4.6009	4.1056	3.7848	3.5580	3.3882	3.2558	3.1494
28	7.6356	5.4529	4.5681	4.0740	3.7539	3.5276	3.3581	3.2259	3.1195
29	7.5976	5.4205	4.5378	4.0449	3.7254	3.4995	3.3302	3.1982	3.0920
30	7.5625	5.3904	4.5097	4.0179	3.6990	3.4735	3.3045	3.1726	3.0665
40	7.3141	5.1785	4.3126	3.8283	3.5138	3.2910	3.1238	2.9930	2.8876
60	7.0771	4.9774	4.1259	3.6491	3.3389	3.1187	2.9530	2.8233	2.7185
120	6.8510	4.7865	3.9493	3.4796	3.1735	2.9559	2.7918	2.6629	2.5586
∞	6.6349	4.6052	3.7816	3.3192	3.0173	2.8020	2.6393	2.5113	2.4073

Table IV (*continued*)

$\alpha = .01$

10	12	15	20	24	30	40	60	120	∞
6055.8	6106.3	6157.3	6208.7	6234.6	6260.7	6286.8	6313.0	6339.4	6366.0
99.399	99.416	99.432	99.449	99.458	99.466	99.474	99.483	99.491	99.501
27.229	27.052	26.872	26.690	26.598	26.505	26.411	26.316	26.221	26.125
14.546	14.374	14.198	14.020	13.929	13.838	13.745	13.652	13.558	13.463
10.051	9.8883	9.7222	9.5527	9.4665	9.3793	9.2912	9.2020	9.1118	9.0204
7.8741	7.7183	7.5590	7.3958	7.3127	7.2285	7.1432	7.0568	6.9690	6.8801
6.6201	6.4691	6.3143	6.1554	6.0743	5.9921	5.9084	5.8236	5.7372	5.6495
5.8143	5.6668	5.5151	5.3591	5.2793	5.1981	5.1156	5.0316	4.9460	4.8588
5.2565	5.1114	4.9621	4.8080	4.7290	4.6486	4.5667	4.4831	4.3978	4.3105
4.8492	4.7059	4.5582	4.4054	4.3269	4.2469	4.1653	4.0819	3.9965	3.9090
4.5393	4.3974	4.2509	4.0990	4.0209	3.9411	3.8596	3.7761	3.6904	3.6025
4.2961	4.1553	4.0096	3.8584	3.7805	3.7008	3.6192	3.5355	3.4494	3.3608
4.1003	3.9603	3.8154	3.6646	3.5868	3.5070	3.4253	3.3413	3.2548	3.1654
3.9394	3.8001	3.6557	3.5052	3.4274	3.3476	3.2656	3.1813	3.0942	3.0040
3.8049	3.6662	3.5222	3.3719	3.2940	3.2141	3.1319	3.0471	2.9595	2.8684
3.6909	3.5527	3.4089	3.2588	3.1808	3.1007	3.0182	2.9330	2.8447	2.7528
3.5931	3.4552	3.3117	3.1615	3.0835	3.0032	2.9205	2.8348	2.7459	2.6530
3.5082	3.3706	3.2273	3.0771	2.9990	2.9185	2.8354	2.7493	2.6597	2.5660
3.4338	3.2965	3.1533	3.0031	2.9249	2.8442	2.7608	2.6742	2.5839	2.4893
3.3682	3.2311	3.0880	2.9377	2.8594	2.7785	2.6947	2.6077	2.5168	2.4212
3.3098	3.1729	3.0299	2.8796	2.8011	2.7200	2.6359	2.5484	2.4568	2.3603
3.2576	3.1209	2.9780	2.8274	2.7488	2.6675	2.5831	2.4951	2.4029	2.3055
3.2106	3.0740	2.9311	2.7805	2.7017	2.6202	2.5355	2.4471	2.3542	2.2559
3.1681	3.0316	2.8887	2.7380	2.6591	2.5773	2.4923	2.4035	2.3099	2.2107
3.1294	2.9931	2.8502	2.6993	2.6203	2.5383	2.4530	2.3637	2.2695	2.1694
3.0941	2.9579	2.8150	2.6640	2.5848	2.5026	2.4170	2.3273	2.2325	2.1315
3.0618	2.9256	2.7827	2.6316	2.5522	2.4699	2.3840	2.2938	2.1984	2.0965
3.0320	2.8959	2.7530	2.6017	2.5223	2.4397	2.3535	2.2629	2.1670	2.0642
3.0045	2.8685	2.7256	2.5742	2.4946	2.4118	2.3253	2.2344	2.1378	2.0342
2.9791	2.8431	2.7002	2.5487	2.4689	2.3860	2.2992	2.2079	2.1107	2.0062
2.8005	2.6648	2.5216	2.3689	2.2880	2.2034	2.1142	2.0194	1.9172	1.8047
2.6318	2.4961	2.3523	2.1978	2.1154	2.0285	1.9360	1.8363	1.7263	1.6006
2.4721	2.3363	2.1915	2.0346	1.9500	1.8600	1.7628	1.6557	1.5330	1.3805
2.3209	2.1848	2.0385	1.8783	1.7908	1.6964	1.5923	1.4730	1.3246	1.0000

Table IV (*continued*)

$\alpha = .005$

m_1 / m_2	1	2	3	4	5	6	7	8	9
1	16211	20000	21615	22500	23056	23437	23715	23925	24091
2	198.50	199.00	199.17	199.25	199.30	199.33	199.36	199.37	199.39
3	55.552	49.799	47.467	46.195	45.392	44.838	44.434	44.126	43.882
4	31.333	26.284	24.259	23.155	22.456	21.975	21.622	21.352	21.139
5	22.785	18.314	16.530	15.556	14.940	14.513	14.200	13.961	13.772
6	18.635	14.544	12.917	12.028	11.464	11.073	10.786	10.566	10.391
7	16.236	12.404	10.882	10.050	9.5221	9.1554	8.8854	8.6781	8.5138
8	14.688	11.042	9.5965	8.8051	8.3018	7.9520	7.6942	7.4960	7.3386
9	13.614	10.107	8.7171	7.9559	7.4711	7.1338	6.8849	6.6933	6.4511
10	12.826	9.4270	8.0807	7.3428	6.8723	6.5446	6.3025	6.1159	5.9676
11	12.226	8.9122	7.6004	6.8809	6.4217	6.1015	5.8648	5.6821	5.5368
12	11.754	8.5096	7.2258	6.5211	6.0711	5.7570	5.5245	5.3451	5.2021
13	11.374	8.1865	6.9257	6.2335	5.7910	5.4819	5.2529	5.0761	4.9351
14	11.060	7.9217	6.6803	5.9984	5.5623	5.2574	5.0313	4.8566	4.7173
15	10.798	7.7008	6.4760	5.8029	5.3721	5.0708	4.8473	4.6743	4.5364
16	10.575	7.5138	6.3034	5.6378	5.2117	4.9134	4.6920	4.5207	4.3838
17	10.384	7.3536	6.1556	5.4967	5.0746	4.7789	4.5594	4.3893	4.2535
18	10.218	7.2148	6.0277	5.3746	4.9560	4.6627	4.4448	4.2759	4.1410
19	10.073	7.0935	5.9161	5.2681	4.8526	4.5614	4.3448	4.1770	4.0428
20	9.9439	6.9865	5.8177	5.1743	4.7616	4.4721	4.2569	4.0900	3.9564
21	9.8295	6.8914	5.7304	5.0911	4.6808	4.3931	4.1789	4.0128	3.8799
22	9.7271	6.8064	5.6524	5.0168	4.6088	4.3225	4.1094	3.9440	3.8116
23	9.6348	6.7300	5.5823	4.9500	4.5441	4.2591	4.0469	3.8822	3.7502
24	9.5513	6.6610	5.5190	4.8898	4.4857	4.2019	3.9905	3.8264	3.6949
25	9.4753	6.5982	5.4615	4.8351	4.4327	4.1500	3.9394	3.7758	3.6447
26	9.4059	6.5409	5.4091	4.7852	4.3844	4.1027	3.8928	3.7297	3.5989
27	9.3423	6.4885	5.3611	4.7396	4.3402	4.0594	3.8501	3.6875	3.5571
28	9.2838	6.4403	5.3170	4.6977	4.2996	4.0197	3.8110	3.6487	3.5186
29	9.2297	6.3958	5.2764	4.6591	4.2622	3.9830	3.7749	3.6130	3.4832
30	9.1797	6.3547	5.2388	4.6233	4.2276	3.9492	3.7416	3.5801	3.4505
40	8.8278	6.0664	4.9759	4.3738	3.9860	3.7129	3.5088	3.3498	3.2220
60	8.4946	5.7950	4.7290	4.1399	3.7600	3.4918	3.2911	3.1344	3.0083
120	8.1790	5.5393	4.4973	3.9207	3.5482	3.2849	3.0874	2.9330	2.8083
∞	7.8794	5.2983	4.2794	3.7151	3.3499	3.0913	2.8968	2.7444	2.6210

Table IV (*continued*)

$\alpha = .005$

10	12	15	20	24	30	40	60	120	∞
24224	24426	24630	24836	24940	25044	25148	25253	25359	25465
199.40	199.42	199.43	199.45	199.46	199.47	199.47	199.48	199.49	199.51
43.686	43.387	43.085	42.778	42.622	42.466	42.308	42.149	41.989	41.829
20.967	20.705	20.438	20.167	20.030	19.892	19.752	19.611	19.468	19.325
13.618	13.384	13.146	12.903	12.780	12.656	12.530	12.402	12.274	12.144
10.250	10.034	9.8140	9.5888	9.4741	9.3583	9.2408	9.1219	9.0015	8.8793
8.3803	8.1764	7.9678	7.7540	7.6450	7.5345	7.4225	7.3088	7.1933	7.0760
7.2107	7.0149	6.8143	6.6082	6.5029	6.3961	6.2875	6.1772	6.0649	5.9505
6.4171	6.2274	6.0325	5.8318	5.7292	5.6248	5.5186	5.4104	5.3001	5.1875
5.8467	5.6613	5.4707	5.2740	5.1732	5.0705	4.9659	4.8592	4.7501	4.6385
5.4182	5.2363	5.0489	4.8552	4.7557	4.6543	4.5508	4.4450	4.3367	4.2256
5.0855	4.9063	4.7214	4.5299	4.4315	4.3309	4.2282	4.1229	4.0149	3.9039
4.8199	4.6429	4.4600	4.2703	4.1726	4.0727	3.9704	3.8655	3.7577	3.6465
4.6034	4.4281	4.2468	4.0585	3.9614	3.8619	3.7600	3.6553	3.5473	3.4359
4.4236	4.2498	4.0698	3.8826	3.7859	3.6867	3.5850	3.4803	3.3722	3.2602
4.2719	4.0994	3.9205	3.7342	3.6378	3.5388	3.4372	3.3324	3.2240	3.1115
4.1423	3.9709	3.7929	3.6073	3.5112	3.4124	3.3107	3.2058	3.0971	2.9839
4.0305	3.8599	3.6827	3.4977	3.4017	3.3030	3.2014	3.0962	2.9871	2.8732
3.9329	3.7631	3.5866	3.4020	3.3062	3.2075	3.1058	3.0004	2.8908	2.7762
3.8470	3.6779	3.5020	3.3178	3.2220	3.1234	3.0215	2.9159	2.8058	2.6904
3.7709	3.6024	3.4270	3.2431	3.1474	3.0488	2.9467	2.8408	2.7302	2.6140
3.7030	3.5350	3.3600	3.1764	3.0807	2.9821	2.8799	2.7736	2.6625	2.5455
3.6420	3.4745	3.2999	3.1165	3.0208	2.9221	2.8198	2.7132	2.6016	2.4837
3.5870	3.4199	3.2456	3.0624	2.9667	2.8679	2.7654	2.6585	2.5463	2.4276
3.5370	3.3704	3.1963	3.0133	2.9176	2.8187	2.7160	2.6088	2.4960	2.3765
3.4916	3.3252	3.1515	2.9685	2.8728	2.7738	2.6709	2.5633	2.4501	2.3297
3.4499	3.2839	3.1104	2.9275	2.8318	2.7327	2.6296	2.5217	2.4078	2.2867
3.4117	3.2460	3.0727	2.8899	2.7941	2.6949	2.5916	2.4834	2.3689	2.2469
3.3765	3.2111	3.0379	2.8551	2.7594	2.6601	2.5565	2.4479	2.3330	2.2102
3.3440	3.1787	3.0057	2.8230	2.7272	2.6278	2.5241	2.4151	2.2997	2.1760
3.1167	2.9531	2.7811	2.5984	2.5020	2.4015	2.2958	2.1838	2.0635	1.9318
2.9042	2.7419	2.5705	2.3872	2.2898	2.1874	2.0789	1.9622	1.8341	1.6885
2.7052	2.5439	2.3727	2.1881	2.0890	1.9839	1.8709	1.7469	1.6055	1.4311
2.5188	2.3583	2.1868	1.9998	1.8983	1.7891	1.6691	1.5325	1.3637	1.0000

* That is, values of $F_{m_1, m_2; \alpha}$, where (m_1, m_2) is the pair of degrees of freedom in F_{m_1, m_2} and

$$\frac{\Gamma((m_1 + m_2)/2)}{\Gamma(m_1/2)\Gamma(m_2/2)} \left(\frac{m_1}{m_2}\right)^{m_1/2} \int_{F_{m_1, m_2; \alpha}}^{\infty} F^{(m_1/2)-1} \left(1 + \frac{m_1}{m_2} F\right)^{-(m_1+m_2)/2} dF = \alpha.$$

† From "Tables of percentage points of the Inverted Beta (F) Distribution," *Biometrika*, Vol. 33 (1943), pp. 73-88, by Maxine Merrington and Catherine M. Thompson; reproduced by permission of E. S. Pearson. If necessary, interpolation should be carried out using the reciprocals of the degrees of freedom.

Index

A

A_2, 147
Ac, see Acceptance number
Acceptable quality level (AQL), 180, 183, 198
Acceptance number, 179, 199
Acceptance region, 113
Acceptance sampling, 1, 7
Algorithm, Yates', 292
Alias, 297–298
Alpha risk, see Hypothesis testing
Alternative, see Hypothesis testing
American National Standards Institute, 197
American Society for Quality Control, 197
Analysis of covariance, 267–271
Analysis of variance, 228, 251, 264
ANOVA, see Analysis of variance
ANSI, see American National Standards Institute
ANSI/ASQC Z1.4, 197
ANSI/ASQC Z1.9, 197, 201
AOQ, see Average outgoing quality (AOQ)
AOQL, see Average outgoing quality limit (AOQL)
Approximations to probability densities:
  binomial approximation to hypergeometric, 40
  normal approximation to binomial, 72
  normal approximation to rank sum, 135
  Poisson approximation to binomial, 182
AQL, see Acceptable quality level (AQL)
Arcsin, 171
Arcsine transformation, 171
Arithmetic moving average chart, 203, 207
ARL, 154. See also Average run lengths (ARL)
Array, see Orthogonal arrays
ASQC, see American Society for Quality Control
Assignable cause, 113, 146, 169, 173
ATI, see Average total inspection (ATI)
Attributes, 29–51
  control charts for, 164–175
Average outgoing quality (AOQ), 184
Average outgoing quality limit (AOQL), 184–186
Average run lengths (ARL), 154, 204, 206
Average total inspection (ATI), 186, 189

B

B_3, B_4, 156, 157
Back-solution, 231
Balance, 274, 342
Bayes theorem, 42
Behnken, D.H., 321–322, 331
Behrens–Fisher problem, 132
Bernoulli trials, 36–37
Beta distribution, see Distributions
Beta risk, see Hypothesis testing

367

Biased estimates, 107
Binomial charts, 164–171
Binomial distribution, *see* Distributions
Bivariate distributions, 44, 66
Blocking, 276
BLUE (best linear unbiased estimate), 106, 248
Box, G.E.P., 7, 284, 321–322, 331, 342
Box–Behnken designs, 321–322
Boxplot, *see* Box-and-whisker plots
Box-and-whisker plots, 22–25, 101
Burman, J.P., *see* Plackett, R.S.

C

c_1, c_2, c_3, 189
c_4, 148, 156, 157
C_p, 159, 332
C_{pk}, 159, 332
Capability of a process, 158–160
Causality, 218
Cause, assignable, *see* Assignable cause
Cause-and-effect diagram, 13–14
c-charts, 172–175
Central limit theorem, 77, 80
CFR (constant failure rate) model, *see* Distributions
Chain, alias, 298
Chain sampling plans, 187
Characteristic life, 60
Charts, 1, 4, 7, 8, 104, 113
    arithmetic average, 203–207
    attributes, 164–175
    binomial, 164–171
    c-charts, 172–175
    cumulative sum, 203, 209–213
    EWMA, 203, 207, 214–215
    geometric average, 203, 207, 213–214
    I charts, 160–161
    np charts, 165–166
    p charts, 167–170
    R charts, 148–151
    s charts, 156
    x-bar charts, 146–156
Chi-square distribution, *see* Distributions
Church, A.C., Jr., 293
Coefficient of variation, 336
Coin tossing model, 33
Concomitant variables, 268

Conditional density functions, 67
Conditional probabilities, 41
Confidence intervals, 88, 90–97, 101, 114, 138, 232
Confirmation experiment, 340
Conforming, 178
Confounded, 297
Constant failure rate, *see* Distributions
Continuous variables, 29, 52–69
Contrast, 288, 292, 295, 297, 311, 314–315, 317, 326. *See also* Defining contrast
Control lines, 113, 146–147
Correlation coefficient, 218, 230, 251
Covariance, 67–69, 218
    analysis of, *see* Analysis of covariance
Covariate, 267
CPL (lower capability level), 159
CPU (upper capability level), 159
Critical region, *see* Hypothesis testing
CSP (continuous sampling plans), 197
Cumulative distribution function, 54
Cumulative sum control chart, 203, 209–213
Cusum chart, *see* Cumulative sum control chart

D

d_1, d_2 (for double sampling), 190
d_2 (for control charts), 104, 147
D_3, D_4, 148–149
Daniel, C., 7, 217, 295–296
Davies, O.L., 342
Deciles, 13
Defect, 164
Defective, 164, 178
Defining contrast, 298, 303
Degrees of freedom, 81, 98, 122, 126, 228, 248, 274, 279–280
Demerit, 174
Deming, W.E., 2–5, 331
DeMoivre, A., 72, 94
Density functions, *see* Probability density functions
Deviation, 10, 98. *See also* Standard deviation
d.f., *see* Degrees of freedom
Discrete variables, 29–51

Distributions:
  beta, 71, 112
  binomial, 30, 33–37, 72, 85, 89, 94, 113, 129, 140, 182
  chi-square, 81–82, 102, 123, 125, 141, 238
  exponential, 52, 56–60, 63–65, 82, 85, 96, 102, 124, 139
  F, 137
  gamma, 52, 61–64, 66, 81, 112
  Gaussian, *see* Distributions, normal
  geometric, 45–46
  hypergeometric, 40, 89
  lognormal, 82–83
  multinomial, 30
  normal, 15, 18, 69, 72–87, 90, 101, 113, 129, 265
  Poisson, 46–49, 71, 172–176, 182
  Student's t, 100
  uniform (continuous), 55, 71, 77, 80, 112
  uniform (discrete), 30
  Weibull, 60–61
Dodge, H.F., 1, 186–187, 189, 197
Dodge–Romig sampling plans, 186, 189
Dotplot, 19, 25, 100, 130
Double sampling, 190
Draper, N.R., 217, 235, 238
Duckworth test, 136
Duncan, A.J., 177, 186, 197, 201

E

Effect, 285–289, 292
EWMA, *see* Exponentially weighted moving average chart
Expectation, 37–38, 46–47, 62, 73, 75, 78, 83, 113
Exponential distribution, *see* Distributions
Exponentially weighted moving average chart, 203, 207, 214–215
Exponential model, fitting by regression, 237–238
Extrapolation, 255

F

Factor, 271, 273
Factorial experiments, 280

Farrell, E., 1
Fence, 23
Finney, D.J., 285, 296
Fishbone diagram, 14
Fisher, Sir Ronald A., 7, 108, 264, 267, 318, 342–343
Foldover designs, 303
Four-level factors, 324–325
Fractional factorial experiments, 285

G

Galton, Sir Francis, 220
Gamma distribution, *see* Distributions
Gaussian distribution, *see* Distributions
Gauss–Markov theorem, 248
Geometric distribution, *see* Distributions
Geometric moving average chart, 203, 207, 213–214
Goodness-of-fit, 125–127, 256–257
Gosset, W.S. ("Student"), 100
Graeco-Latin squares, 323–324
Graphs, 13

H

Half-normal plot, 295
Hazard rate, 59
Hinge, 22–23
Histogram, 13–21, 25
Homoscedasticity, 227
Horizontal V-mask, 212
Houghton, J.R., 3, 6
H-spread, *see* Interquartile range
Hunter, J.S., 214–215, 284
Hunter, W.G., 5, 284
Hurdle, minimum, 293
Hyperbola, fitting by regression, 235–237
Hypergeometric distribution, *see* Distributions
Hyper-Graeco-Latin squares, 324
Hypothesis testing, 31, 113–127. *See also* Tests
  acceptance region, 115–116
  alpha risk, 115, 118, 147, 181
  alternative hypothesis, 114, 116
  beta risk, 115, 120, 181
  critical region, 115–116

Hypothesis testing (*Continued*)
  null hypothesis, 116–117
  power, 118, 120
  rejection region, 115–116
  sample size, 121
  type I error, 115
  type II error, 115, 120

I

I charts, 160–161
Independent events, 41
Independent variables, 45, 68, 77, 81
Indifference quality, 198
Inner arrays, 337
Inspection level, 199
  normal, 199
  reduced, 199–200
  switching rules, 201
  tightened, 199–200
Interaction, 276, 287, 289, 303
Interaction plot, 287
Interquartile range, 22
Interval estimates, *see* Confidence intervals
Ishikawa, K., 13

J

John, P.W.M., 261, 284–285
Joint density function, 67
Juran, J.M., 2, 6, 24

K

Kaizen, 4
Kelvin, Lord, 4
Kurtosis, 76–77, 87

L

Latin squares, 323
Lattices, 297, 299, 300–301, 304, 312, 318–319
  $L(8)$, 299
  $L(9)$, 312
  $L(12)$, 304, 320
  $L(16)$, 300–301, 327
  $L(18)$, 319, 337–338
  $L(27)$, 318
  $L(36)$, 319

LCL (lower control limit), 156, 158, 166, 170–171
Leaf, *see* Stem-and-leaf diagrams
Least squares, method of, 223, 243, 245
Lifetime, 56–57, 59–61, 83, 87, 89
Likelihood ratio, 191
Limiting quality (LQ), 198
Limits, 74
Linear combinations of random variables, 78, 80, 174
Linear estimator, 105
Lot, 177
  disposition action, 179
  sentencing, 179
Lot tolerance percent defective (LTPD), 180
LQ, *see* Limiting quality (LQ)
LSD (least significant difference), 267
LSL (lower specification limit), 159, 183
LTPD, *see* Lot tolerance percent defective (LTPD)
Lucas, J.M., 209

M

Mann, H., *see* Tests, Wilcoxon's test
Marginal probabilities, 45, 67–69, 71
Mathematical expectation, *see* Expectation
Maximum likelihood estimates, 108–110
Mean, 9–10, 18, 71–73, 75, 78, 101
Mean square, 251, 256
Mean square error, 332
Median, 11, 15, 18, 22–23, 58, 70, 72, 134, 332
MIL-STD-105D, 177, 179, 189–201
MIL-STD-414, 197, 201
MIL-STD-1235B, 187, 197
Minitab, 12, 14–15, 101, 122, 132, 218, 249
Modal interval, 15
Moment estimators, 106–107
Moment generating function, 64–66, 71, 76–78
Moments, 63–65, 70, 75
  estimators, 106–107
  generating function, 64–66, 71, 76–78
Moving average charts, 203
MS, *see* Mean square

INDEX   **371**

Multinomial distribution, *see* Distributions
Multiple sampling plans, 177
MVUE (minimum variance unbiased estimate), 105

N

Nelson, L.S., 171, 209
Nelson, P.R, 158
Neyman, J., 91
Nonadditivity, 277
Nonconforming, 164, 178
Nonconformity, 164, 172
Nondefective, 178
Nonorthogonal, 285
Nonparametric tests, 129
Normal equations, 224, 247
Normal distribution, *see* Distributions
Normal plots, 295-296
Normal scores, 83-85
np-charts, 165-166
Null hypothesis, *see* Hypothesis testing

O

OC, *see* Operating Characteristic curve
Octane blending, 258-259
Ogive, 83, 121
One-at-a-time design, 342
One-sided intervals, 93
One-sided tests, 117
One-way layout, 265
Operating Characteristic curve, 120, 198
Orthogonal arrays, 297. *See also* Lattices
Orthogonality, 274
Ott, E.R., 277
Outer arrays, 337
Outliers, 16, 18, 169, 296

P

Page, E.S., 204, 209
Paired *t*-test, 133
Parabolic masks, 209
Parameter, 53, 73, 88-90, 113
Parameter design, 333-340
Pareto diagrams, 13, 24
p charts, 167-170
Pearson, E.S., 91

Pearson, K., 125
Percentiles, 13
Performance characteristic, 333
Plackett, R.S., 304, 320
Plot:
    as an experimental unit, 264
    interaction, 287
Point estimates, 88, 90
Poisson distribution, *see* Distributions
Polynomials:
    fitting by regression, 253, 271
    orthogonal, 255, 271
Pooled estimate of variance, 131, 294
Population, 10, 74
Power, *see* Hypothesis testing
Probability density functions, 52
    conditional density functions, 67
Process capability, *see* Capability of a process

Q

Quadratic blending model, 259
Quadratic model, fitting by regression, 253
Qualitative factor, 271
Quantitative factor, 271
Quartiles, 11-12, 22, 58

R

Radians, 171
Random samples, 79
Random variable, 30
Range, 104, 147. *See also* Interquartile range
Rank tests, *see* Tests
Rational subgroup, 146-147
R charts, 148-151
Re, *see* Rejection number
Rectification, 179, 184, 186
Regression, 7, 220, 270, 290
Rejection number, 179
Rejection region, *see* Hypothesis testing
Reliability function, 59
Residual plots, 233
Residuals, 223, 231
Resolution of a fractional factorial, 303-304, 340
Roberts, S.W., 213

Robust design, 332–333
Robust statistics, 18
Romig, H.G., 187, 189
Rule, stopping, see Stopping rules for control charts
Rules for switching between levels of inspection, 201
Rule of thumb, 74, 165, 229
Runs:
  experimental points, 265
  of points on quality control charts, 204

S

Sample, 10
Sample size, 122, 183, 190
Sampling:
  without replacement, 39
  with replacement, 39
Sampling plans:
  ANSI/ASQC Z1.4, 197
  ANSI/ASQC Z1.9, 197, 201
  chain sampling plans, 187
  CSP (continuous sampling plans), 197
  Dodge–Romig sampling plans, 186, 189
  MIL-STD-105D, 177, 179, 189–201
  MIL-STD-414, 197, 201
  MIL-STD-1235B, 187, 197
  skip-lot sampling plans, 187
Scatter plots, 13, 218
s charts, 156
Scheffe, H., 258
Schilling, E.G., 158, 177–178, 186
Scores, normal, see Normal scores
Screening experiments, 301–302
Sequential probability ratio test, see Tests
Sequential sampling, 8, 177, 191–197
Seven tools, Ishikawa's, 13
Shewhart, W., 1, 4, 11, 145, 203, 206, 214
Signal-to-noise ratios, 335
Single sampling plans, 179–187, 190
Skewness, 76, 87
Skip-lot sampling plans, 187
Smith, H., 217, 235, 238
Snee, R.D., 258, 277
Standard deviation, 9–10, 101, 104, 114

Statgraphics, 15
Statistic, 9, 12
Stem-and-leaf diagrams, 20–22, 25
Stopping rules for control charts, 154, 174, 203
Stratification, 13
Student's distribution, see Distributions
Sum of squares, 98, 229, 264
Symmetric, 17, 76
System design, 332–333

T

Taguchi, G., 5, 7, 284–286, 297, 331–343
Tally sheets, 13
Target value, 159, 332
Tests, 31
  chi-square test, 123, 125–127, 129
  Duckworth test, 136
  F-test, 137, 230, 251, 264
  Mann–Whitney test, see Tests, Wilcoxon's test
  paired t-test, 133
  rank tests, see Tests, Wilcoxon's test
  sequential probability ratio test, 191–193, 209
  t-test, 122, 131, 133, 228, 264
  Wilcoxon's test, 129, 134–136, 332
  z-test, 119–120, 126
Ties in rank tests, 134
TINT, 101
Tolerance design, 333, 340
Torrey, M.N., 187, 197
Transformations, 171
Treatment, 264
Trials, see Bernoulli trials
Trimmed mean, 12, 18
TRMEAN, 12
t-test, see Tests
Tukey, J.W., 21–22, 136
2 × 2 tables, chi-square test for, 129, 141
Two-sample t-test, 131
Two-sided intervals, 92
Two-way layout, 273
Type I error, see Hypothesis testing
Type II error, see Hypothesis testing

U

UCL (upper control limit), 156, 158, 165, 171

Unbiased estimators, 105
Unequal variances, 132
    Behrens–Fisher test, 132
    F-tests for, 137
Uniform distribution, *see* Distributions
Upper confidence intervals, 93, 103
USL (upper specification limit), 159

V

Variables:
    concomitant, 268
    continuous, 29, 32–69
    discrete, 29–51
Variance, 10, 38, 64, 71, 75, 78, 97–102, 123, 131, 137, 294
    estimation of, 97–102, 131, 294
Variation, coefficient of, 336
    use as SN ratio, 336
V-mask, 209

W

Wald, A., 189
Warning lines, 154, 203
Weibull distribution, *see* Distributions
Whiskers, *see* Box-and-whisker plots
Whitney, D.R., *see* Tests, Wilcoxon's test
Wilcoxon's test, *see* Tests
Wood, F., 217

X

x-bar, 9, 90
x-bar-bar, 147
x-bar charts, 146–156

Y

Yates, F., 7, 284, 292, 311, 342
Yates' algorithm, 292

Z

Zero-defects, 178
z-test, *see* Tests